Henry Haversham Godwin-Austen (1834-1923)

THE K2 MAN (AND HIS MOLLUSCS)

THE EXTRAORDINARY LIFE OF HAVERSHAM GODWIN-AUSTEN

For Chris,
With my thanks,
and every good wish
from Catherine

RGS Guildford
November '13

For my friends
Alison and Mark, Pamela, Richard and Annette,
Ritchie and Katharine, Simon and Katie

I count myself in nothing else so happy
As in a soul rememb'ring my good friends.
Shakespeare: *Richard II*, 2.3

THE K2 MAN (AND HIS MOLLUSCS)

THE EXTRAORDINARY LIFE OF HAVERSHAM GODWIN-AUSTEN

CATHERINE MOOREHEAD

FOREWORD BY JIM PERRIN

THE IN PINN

is an imprint of
Neil Wilson Publishing
www.nwp.co.uk

First published in 2013.

All maps were produced by Rob & Rhoda Burns (robrhoda@drawingattention.co.uk)

A catalogue record for this book is available from the British Library.

Print edition ISBN: 978-1-906000-58-5
Ebook ISBN: 978-1-906000-60-8

Printed and bound in the EU.

CONTENTS

Part Five: The Shadows Lengthen

ACKNOWLEDGEMENTS

MY THANKS ARE DIVIDED YET overlap between those who made this book possible, and those who made researching, writing and publishing it such an enjoyable experience along the way. The driving force in helping me from the outset is Ritchie Robertson, who has fairly recently taken up his appointment as Taylor Professor of German at Oxford University. He introduced me to Dr Mike Riordan, Archivist at St John's College Oxford, the source of much genial and professional advice on how to handle Godwin-Austen's private papers, and Dr William Whyte, also of St John's College, who provided the historical foundations on which most of my research was built. Ritchie has looked at the evolution of the text, invariably with kindly understanding, remarkable critical acumen and an astonishing depth of scholarly knowledge. Dr Elizabeth Baigent, University Reader in the History of Geography and Senior Tutor at Wycliffe Hall, Oxford, who revised Kenneth Mason's original entry for Haversham in the *Dictionary of National Biography*, patiently answered my emails, pointed me in correct research directions, and provided encouragement when the tangle was thick. The staff at the Bodleian, in particular Sue Bird and Mike Webb, went out of their way to be helpful.

This book would never have been written without the very specific help of the Royal Grammar School, Guildford. The previous headmaster, Tim Young, appointed me to this job at a difficult time in my life; the present headmaster, Dr Jon Cox, and the Governors, particularly the then chairman of their Academic Committee, His Honour Judge Critchlow, provided me with a sabbatical term at exactly the right time, during which the bulk of the book was researched and written: I am most grateful to them all.

One of the joys about working at the RGS Guildford is its Parents' Association. For this project, RGSPA not only produced hard cash, but also ran an entertaining Gurkha Curry Evening, half the proceeds of which went towards assisting me. To all of them, particularly the dynamic Lou Montagu, the incisive Simon Page and the ever-constructive Mark Hoskins, I offer my most grateful thanks.

Among my colleagues, John Simpson, lately Director of Studies, John Saxton, Head of History and our Librarian, Tamsin Farthing, have all helped in practical and other supportive ways. My thanks also to Phil Griffith for his enthusiastic and skilful photographic help. And I would also thank Jenny Hipwell, its former director, and Jenny Rothwell and Kathryn Sweet at the RGS Foundation Office for their unfailingly sunny help in providing access to ancient copies of the school magazine *The Guildfordian* and other useful database material. Jenny Hipwell's son Sam also deserves my enthusiastic thanks for his unflappable and talented expertise in producing the front cover design.

At the Royal Grammar School, we do teach some of the brightest boys in the country. It was a form of that brightness which led Sam Jelley alertly to exclaim that his father had a friend called Godwin-Austen, the remark which fired the starting-gun. I thank him for using his brightness so well.

Having been led to the Godwin-Austen family, I should like to say how very much I have appreciated their help and enjoyed their company over the last three years. It has been the forthcoming, trusting and invariably positive nature of Robin and Richard Godwin-Austen which has been such a help and encouragement. Both kindly offered a critical response to my chapters, factually and stylistically; I have been most fortunate that they have been so open about Haversham's life and the material he (and this most resilient family) produced. One of the pleasures of researching this work has been Godwin-Austen hospitality, produced with effortless kindness by Kate at Shalford, Sally in Southwell and Mary at Papplewick; this particular writer appreciates that kindliness very much. But it was also their moral support which has been so telling. And I also thank Robin and Kate's son Stephen, and Richard and Jane's son, Jonathan, for answering my queries, particularly about their K2 trek in the footsteps of their astonishing ancestor.

Although I met them only briefly, I should much like to thank Gilly and Aline Marsh for filling me in on a number of interesting family details.

Keeping track of extended families is a tricky process. In that context, I should very much like to thank family historian Robin Pearce (and through him, Patrick and Catherine Conquest) for all the comprehensively provided, quickly sent and genially discussed material concerning Haversham's descendants through Edward Milner. Our correspondence has made my job much easier, and generated a pleasant exchange of information and discussion.

Among those professional bodies who helped me along the way, pride of place in my thanks should go to the Natural History Museum, South Kensington. Professor Fred Naggs talked me through the scientific research side of the project and provided me with the research tools to do the job (one of which, the Mollusca Departmental Library, is so helpfully run by Kathie Way). And he introduced me to John Mathew, who kindly agreed to take time out from a very busy Wellcome Foundation scholar's schedule to write the illuminating essay in Chapter 21. Polly Parry, the NHM Archivist, went out of her way on several occasions to follow up some of my queries; I am most grateful.

We are fortunate in this country that our great public institutions offer such dedicated public service, in a way that transcends economic and political realities. St John Simpson and Helen Parkin at the British Museum epitomise such high standards. My thanks to them for their friendly help, interest and disinterested hard work, and also to Alison Harding, at the British Musuem's fascinating country branch in Tring.

Despite an at times slightly peculiar institutional policy towards research help for out-of-towners, I found the desk staff of the Foyle Reading Room at the Royal Geographical Society to be enthusiastic, highly knowledgeable and most helpful. My thanks to them.

Among that esoteric but, in Godwin-Austen terms, most important group, the malacologists, I would particularly like to thank Dr J Reid of the Malacological Society, and Professor George Dussart, University of Kent at Canterbury, the keeper of the society's archives, for making so much valuable research material available to me so readily.

Librarians have in general come out of my experiences very well. Apart from those mentioned above, I should very much like to thank the staff I have met at the British Library, particularly the India Reading Room, and at the V&A for their help in steering me through what at times seemed quite a maze.

Nearer home, various people in and around Guildford have been keen and helpful in supporting my attempt at creating a wider knowledge of Haversham. Jackie Browning at the Guildford Institute, Mary Alexander and Carol Brown at Guildford Museum, Hannah Jeffrey at Surrey Archaeological Institute and local historian Ann Laver have all helped me with my enquiries admirably.

And in some respects, the person who is *the* local historian is Margaret Dierden. Her output is prodigious. But it is the high quality of her work, its attention to detail and accuracy, as well as its impressively wide range and utter engagement, which has been such a help. She has provided insight and comment on such a variety of topics that it is hard to know where to begin in thanking her. Guildford, and Shalford in particular, is most fortunate to be served by such a historian.

Still quite close to home, I should very much like to thank Graham and Jane Kemp at Nore, for their welcome, hospitality and interest, as well as for so much information on the history of their beautiful country home.

To write the book has of course been one thing; to get it published quite another. Jeremy Butterfield, perhaps the most adroitly lively user of the English language (and others) I have ever come across has produced connections aplenty in the publishing world. And even where these publishers have not been able to say yes, they have sent the kind of positive comments and advice which this tyro writer greatly appreciates. In this respect I would particularly like to thank Ben Harris, Ken Wilson at Batonwicks, Sir Chris Bonington, Michael Dover and Margaret Body, who, I am glad to say, first connected me with my publisher, Neil Wilson. And Diana Cave has also pitched in; my thanks to her.

In the USA, a footnote in that magisterial Himalayan history, *Fallen Giants*, stating that, 'remarkably, Godwin-Austen still awaits his biographer' has led to a most enjoyable and encouraging correspondence with its co-author, Professor Stewart Weaver of the History Department, Rochester University, New York.

Of course, a mountaineering biography would be incomplete without the assistance of my fellow members at the Alpine Club. Peter Mallalieu's kindness and informativeness have been of immense help, as well as his most appreciative chapter on Haversham as an artist. From John Cleare I have been given lessons in photo-scholarship. His investigations into the siting of Haversham's watercolours in the K2 and Nun-Kun areas were carried out with the most meticulous accuracy and lively good humour, materially extending my understanding of where this great explorer had been. I am deeply grateful to him. My thanks also to Jerry Lovatt for his support in relation to Haversham's paintings, and for his moral support generally.

Apart from the mention of Ritchie Robertson above, this leaves my friends, without whose encouragement, reading, criticism and general bonhomie everything would have been so much harder. Green-fingered and critical comments from Pamela MacGregor; the readings and advice of a published fellow-writer from Simon Watson (with further assistance from Bruce Phillips); the encouragement of David Dunmur, Barney Rosedale, Richard Howlett Jones, CJ and Zola Meadows; the generosity of Alison Raeside and the enthusiastic help both she and Mark Raeside gave in producing my publisher's proposal – as well as immense moral support as the writing evolved – are all greater than the sum of their parts.

I very much appreciate Jim Perrin stepping out of a busy life to write the Foreword in his characteristically lively way. Finally, I should like to thank my publisher, Neil Wilson, for his adroit, canny and highly professional help in steering me through the choppy waters of getting this sizeable work into the public domain where, after a wait of more than 90 years since Godwin-Austen's death, it rightly belongs.

FOREWORD

THE NAME IS ONE REMEMBERED from the atlases and geography text-books current in my schooldays: 'Mount Godwin-Austen', a climber's peak, second-highest in the world and far more difficult (and dangerous too) than that other relic of colonialism and militarism and focus for dependent ambition, 'Mount Everest'. The spring-clean of self-determinism has swept the better mountain's former name away. Perversely, it has had fixed to it and accepted another relict colonial tag. 'K2' it now is, though 'Everest' (in the proper pronunciation of which Catherine Moorehead kindly instructs us), despite being possessed of perfectly good variant indigenous names, is still branded with the appellation of a notably unpleasant and singularly uninteresting servant of the British Raj, who had never even seen the mountain that now suffers the imposition of this association. It would be hard to feel any enthusiasm for a biography of Sir George Everest of still-far-too-current name. It's impossible to respond with anything but interest and pleasure to Catherine's account of the extraordinary life and milieux of the all-but-forgotten Haversham Godwin-Austen. That defunct mountain name apart, I knew nothing of the man or his life before reading this exhaustively researched, scrupulously written, witty and authentically knowledgeable book. We have a great deal to thank her for. She has trawled the literary remains and travelled the landscapes of her subject's life, and in doing so has provided a fresh and comprehensive view and perspective on one of the great sagas and periods of mountain exploration.

It was not just a saga of discovery, but one of military conquest, record and survey. The 1600-mile ruckle across the surface of the globe that is the Himalaya and the Karakoram, within which ranges lie all the world's highest peaks, provided the venue for it. This was the halfway line across the pitch on which The Great Game was played out between the empires of Britain and Russia – and still is being played out, the oropoliticians would argue, albeit between re-grouped adversaries.

The notion of the 'English Victorian Gentleman' trails elements of the pejorative in its wake these days. Catherine's lively sympathy is always with her subject as she follows him in minutest detail through his Surrey-squirearchy upbringing; his Sandhurst-and-colonial career; his marriages with an Afghan woman, an English socialite, and finally to a woman of true mind with whom he found domestic happiness; his travels among unexplored mountains and passes; the development of his obvious surveying brilliance and artistic distinction; his very 19th-century Darwinian passions for malacology and palaeontology and his spiritual progress from an orthodox Anglicanism through Islam to a late immersion in Buddhism. Rather than dry objectivity, she brings to her account of Haversham's life a biographer's necessary qualities of compassionate understanding and genuine affection, thus enabling us clearly to see the man among the wreathing mists of his time.

The central interest of her book, for most readers, will be in the chapters concerning expeditions Haversham undertook to the Karakoram, Ladakh, Zanskar and Bhutan between 1860 and 1865, in the course of his work on the Great Trigonometric Survey. This massive project, initiated by the East India Company but brought under Crown administration after the Great Rebellion of 1857, established with a meticulous accuracy (upon which satellite technology and digitization have failed significantly to improve) the heights of all the giant peaks of the Himalaya and the Karakoram. Haversham's journeys particularly of 1860 and 1861 to the latter, as Catherine states, 'rank among the very highest achievements by any British mountaineer-explorer,' and prefigure the later mountain journeyings of Martin Conway, Francis Younghusband, Tom

Longstaff, Eric Shipton and HW Tilman. Haversham was the first Westerner to broach the immense Baltoro glacier system; the first to set eyes on the world's second-highest peak, that for so long bore his name. His ascent of Mata Peak (6283m/20,613ft) above the Rupshu valley in Ladakh stood as a summit altitude record for the best part of three decades. The quality of his surveying and mapping was exemplary. And yet, until now, he was the forgotten man of early British greater-ranges mountain activity. I'm delighted that Catherine Moorehead's painstaking researches have set the record straight, and am honoured to be asked to provide the Foreword to her fine and quirky book.

Ariège, June 2013

INTRODUCTION

WHEN I MOVED TO SURREY in 2001, I carried with me round the M25 the emotional baggage of 25 years' English teaching, 46 drama productions, four expeditions to very obscure parts of Central Asia, my name somewhat improbably in *Munro's Tables*, and an interesting set of personal suitcases of varying degrees of bulkiness. Shortly after taking up my teaching appointment at Godwin-Austen's old school, during a casual conversation about my mountaineering interests, a new colleague suggested that Godwin-Austen – about whom I knew nothing other than that he was connected by name with K2 – was somehow associated with the school (an 'Austen' having been one of its 16th-century co-founders) and that his descendants lived 'somewhere in Surrey'.

Not much to go on, and I gave the matter little further thought, though the conversation did recall Godwin-Austen's name, bracketed after K2, in one of my earliest atlases. At the time, it had seemed rather odd – why should an English name, and a pararhyming, hyphenated one at that, be attached to the world's second-highest mountain lying in the very depths of Asia? For most of the rest of the decade, however, I was preoccupied with two expeditions to quite different parts of Central Asia: one to the westernmost extremity of Outer Mongolia, the other to just about everywhere in Tibet that was worth going to,[1] followed by setting up a school hillwalking club, the RumDoodle Society (the author, WE Bowman, having lived in Guildford for 26 years).[2]

In 2009, however, Godwin-Austen turned up again, very suddenly indeed. I was introducing a play called *Burning Everest*, by Adrian Flynn and Michele Celeste, to my youngest pupils (11-year-olds). It concerned a boy with domestic problems – mainly a wayward mother – who took psychological refuge in fantasising about himself as George Mallory, his grandfather having fictionally been a steward on the ship which brought Mallory and the other members of the 1924 Everest Expedition to India. (I subsequently discovered by chance that one of the first accounts in Britain of Mallory and Irvine's last climb was given at the RGS Guildford[3] in January 1925 by Noel Odell.) With a flagrant and enjoyable disregard for keeping to Assessment Objectives, the kind of formulaic nonsense which bedevils real teaching these days, my introduction rambled on beyond the history of Everest, and Mallory and Irvine's famous disappearance (and in Mallory's case, his sad re-appearance as a bleached corpse, perhaps the most interesting part of the lesson as far as the young were concerned) to K2. I mentioned its alternative, unofficial name. At that point, one of the boys not yet lulled into a torpor (see Acknowledgements on p.xi) piped up and declared that his father, 'had a friend called Godwin-Austen.'

---

[1] The 2003 Godong Kangri Expedition travelled from Kathmandu to Lhasa, then westwards for five days through Xigatse and Saga to Mount Kailas. Its circumambulation, which rises to 5800m/19,029ft plus, was undertaken as an acclimatisation trek. From there, we returned east, half way to Lhasa, then struck north onto the Chang Tang (northern plateau) where we put up three first ascents between 6000m/19,685ft and 6200m/20,341ft. The return journey adventurously crossed the Tsangpo at Saga on the old non-motorised ferry then went overland via Tingri and the Pang La to Rongbuk and Everest base camp, above which a couple of treks were enjoyed. The homeward journey continued over the Lamna La to Tingri then out by the Tang La, the Nyalam Gorges, Zhangmu and (after negotiating a vast landslide) Kathmandu. The duration was seven weeks; the cost was astronomical; the experience life-altering – the grandest school expedition ever? An *Expedition Report*, with good maps, detailed text and numerous photographs is obtainable from the author, c/o The Royal Grammar School Guildford, GU1 3BB (£25).

[2] The achievements of this Society can be seen on the school website at www.rgs-guildford.co.uk/About the RGS/Activities/Clubs and Societies/RumDoodle Society.

[3] I am obliged to give the school its full title in order for it not to be confused with the Royal Geographical Society.

I was startled: here was a link to the great man being offered to me. An introduction to Stephen Godwin-Austen produced an introduction to his father, Lt Col Robin Godwin-Austen, and his mother, Kate. An enjoyable lunch ensued: it turned out that these descendants of the K2 man lived not far from my own house. Robin kindly invited me to take a look at the surviving (there was a big clear-out in the late 1960s) papers. These turned out to be a treasure trove of personal memorabilia.

The archive (for want of a better term) was in need of some ordering. With Robin and Kate's permission, and Kate's kindly and skilful attentiveness in seeing to the needs of the inner woman, I set to work in the summer holidays and managed to classify some boxes of letters, newspaper cuttings and so on. Much spidery Victorian handwriting was arduously transcribed: Haversham's handwriting, particularly when a young man, was rather scrawly and idiosyncratic. This challenge was considerably extended when it came to reading Haversham's four surviving diaries.

This enjoyable if eyeball-stretching work fortunately coincided with the possibility of a sabbatical from school. To write Haversham's biography (which for obvious reasons does not exist outside the *Dictionary of National Biography*) was an ideal project. The Headmaster and Governors' Academic Committee were won over, and the work begun.

This was mostly engrossing, occasionally frustrating. After meeting Robin and Kate, followed by the excitement of seeing the papers – when I first looked through them, I could hardly speak for a couple of days – the following best part was getting to know Dr Richard Godwin-Austen, Robin's brother and fellow great-nephew of Haversham, and his wife Sally, who now live in a perfect town house in Southwell, Notts, by the Minster. Sally produced an outstanding lunch on my first visit; Richard provided the last of the old century's wine. Ready for anything, we then drove out to Papplewick Hall (just outside Nottingham) to meet son Jonathan and daughter-in-law Mary, who with their children now occupy this well manicured, beautifully restored, Adamesque country house. (Visiting hours are posted on its website: www.papplewickhall.co.uk) It, too, turned out to have its share of Haversham memorabilia: watercolours, the first geological account of Kashmir, Buddhist mandalas probably painted by Haversham, family portraits (including an amazingly impressive oil of Haversham's mother, the general's daughter, she who, with Queen Victoria, was responsible for converting the family name to Godwin-Austen in 1854).

Indeed, almost everyone I have come across on this quest, which has taken me out of Guildford to rural Surrey, Ashford (Kent), London, Oxford, Cambridge, Nottingham, Bampton (Oxon) and Brecon, has been almost unfailingly constructive. Local institutions such as the Surrey Archaeological Society and Guildford Museum have naturally been keen to promote their local hero; the great London institutions such as the Natural History Museum, the British Museum, the V&A and the Royal Geographical Society (where, usefully enough I have been a Fellow for 20 years) have all taken a real interest in the project, as have the great libraries: the Bodleian and the British. Friends helped in all sorts of ways, by way of reading text, securing contacts (particularly in Oxford), offering opinions and advice. And of course several members of the Alpine Club pitched in with their quite extraordinary and indeed very humbling expertise.

Rather less enjoyable was trying to fund the whole process. A trawl of the Royal Geographical Society's grants, and one subsequent application, produced nothing – it seems difficult to acquire funds if you don't have a PhD or are attached to a university, these days. Other sources where some contribution might reasonably have been expected also failed to be positive. To the rescue came a friend whose income and natural generosity allow her to do this sort of thing, and the school Parents' Association (already generous in their support for some of my previous projects), not only altered their constitution in order that a teacher should benefit, but gave me 50% of the proceeds from a most convivial Gurkha Evening to which rather more than a couple of hundred loyal parents turned up and, after an authentic Nepalese curry, listened to my talk, which was probably more humorous than originally intended, on the great man. The expense of travelling, of buying into academic and other websites, of paying the anti-scholarly fees for images and licences, and even of stationery and printer ink, mounts up to something more substantial than one might think, but thanks to the generosity of my friend and the Association, plus a bit of belt-tightening, I gained the funds to see me through.

Several further difficulties presented themselves. The greatest, in terms of actually describing Haversham's life, has been the unevenness of the available material. His diaries for the best part of four years of his professional life, a very detailed memoir of his three years at Sandhurst and a memoir of his third wife, Jessie Robinson, written after her death in 1913 by Gertrude Jekyll (and a local artistic friend now identified as Margaret Watts) all provide fairly minute descriptions which cover 39 years of the near 90 he lived through. After that, every time one reads his correspondence which has survived in the natural history museums at South Kensington and Tring, the Royal Geographical Society, the British Library, the Bodleian and the V&A, one is reminded, with a sense of frustration, that other diaries and travel journals undoubtedly existed. Apart from the archives of the Malacological Society lurking dustily in a couple of wooden chests in deepest Kent, and the many scholarly articles he wrote, the remaining years are a matter for careful jigsaw-puzzle work, with the imagination and an estimate of probabilities called on to fill in the missing pieces. Fortunately, some of the papers, such as the seminal 1864 work on Haversham's exploration of the Karakoram and close encounter with K2, do contain blow-by-blow accounts of what went on.

And like any human being with a rich inner life, Haversham kept much to himself that he regarded as private. His affair with his Afghan lover, Kudikji, his subsequent lack of relationship with their (illegitimate) son, the dark side of the character of Arthur, the son and heir he had by his first English wife; her extravagance, and Haversham's probable conversion to Buddhism very late in his life are understood only through piecemeal evidence, some of it second- or third-hand.

And such uncertainty did not just apply to the verbal evidence. The delight which I experienced when I first saw Haversham's watercolours (with which I fell very quickly in love) was tempered by their lack of dates, or unclear dates, or dates which did not seem to fit the rest of the available evidence. Much careful study was required in order to resolve a whole variety of problems in this area.

Apart from unevenness of source material, two opposing problems then arose. The first was the vastness of the main Godwin-Austen family archive. It runs to several thousand pages, dating back several centuries. Fortunately, General Sir Alfred Bingley had already trawled this shoal and landed a fine family history which, if it occasionally confuses the odd George or John or Robert or Henry Austen, remains nonetheless a meticulous piece of work with much enjoyable detail in it; frustratingly, however – and slightly puzzlingly, given that it was published in 1936 – it stops at about 1797. I was also greatly assisted by Guildford local historian (particularly of Shalford) Margaret Dierden. She, too, had processed this vast collection of documents and produced most helpful, precisely expressed and very readable articles from it. Reading the *Surrey Advertiser* from 1899 to 1902, the period of the Godwin-Austen bankruptcy, on badly-produced microfilm at the Surrey History Centre was, on the other hand, torture.

My other cornucopic difficulty centred on Haversham's abilities as a polymath. I had taken up my interest in Haversham because I naively presumed that he was merely a mountaineer. He was, as Kenneth Mason said, 'probably the greatest mountaineer of his day', having several times climbed above 6000m/19,685ft, and explored the greatest glaciers on the planet before Whymper had made it to the top of the Matterhorn. Unfortunately from this non-scientist's point of view, he also turned out to be a highly productive geologist, ornithologist, zoologist, malacologist, conchologist and ethnologist, in the best Victorian polymathic tradition. While I have a little basic grounding in geology and ornithology, both content and method of the other disciplines remain a mystery to me. The same applies to Haversham's vast collection of watercolours, drawings and sketches. A scrape 'O' level in Art, even when acquired in a serious exam in 1967, is hardly sufficient for a sophisticated understanding of artistic evaluation. As when something goes awry on my Central Asian expeditions, however, I am very much a believer in asking the locals. To that end, John Mathew of Harvard, Peter Mallalieu, Keeper of Pictures at the Alpine Club, and Jerry Lovatt, Vice-President and Honorary Librarian of the Alpine Club, have kindly contributed expert essays which assess Haversham's most substantial contribution to the exciting history of 19th-century science (his initial natural history training having been carried out by 'Darwin's bulldog', TH Huxley) as well as his character and abilities as an artist. I am very grateful to them for undertaking these tasks, without which the whole work would be much less well informed and a great deal more unbalanced.

It was both a surprise and a pleasure to realise that Godwin-Austen was an extraordinary polymath as well as one of our greatest explorers. What puzzled me, given these distinctions, was: why was Haversham not better known? There are several answers. The most telling reason is that his papers and memorabilia have been inaccessible since his death 90 years ago. He has an entry in the *Dictionary of National Biography* and some other specialist encyclopædias to do with mountaineering-exploration and his branches of science, but these are neither popularly available nor expressed, in many cases, in the most accessible of language.

Second, Haversham was no populist; he was also quite a modest, and certainly a private, man although, where matters of scholarly rigour were concerned, he could be tetchy and acerbic. Consequently, while also perhaps fearing public exposure on matters which he would prefer to remain private, he never wrote a work for general consumption other than letters and articles for the *Surrey Advertiser*. Even here, however, in an article on the composition of the first Mount Everest expedition, in which there was widespread public interest, he spoiled his populist chances in a way for which he was rightly, if not very competently, satirised in *Punch* by devoting a paragraph to the delights of 'a fossiliferous limestone on the Donkia La containing doubtful nummulites.'

His books are highly specialised and, while in themselves they are some of the most important works in the history of their subjects, they have no popular appeal compared to, say, Captain Cook's *Journals*, Darwin's *The Voyage of the Beagle*, Stanley's *How I found Livingstone* (and other works), Captain Scott's *Diaries and Journal* (the British reading public do enjoy a good martyrdom), PH Fawcett's *Exploration Fawcett* or Bonington's *Everest Expeditions*. (These works have all generated speedy biographies or autobiographies, out of which the subject's popular appeal has been extended. Bonington, of course, has also enjoyed the immediate bonus of film and television, although the other authors have also been the subject of biopics.) Given his energy, one can only regret that *My Karakoram Adventures*, or *My Life among the Headhunters* were never forthcoming. The Victorian appetite for such works was considerable; if he had published such works, his fame would have been assured, and permanent. And one can also only regret that, until now, no one has known enough about Godwin-Austen to make a film of his highly cinematic life (a matter which is in the process of changing).

A corollary of this lack of a popular account of his spectacular life's work arises in considering what he did write. His legacy of eight books and 174 scholarly papers is just that. The books and papers are there for the advancement of science, nothing else. They were designed to be read to a small number of scholars, anything from a handful to a few dozen, usually in London or Guildford. And it was not just the oral delivery. The scholarly journals for which Haversham wrote, *The Geographical Journal*, or the *Journal of East Bengal Studies*, for example, all had quite restricted circulations. In the societies of which he was a member (and in most cases a high office-bearer), his work is still much revered.

It would be fair to say that I was shocked by the discrepancy between the colossal substance and variety of Godwin-Austen's life and the disproportionate contemporary ignorance – my own included – of his achievements. The aim of this first biography of Haversham Godwin-Austen, therefore, has been to give a great British explorer and scientist his proper place in our national history and to offer to today's reading public the opportunity to enjoy the life of a most extraordinarily talented human being.

PART ONE

AUSTENS, GODWINS AND GODWIN-AUSTENS

CHAPTER ONE

ORIGINS

The Surrey Austens

THE SURREY FAMILY INTO WHICH Henry Haversham Austen was born on 6 July 1834 was ancient, noble to the point of several royal connections, and interesting. Merchant adventurers, eclectic collectors, estate-improvers, scholars and soldiers throughout the Austens' long history played a complex and insistent role in forming Haversham's scholarly yet daring character as well as helping to shape the unusual plot of his adventurous, scientific, artistic and emotional life.

Much of the colour in Austen family history stems from their marriages. The most controversial dynastic shift among the Surrey Austens took place after George Austen (d.1728), disinherited his childless elder son John in favour of the second son, Robert (d.1759); Robert, in turn, had to adopt his nephews by Act of Parliament (*Austen Name Act, thirty-third of George II*) because either he was childless or had had his 'nephews', Henry and Robert Stoffold, out of wedlock. Haversham's father claimed that the marriage between Robert and his wife, Joan – ostensibly childless – had been kept secret because both Henry and Robert had been born to them before the marriage took place. There are several pieces of evidence – wills, deposed oaths and so on – which contradict this assertion, however. In the main family history Sir Alfred Bingley ascribes Haversham's father's opinion to wishful thinking because an unbroken line would have improved the length and status of the Austen pedigree to the time of William II Rufus, 1056-1100. This question is unlikely to be resolved.

Not very long after this 'nephews' event, what was perhaps the most 'imprudent' of the Austen marriages occurred when Sir Henry Edmund (1785-1871) married while under the age of 21, with awkward consequences for the settlement of the estate extending as far as bankruptcy in 1899. This consequential event can be linked to Haversham's impetuous first marriage, to the Sudhun landowner's daughter, Kudikji. She and Haversham enjoyed a love affair, resulting in a love-child (see Chapter 9), the consequences of which, off and on, haunted Haversham throughout his long life and were probably the principal impediment to him gaining much earlier the public recognition and fame that should have been rightfully his. Haversham's second wife, Pauline (see Chapter 14), was an injudicious choice in a different way: she appears to have been something of a spendthrift. These first two marriages were balanced, however, by the 27 years which Haversham and his third wife, Jessie, spent together in a union of love and mutual support. And in Haversham's old age, in the early 20th century, his nephew Annesley quarrelled violently with his father when he discovered 'Old Godwin Godwin' carrying on with the governess (see Chapter 3). Other than that, and the various liaisons discussed above, there seems to be little further evidence of that kind of waywardness in a family history which has been reasonably well documented for 13 generations and about 900 years.

The 'ancient' character of the Austens comes across somewhat self-consciously in two pedigrees and a family history.[1] In one pedigree, which begins convincingly then oversimplifies as it regresses in time, an 18th-century marriage between Robert Austen (d.1797) and Lady Frances Annesley,[2] the great-granddaughter of Sir John Thompson, first Baron Haversham, whose wife was the daughter of Arthur Annesley, the first Earl of Anglesey,[3] created a Whig branch which lasted from the Puritan Austens of the 17th century to Haversham's father, Robert Alfred Cloyne Godwin-Austen, in the 19th. This line had meandered down from Cromwell and Hampden (also the friend of an earlier Austen), then from Edward I, *malleus Scotorum*, and to Robert Alfred Cloyne from William the Conqueror. Before this pivotal ancestor, it can be followed through the daughter of Alfred the Great and Baldwin I, Count of Flanders, then across several generations to the original progenitor, the mighty Charlemagne himself.

The second and rather more credible family tree, through the males, shows the Austens enjoying a direct lineage to Stephen Austen, of Stapleford, Cheshire, whose precise dates are uncertain but who lived during the reign of Richard II, from 1377 to 1399. This line may or may not – see page 3 – have suffered a break in the 18th century after Robert Austen's father probably died childless and adopted his nephews, Robert and his brother Henry, as his heirs.

The Austens (from the Latin *augustus*, meaning majestic, revered) began their long English career in Kent, the name possibly surfacing by association with Saint Augustine of Canterbury. The branch from whom the present family has grown moved to Shropshire and Cheshire in the 11th or 12th centuries, though it is not known why. With characteristic Austen enterprise, they had become a landowning family by the time of Richard I (1157-99). A pedigree of the ancient Cotgreave family of Cheshire confirms their presence through a quartering of them in its arms.

In the 13th or 14th centuries, the family probably lost much of its land, but a knight named Austen – the first mention in the family history of several highly able soldiers of whom Haversham was one – somehow distinguished himself on 22 January 1351, three years after the capture of Calais by Edward III, when the nearby castle of Guînes was delivered up to the English. The presence of a knight suggests that Austen social circumstances were again on the rise: it is an undeniable theme of this family's history that their fortunes have rollercoasted through the various phases of English then British history. Haversham's bankruptcy in 1899 (although, as far as I can see – the reasons are examined in Chapter 18 – this was hardly the fault of Haversham himself) takes its place in this undulating historical sequence.

During the Wars of the Roses the Austens' rising fortunes were reversed when they chose the wrong side to support and were made to forfeit their Shropshire properties once more. This 'much impoverished' family then shifted their ground to Toddington in Bedfordshire (bathetically, now better known as an M1 service station, although the village is still close by), presumably named thus in memory of their lost estates in Shropshire.

---

[1] I am indebted for much of the information in this chapter to *The Austens of Shalford and their Kinsfolk*, by Sir Alfred Bingley; privately published, 1936. This history, closely based on Godwin-Austen family papers, the majority of which are now lodged in the Surrey History Centre, Woking, does, however, only go up to the end of the 18th century.

Lt Gen Sir Alfred Horsford Bingley KCIE, CB, K St J, (1865-1944) was the son of Alice Austen, daughter of John Wentworth Austen (1820-75), the youngest son of Sir Henry Edmund Austen (1785-1871), Haversham's grandfather (see Chapter 3).

[2] She died aged 97 and may have been responsible for introducing the longevity gene into the family.

[3] Arthur Annesley, 1st Earl of Anglesey PC (1614-86) was an Anglo-Irish royalist statesman. He rose to the position of Lord Privy Seal, and had several styles throughout his lifetime: the Hon Arthur Annesley (until 22 November 1660), the Viscount Valentia (1660-1), and the Earl of Anglesey (after 20 April 1661).

THE MODERN FAMILY

The family had already completed its first half-millennium of recorded history when the Surrey dynasty was founded by John, the son of a John Austen of Toddington who died in 1478. This uncelibate priest became rector of Chiddingfold (near Godalming) but probably married before he took holy orders. His wife, the daughter of a Roger Elyot of Shalford,[4] brought this first mention of the family seat's name into the Austen story. John the priest died in 1542.[5] (Roger Elyot was a collector of tithes for St Mary, Shalford, which at that time was owned by the Augustinian Canons of St Mary of Bethlehem, a London hospital. St Mary's became the Austen family church. Most of the leading members of the family are buried or commemorated there.)

In the good years, John the priest probably did well out of the agricultural produce and woolpacks he received in tithes. His son, also called John[6] (buried in Holy Trinity church, Guildford, 16 April 1572[7]), became a prosperous Guildford wool-stapler who bought wool from farmers, sorted and treated it and then sold it on to manufacturers. John had made his professional move at the right time – the well-situated and financially robust town of Guildford had become an important centre of the wool trade; woolsacks appear on the town's coat of arms. The durable part of the family fortune was founded here; the Austens' wealth was and is closely tied to agricultural conditions in Surrey. Indeed, the Great Agricultural Depression of the late 19th century was the main culprit in bringing about the failure of the Austen estates after Haversham's return from India.

It was John, the son of John the rector of Chiddingfold and therefore the second Surrey Austen, who redefined the family's social status. He was ambitious: he became an MP (from 1563 to 1567) and was appointed one of the *Probi Homines* or Approved Men of Guildford. (This assertion of small town morality effectively means a town councillor, although local council executive power was much greater then than now.) He progressed to become Mayor of Guildford in 1566. He married into local politics; his wife, Joan, was related to a Snelling who was twice Mayor of Guildford.

John the Approved Man was evidently someone of enterprise and intellect: he co-founded the Royal Grammar School Guildford[8] (his name appears, with the other founders and major benefactors, on the crossbeams of one of the school's oldest rooms). To achieve this feat, he had locked himself into a legal battle with the Archbishop of York over possession of, and rents from, the chantry lands on which the school was to be built, it having enjoyed an uncertain existence since the time of its founder, Robert Beckingham, in about 1509. The legal judgement went against the archbishop; John secured an Act of Parliament in 1563, at considerable personal expense, to consolidate his victory, then raised money in 1569 from the local gentry and the City of London (a further testimony to his importance in both Guildford and, now, the capital) to further the building of the schoolmaster's house which had been begun by the mayor and Approved Men at their own expense in 1557, although the work was still nowhere near complete when he died three years later. John's younger son George appears to have consolidated much of the legal and building work undertaken by his father: he clarified the endowments, completed the

---

[4] Shalford is (by Surrey standards) quite a well-preserved village, lying just south of Guildford on the A281. It was mentioned in the Domesday Book in 1086. The name probably derives from 'shallow ford', although other possibilities exist.

[5] The family pedigree says 1553, which is probably wrong, because of a muddled transcription of the dates 1553-8, which were those of the marriage of John Austen's son.

[6] The Austens are not imaginative when it comes to Christian names. John, George, Henry and Robert abound. A sprinkling of Charles, James or William would have helped lessen the possible confusions. The women's names are more varied.

[7] Bingley wrongly states 1571.

[8] The school's origins appear to lie in a deed of gift of land to Robert Polsted by Robert Beckingham in 1507. A 'licence in mortmain' having failed to be obtained in two years by the villagers who might have benefited, a deed for the founding of the 'free school' was drawn up in 1509. During the academic year 2009-10, the school celebrated its 500th anniversary.

building and furnishing of the schoolmaster's house and secured the Bishop of Winchester's approval for the school's statutes. The Austen then Godwin-Austen families have been connected with the school ever since its co-founding. Many of them, including, almost certainly, Haversham, from 1845 to 1848 (see Chapter 4) have been educated there. Austen money, where public benefaction is concerned, has tended to go into education rather than, say, medicine.

John the Approved Man was evidently dissatisfied with small town wool-trading: having extended his business to London, he joined the Merchant Adventurers, founded the previous century in 1407, a thriving regulated company which had prospered thanks to a monopoly on cloth export mainly to the Mediterranean. Records show that for example he was on Crete in 1523. He must have been one of the earliest English traders in the Mediterranean. His voyages, which took in Genoa, Venice and the Levant, were 'accounted difficult and dangerous', and might equally describe Haversham's work in the icy Karakoram and humid jungles of northern Burma, probably motivated by similar reasons of trade expansion three and a bit centuries later. John's social skills were also evidently remarkable, as was his perceptiveness: he established some sort of connection with the Doge of Venice, Antonio Grimani, in 1521. This exotic link gave rise to one of the best Austen family legends (see p7).

POLITICS AND RELIGION

This same John Austen cleverly survived one of the most turbulent periods in English religious history. Having begun life as a Roman Catholic, he is said to have approved of the spread of the Protestant Reformation under Edward VI (1537-53) and no doubt watched with consternation the reaction under Mary (1516-58), especially as the Abbots, Guildford friends whose sons became Archbishop of Canterbury and Bishop of Salisbury, were 'nearly brought to the stake'.[9] (That John Austen was friendly with a couple who would produce a bishop and the Primate of All England shows how rapidly he and his family had risen through the social ranks.) John Austen, however, shrewdly went along with England's religious swings and so weathered the Marian storm. By accepting the Anglican compromise, he lived to prosper through a good part of Elizabeth's reign.

Other religious dalliances are also apparent. Methodism appears in part of the family in the 18th century, for example, and other significant family members have oscillated between puritan and cavalier tendencies (with the religious practices thereby implied), rather like the nation at large.

The most spectacular exponent of religious change of identity is, however, Haversham himself. As a young man in the Punjab and Kashmir, when in love with his Afghan, and therefore Muslim paramour Kudikji, and preparing to go through a form of Islamic marriage service with her, he declared in a letter (though probably with the intent of giving a placebo to his correspondent) that he was 'at that time to all intents and purposes a follower of the Prophet'. (The truth of this statement is examined in Chapter 9.) And towards the end of his life, although it may be simply a stronger statement of a latent interest since first arriving in Burma early in 1853, Haversham, despite his regular observation of his Christian duties as local squire, undoubtedly enjoyed a leaning towards Buddhism (the substantial evidence for which is examined in Chapter 24).

There have been just as many swings in the Austens' political positions, which remained, however, closely allied to the notion of public service: over the centuries they have generally been Royalists, Whigs and Liberals, while the exceptions have been the 17th-century Puritans and Haversham himself who, with his third wife Jessie, a co-founder of the right-wing post-Disraelian Primrose League, became the first modern Austen Conservative. Robert Alfred Cloyne, Haversham's father, in contrast to his son, stood for Parliament in 1835 on an 'Austen for religious and civil liberty' and, topically enough, on an 'Austen Electoral Purity' platform. He spoke in favour of the 1866 Reform Bill – he was not only defeated but came third. In his obituary, the

[9] Bingley, *Ibid*, p5.

Surrey Advertiser describes him taking 'a prominent part for years in Public Matters', especially county business, and he was, 'a regular attender of the Court of Quarter Sessions, where he frequently took part in the debates, and generally figured on the side of economy, as against what he deemed to be unnecessary or inflated expenditure. The learned Gentleman was also a member of Guildford County Bench of Magistrates until failing health compelled him to give up. In politics, Mr Godwin-Austen was a consistent Liberal ... '[10]

THE RISE OF THE ESTATE

John Austen, the second Surrey Austen and its first Merchant Adventurer, consolidated the rising social position of the Austens through his two sons: John (d.1613) and George (d.1621). These Jacobethans form the real basis of the modern family. John, like his father, became a Merchant Adventurer; he was also a Citizen of London and a member of the Haberdashers' Guild. The connections with Haversham the adventurer are not hard to make: although Haversham's 'adventures' occurred within the well-structured framework of Army life, time and again he disregarded instructions not to go above 16,000 feet and thereby established his reputation as one of the greatest mountaineers of his day. On his return to England he became a leading, if supposedly rather grumpy, member of a number of societies and institutions, such as the Royal Geographical Society, the Natural History Museum and the Malacological Society, which could be seen as the modern equivalent of John and George Austen's membership of guilds at the time when the country was moving towards an economy dependent on scientific and technological advance.

John's financial acumen led to his success in the import-export business (cloth and hats for export; pins, luxury items such as pictures, tapestries, silks and jewellery and exotica being imported). His trading interests naturally led him to the hub of the Western world's trade, Venice, a boom-city for craftsmen after the fall of Constantinople in 1453. Here, presumably thanks to the connection established by his father with Antonio Grimani[11] in 1521, he befriended the Doge, Marino Grimani, around 1595. There is no documentary evidence, but family tradition has it that he fell in love with and married Giuliana, a kinswoman of the Doge. The one quite strong piece of evidence for this story is that her portrait, of the school of Veronese, is still in the family's possession. Her clothing and jewellery show the right fashion for the period and their richness certainly suggests someone of high social standing. The marriage (if it got that far, which is unlikely as marriage between Merchant Adventurers and foreigners was forbidden at the time) must have been a mixed one: the Catholic Grimani and the Protestant Austen. Again, the link with Haversham can readily be seen: one of Haversham's first acts when not directly governed by his family, i.e. when alone after the death of General Godwin in Simla in 1853, was to fall in love with, have a child by and eventually marry Kudikji in the romantically wild, semi-explored border country between Kashmir and the Punjab. This liaison, which would have been much frowned upon by Haversham's regiment, strongly resembles John Austen and Giuliana Grimani's earlier romance, objected to in their case by ecclesiastical and trade officialdom. In Haversham and Kudikji's case, fortunately, we can look at the matter (Chapters 9 and 25) in rather greater detail.

Although Elizabethan exuberance faded into the darker years of the Jacobean period, the Austens continued to rise in wealth and status. John, now the third Surrey Austen and grandson

[10] *Surrey Advertiser*, 29 November 1884. I am indebted to Margaret Dierden for this helpful and illuminating reference.

[11] Marino Grimani (b.Venice 1532, d.Venice 1605) was the 89th Doge of Venice, reigning from 26 April 1595 until his death. He was a member of the same noble family to which the earlier Doge, Antonio Grimani, who had met John Austen, second of Surrey, in 1521 belonged. Grimani's reign as Doge was principally remembered for (1) the splendid celebrations for the coronation of his wife, Morosina Morosini and (2) the beginning of the quarrel with the papacy that resulted in Pope Paul V placing the Republic of Venice under papal interdict in the reign of Grimani's successor, Leonardo Donato (1606-7).

of the founding rector of Chiddingfold, continued to strengthen the family fortune until throwing over his adventurous Mediterranean career in 1598; he then settled in Guildford (although not with Giuliana – there is no record of her in England). His stay-at-home brother George, continuing his trade as a successful wool-stapler and as the Member of Parliament for Haslemere in 1597 then Guildford in 1603, had also prospered but had spent the bulk of his time in Guildford. He married three times and produced at least 14 children. This early Jacobean increase in the Austen fortunes was not, therefore, motivated just by acquisitiveness: George's three families naturally required quite a lot of space.

Probably the first addition to the Austen property portfolio at this time was Nore, a substantial farmhouse brought to the Austens in the 1580s through George's marriage dowry from his first wife, Ann Mellersh, her family having built Nore during the reign of Henry VIII (1519-47). Some of its timbers, still in the roof, are said to date from the time of the Armada. It is a beautiful place, enjoying, southward to the (South) Downs, one of the finest views in a county of fine views, enhanced in its foreground by a garden designed and part-planted by Gertrude Jekyll. Nore lies on the face of a steepish wooded hillside below the hamlet of Hascombe, a little to the south-west of Godalming. It remained in Austen ownership for over 300 years and was still a working farm until late in Haversham's life: he moved there permanently, having decamped from Shalford House in about 1907 and remained, working with Jessie on his molluscs and shells in the Garden Room until her death in 1913 and his own in 1923. His letters are peppered with references to haymaking, the problem of feeding the hens in wartime and other agricultural matters. Nore's curious modern history, which is described in greater detail in Chapter 19, does not, however, end with Jekyll's designs: Captain Scott spent his last night in England there and Dirk Bogarde bought it in the 1950s.

Even before retirement, John and George had already worked together to buy substantial amounts of land and property, such as the Manor of Smithbrook, acquired in 1588. Separately, George bought property around the villages of Bramley and Dunsfold, while John had begun to buy agricultural and urban sites around Guildford and Shalford. George and John (who was childless) used their property-buying not just to accommodate George's growing family but to elevate their social standing. While their rise in status was mostly local, they began at this time to connect to people of national significance: for example, George's daughter by his first marriage married the brother of George Abbot, Archbishop of Canterbury[12] (whose statue occupies a focal position at the top of Guildford High Street). The Archbishop's brother, Maurice, was one of the earlier governors of the East India Company, the first family connection with the body which was to play such an influential part in Haversham's professional life.

---

[12] Born at Guildford where his father Maurice Abbot (d.1606) was a clothworker, he was taught at the Royal Grammar School Guildford. He later studied, and then taught, at Balliol College Oxford, was chosen Master of University College in 1597, and appointed Dean of Winchester in 1600. He was three times Vice-Chancellor of the University, and took a leading part in preparing the Authorised Version of the New Testament. In 1608, he went to Scotland with George Home, 1st Earl of Dunbar, to arrange for a union between the churches of England and Scotland. He so pleased King James in this affair that he was made Bishop of Lichfield and Coventry in 1609 and was translated to the see of London a month afterwards.

On 4 March 1611, Abbot was raised to the archbishopric of Canterbury. He defended the Apostolic Succession of the Anglican archbishops and bishops and the validity of the church's priesthood. In spite of this defence, his puritan instincts frequently led him not only into harsh treatment of Roman Catholics, but also into courageous resistance to the royal will, e.g. when he opposed the scandalous divorce suit of the Lady Frances Howard against the Earl of Essex, and again in 1618 when, at Croydon, he forbade the reading of the declaration permitting Sunday sports.

In 1622, while hunting in Lord Zouch's park at Bramshill in Hampshire, a bolt from his cross-bow aimed at a deer happened to kill one of the keepers; Abbot was so distressed by the event that he fell into a state of deep depression. His enemies maintained that the fatal issue of this accident disqualified him for his office, and argued that, though the homicide was involuntary, the sport of hunting which had led to it was one in which no clerical person could lawfully indulge. The king had to refer the matter to a commission

SHALFORD HOUSE AND PARK

In 1599, John and George succeeded in raising the family social status further by buying the rectory and manor of Shalford, another important local estate, from Sir George More of Loseley, the first Austen liaison in English law with titled gentry. (The rectory of Bramley was also included in the sale.) The manor came to be called Shalford House and was jointly rebuilt by these dynamic brothers between 1601 and 1611.

The rise continued. In the 18th century, Sir Henry Austen (1735-86), the first knighted Austen since the founding of the modern family, enlarged and converted the house and estate into an Adamesque manor with much improvement to the grounds; it then became known as Shalford Park. True to the Austen form of economic ups and downs, however, it had to be sold in 1899, with its magnificent contents (see Chapter 18), to settle Haversham's bankruptcy debts. Thereafter, it was rented out (as it had been by Haversham on occasion before 1899) before declining through stages of increasing indignity, from becoming the wartime headquarters of the Cornhill Insurance Co to eventually being pulled down by Guildford Borough Council in 1967-8, to make way for the River Wey water-extraction plant. (A small plaque behind St Mary's Shalford marks the spot.)

In the best Jane Austen[13] tradition, where those who improve the estate, for example in *Mansfield Park*, are the morally worthy, a look at the two main phases of improvements to Shalford House in the early 17th century and the late 18th century will help to define rather more precisely this theme of estate improvement, varied by downturn, which runs throughout post-Elizabethan Austen history and which caught Haversham so fiercely in its grip.

Shalford House enjoyed its first serious rebuilding between 1601 and 1611. It was 'a moderate sized gentleman's country house in the ordinary late Tudor or early Jacobean style' (Bingley). An armorial mantel was created in the north-west corner, known as the Dining Room, while an over-mantel, containing carved figures of saints, in the north-east corner, became known as the Oak Room. This over-mantel appears to have been removed either from the Dominican friary which George pulled down in Guildford in 1606 or from the vicar of Shalford's parsonage (built by the prior of the Augustinian canons, the lay rectorship of Shalford having been purchased by the Austens). The carved figures are Roman Catholic, i.e. pre-Reformation in origin and family historian Bingley speculates that they might have been John Austen's recognition of his relationship with Giuliana Grimani. Also notable were the great fireplaces, some with Dutch tiles, their magnificence and craftsmanship being visible in photographs taken before Shalford House was demolished. (Guildford Town Council gave one of the chalk-stone fireplaces, dating from 1609, and probably originating from the ancient Tyting Farm near the base of St Martha's Hill, to Annesley Godwin-Austen. Four carved corner wood panels can still be seen at Papplewick Hall, Notts.) Shalford House and its furnishings confirmed the Austens as members of the landed gentry

---

of ten, though he said that 'an angel might have miscarried after this sort'. The commission was equally divided, and the king gave a casting vote in the archbishop's favour, while also signing a formal pardon or dispensation.

After this the archbishop seldom appeared at the council, chiefly on account of his infirmities. In 1625 he attended the king constantly, in his last illness, and performed the ceremony of the coronation of King Charles I. His refusal to licence the assize sermon preached by Dr Robert Sibthorp at Northampton on 22 February 1627, in which cheerful obedience was urged to the king's demand for a general loan, and the duty proclaimed of absolute non-resistance even to the most arbitrary royal commands, led Charles to deprive him of his functions as primate. The need of summoning parliament, however, soon brought about a nominal restoration of the archbishop's powers. His presence being unwelcome at court, he lived from that time in retirement, leaving Laud and his party in undisputed ascendancy. He died at Croydon on 5 August 1633, and was buried at Guildford, where he had endowed a hospital with lands to the value of £300 a year.

[13] There appears to be no direct blood connection between Jane Austen's branch of the family, from Hampshire, and the Surrey Austens. The only link between the families is that Jane Austen's brother was rear-admiral of the fleet sent in support of Haversham's grandfather, General Godwin, at the outset of the Second Anglo-Burmese War. When Charles Austen died suddenly of cholera at Prome, Burma in 1852, General Godwin used the opportunity of him being replaced by only a commodore to turn the Burmese War (successfully) into a combined, proto-Marine, Army and Navy campaign.

with an established family history: Bingley, no doubt rightly, gives the heads of the family the rank of squire from this point.

The changes to Shalford Park made in the 18th century by Henry Austen (formerly Stoffold), (1735-86), and his brother Robert Austen (1739-97) after 1786 (when he became sole owner) and then by Sir Henry Edmund Austen (1785-1871) were even more extensive. Pevsner[14] describes the building as, 'heavy stuccoed Classical of 1797. The main (N) front of nine by three bays had an elegant porch of four Ionic columns shielding a big Adamesque fanlight, up to the best Dublin standard.' The house was expanded to 71 (possibly 73) rooms on three storeys (the 17th-century version probably having had two). Most notably, a Drawing Room, which looked out onto a large cedar of Lebanon and had a Carrara and red Scottish marble fireplace, was added; this was installed in Papplewick Hall in 1986. A Music Room was also created. Some parts survived demolition: the unique fanlight, over six feet across, above the front door was salvaged and stored by Guildford Borough Council. It was eventually donated to the Brooking Collection of Architectural Artefacts, in Greenwich.[15]

In the grounds, true to the neo-classical principle of Alexander Pope's 'nature methodis'd', Sir Henry built stables, entrance lodges and a summerhouse in the Monks' Walk. From the summerhouse, a picturesque prospect of the deliberately preserved ruins of St Catherine's Chapel could be enjoyed. The garden was laid out and an avenue of limes was planted. Inside, Robert Austen had the aesthetic sense to preserve the Jacobean Dining and Oak Rooms. (Some confusion remains about whether the north-east corner was in fact the Oak Room or the Library.) The 'Ancient Dining Room' survives in an engraving presented by Sir Henry Edmund (see plate 7). Above the fireplace, the family coat of arms can be seen; the ceiling is beautifully panelled; a table and chairs evinces very high quality craftsmanship; the stained glass windows are studded with coats of arms, the curtains deep and grand; a sizeable portrait suggests an ancestor in high-ranking ecclesiastical robes.[16]

Robert Austen's acquisitions extended to purchasing Shalford Manor Farm from Lord Onslow in 1779. It is a very handsome, beautifully proportioned house with nicely proportionate outbuildings and elegant garden. And Robert's tree-planting extended to the well-known nearby feature of Chantry Wood, sold in the 1930s to Guildford Borough Council as a part of Green Belt development.

These properties and lands were what Haversham would have inherited when his father, Robert Alfred Cloyne, died in 1884 – splendid residences, necessary to accommodate all his many brothers and sisters (particularly the unmarried ones), valuable possessions which were both tasteful, if conventional, beautiful pieces of parkland and fields which would have kept him in touch with his natural history interests and a strong sense of local prominence. He would have no doubt felt the weight of his responsibilities in preserving and developing this estate which of course extended far beyond the confines of Shalford Park; the only reliable estimate of the estate's size at that time suggests it ran to about 5,000 acres, the value of which, today, given that this corner of Surrey has some of the highest land and property prices in the UK outside London, would be vast. The vigour with which Haversham tried to protect the estate against bankruptcy as the great agricultural depression developed – and the mistakes he made – are described in Chapter 18. Selling off Shalford Park and its regal contents undoubtedly marked a turning-point in the modern family's fortunes.

Haversham's two senior descendants – Robin and his brother Richard – have also done their bit as 'improvers of the estate'. After a distinguished army career, culminating in becoming Commanding Officer of the 1st Battalion, Royal Regiment of Wales (24th/41st Regiment), among

---

[14] Ian Nairn and Nikolaus Pevsner, revised by Bridget Cherry: *Surrey*, in *The Buildings of England*, Yale University Press, Penguin, 1962, p452. (The opinion is largely Ian Nairn's: see Foreword, p11.)
[15] The Brooking Collection, which deserves wider support, can currently be found at Fawley Hill, Fawley Green, Henley on Thames, RG9 6JA. See also at www.thebrookingcollection.com.
[16] For the details of the size and contents of Shalford House when it was sold in 1899, see Chapter 18.

other administrative posts (as part of the Commonwealth Parliamentary Association in the Houses of Parliament for eight years, for example), Robin (b.1932) became chairman of the Surrey Country Landowners' Association, a natural outcome of his strenuous efforts to turn round the economic fortunes of the remaining Godwin-Austen estates (much reduced after Haversham's bankruptcy in 1899, and particularly after the recession of the late 1980s and early 1990s – the cyclical nature of Godwin-Austen estate history is unsurprising but also very insistent). In 1996, he followed in ancestral footsteps as a Deputy Lieutenant of Surrey. Dr Richard Godwin-Austen, having risen to become Secretary Treasurer General of the World Federation of Neurologists,[17] in 1994, for the statutory period of one year, was also honoured by being nominated High Sheriff of Nottinghamshire. He bought *his* estate, Papplewick Hall, just outside Nottingham, in 1985. With the drive and skill of his first wife, Jane (d.1996), the Hall and its grounds have been completely, expertly and very beautifully restored.

'The improvement of the estate' has been, therefore, for the Surrey Godwin-Austens over the past five-and-a-half centuries, a very movable feast, largely thanks to historical forces over which they could have exercised little or no control. Many individual Austens, Godwins and Godwin-Austens possessed unusual talents, however, such abilities playing their own part in influencing Haversham's exceptional character, as the next chapter shows.

[17] When I first visited Richard and Sally at their beautiful Southwell home, I was, after a thoroughly enjoyable lunch, taken out to Papplewick Hall. Richard then kindly ran me back to Newark Station. I was not a little delighted to become aware during that drive that I was being conducted through Sherwood Forest by the (then ex-) Sheriff of Nottingham!

CHAPTER TWO

HISTORICAL STRANDS

Thinkers and Men of Action

IN RETURNING TO THE DIVERSE character of Robert Austen, the late 18th-century improver of Shalford Park, we can see the beginnings of a second major trait in the family after the inclination towards estate development: collecting. Robert's coin-collection was said to have formed the nucleus of the Royal Mint; his library contained many works on coins and medals. He was also an antiquarian, an occupation he appears to have inherited from the founding George Austen, who had compiled a 'family book', a work developed by Robert Alfred Cloyne and his eldest son and which has been kept up to date by the present family. He also collected pictures, by all accounts very fine acquisitions subsequently dissipated by Robert Alfred Cloyne then by Haversham when the Shalford House contents had to be sold by auction at the time of the bankruptcy. He was even said to have received the mysterious gift of a Canaletto – left one night on the front lawn, if a family story is to be believed – although no trace of this painting (or who its donor was) now remains.

The landed interests having been consolidated, the Austens now turned to making a distinction in science. One of Robert Alfred Cloyne's friends was Dr Daniel Carlsson Solander (1733-82), a Swedish botanist who accompanied Captain Cook and whose extensive botanical work included the study of land slugs. The connections with Haversham are, again, not hard to find. Haversham was probably one of the greatest collectors of natural history objects, principally freshwater molluscs, bird-skins and geological specimens, in British history. (For a closer description of his collecting, see Chapter 21.) Haversham also collected cornelians and other semi-precious stones, stamp seals and a few other oriental *objets d'art*, mostly while on his Asiatic travels, which were sold and/or donated to the British Museum, where, sadly, they have never been publicly exhibited. (But for the British Museum's redemptive action, see Chapter 12.) Investigations at Surrey Archaeological Society and Guildford Museum also show that, locally, Haversham collected flint arrowheads and Roman remains, particularly pottery, several examples of which contain fragments of the charred bones of luckless Romans.

CARTOGRAPHERS

If collecting was in the blood, so too was surveying and draughtsmanship. John Austen in the 17th century, Robert Austen in the 18th century, Haversham, *par excellence* (see Chapter 8) developed the tradition. It was furthered by Annesley Godwin-Austen (1885-1977), who was the first surveyor of the Queen Charlotte Islands off Canada's west coast; during the First World War he surveyed trench tunnels with the Royal Engineers and was mentioned in despatches. After the First World

War and a formal training in surveying, he joined the Colonial Office, as it was known, and began work, first in Persia (then under a British mandate) followed by East Africa (in Lusaka, at that time in Northern Rhodesia, now the capital of Zambia). From 1931 onwards, his work took him to Tanganyika (mainly modern Rwanda and Burundi) then Cyprus. He rose to become the Colonial Office's Deputy Director of Surveys. Annesley's talent for and knowledge of surveying therefore transcended its only being a method required for defining the nature of his lands and, like Haversham's, was used in public service, mostly in exotic and difficult locations. Happily, now, Stephen Godwin-Austen (b.1962), who in due course will take on the running of the estate, apart from demonstrating technical and aesthetic talents in such diverse interests as film-making and property restoration, is described as an outstanding draughtsman and carries on this well-established family tradition.

Nor was it purely a question of just technical abilities, either: there is an artistic sub-plot, dealt with in greater detail in Chapter 22, where Haversham's considerable talents as a watercolourist are set out. Cicely and Marcia, Annesley's younger sisters and therefore Haversham's nieces, were both accomplished watercolourists, Marcia in a kind of neo-Impressionist style; and it appears that a similar talent has surfaced in Stephen's sister Diana.

SCHOLARS AND ACADEMICS

These antiquarians, collectors and surveyors lead us to one of the most important strands in this family's history: the academics and scholars. The root might be said to have taken hold with John Austen's 1559 co-founding of the Royal Grammar School Guildford,[1] now one of the leading academic independent schools in the country and the *alma mater* of George Abbot, Archbishop of Canterbury. Robert, the son of family co-founder George Austen, became an MA and Fellow of King's College Cambridge while his brother Ralph (sometimes Raffe) also gained his MA and became a Fellow of Magdalen College Oxford. Ralph's unmemorable contribution to scholarship is remembered, according to Bingley, through a commentary on the Book of Job which was 'both dull and obscure'.

The 18th-century antiquarian Robert Austen was known as a lawyer and classical scholar. He was called to the Bar at Lincoln's Inn in 1767 (the letter of appointment still survives in the Godwin-Austen papers) and became one of the six clerks of the King's Bench in Chancery. (He thereby acquired documents from important state trials such as the impeachment of Robert Clive and the quasi-judicial murder of Admiral Byng.)[2] He was elected FSA in 1779. From there, we go forward for about half a century to Haversham's father, Robert Alfred Cloyne. His important contributions to geology and palaeontology are described in Chapter 3. After becoming a Fellow of Oriel College Oxford (from where he had graduated) he became friends with some of the leading geologists of the day – Lyell, Horner and Murchison, and was a pupil of Buckland; one of his papers was remarked on by Darwin himself; he won the Wollaston Medal, the highest honour for a geologist, in 1862,[3] and was elected FRS as early as 1849. His contributions included *The Tertiary-Fluvio-Marine Formation of the Isle of Wight* (1856) and *Outlines of the Natural History of Europe, the Natural History of the European Seas* (1859).

---

[1] An idea of the present and past history of the school can be obtained from its labyrinthine website at www.rgs-guildford.co.uk.

[2] Admiral John Byng (1704-57) built up a reputation as a solid naval officer and received promotion to Vice-Admiral in 1747. Byng is best known for the loss of Minorca in 1756 at the beginning of the Seven Years' War. He was court-martialled and found guilty of failing 'to do his utmost' to prevent Minorca falling to the French. He was sentenced to death and shot by firing squad on 14 March 1757. Voltaire satirised the political (rather than strictly correct judicial) process which led to Byng's execution in his novel, *Candide*, where he famously declared that the execution was 'pour encourager les autres'.

[3] Darwin won this medal in 1859. Murchison in 1864, Lyell in 1866.

Despite some genuine originality of thought (particularly concerning the probability of coalfields in Kent) and a considerable scholarly output, Robert Alfred Cloyne was surpassed, in productivity if not necessarily in major insights, by his eldest son, Haversham, whose extensive bibliography (see the Appendices), running to eight books and about 170 scholarly papers, as well as his extensive natural history collections in geology, ornithology, malacology, conchology and ethnography, remain an awe-inspiring example of scholarly range and applied self-discipline.

Haversham's vast contribution is, however, only one line in the ideas and endeavours running through this talented family.

SOLDIERS

If some of the Austens have distinguished themselves as scholars, many of them have made a historic name for themselves as soldiers. Although remembered as a surveyor-mountaineer-explorer, Haversham was actually in the army for 26 years. The military line appears to start with the unspecified knight who distinguished himself at Guînes (see Chapter 1). From there, we have to move forward several centuries to John Austen (1640-80) who retired from the Parliament Army as a colonel. During the Civil War he fought in the Onslow Surrey Militia as captain, then major despite, according to Bingley, being a natural compromiser by nature: he is said to have expressed an aversion to King Charles's trial and execution. In the post-execution struggles, the Surrey Regiment turned up late for the Battle of Worcester (1651), perhaps on purpose. John Austen also fought in Cromwell's campaign to subjugate the Irish. His second son, Robert, also rose to the rank of colonel (of the militia) and became a Lord Commissioner of the Admiralty (painted as such by Kneller[4]) from 1690 to 97.[5]

GENERAL GODWIN

The Austens then seem to have settled into peaceability for over a century, until we reach General Godwin (1784-1853), one of the most controversial yet able figures in the family's history, and the greatest single influence, after his father, on young Haversham. The controversy begins with his birth: was he the illegitimate son of George IV? Since at the date of his birth the then Prince Regent would have been 22, this seems possible.

The claim has been subjected to much scrutiny, but the results are still inconclusive. The most tantalizing evidence comes in a letter from Alfred, Haversham's younger brother, to Haversham's son and heir Arthur, dated 4 May 1937. In it, he refers to the letter from 'your uncle Godwin', who, at that date, can only have been Haversham's youngest sibling, Harold (1856-1943, therefore about 81 when he wrote). A letter to Arthur from Harold (though signed only 'uncle Godwin') concerning General Godwin's parentage survives, and is dated 29 April 1937. Alfred's letter suggests he has taken some time to reply ('No displeasure, but very busy with business & social engagements ... ') – an intervening letter has probably been lost. Harold's letter is not very accurate: he suggests that General Godwin's mother was Mary Godwin,[6] daughter of the celebrated 'free thinker'.[7] This cannot be right, as Mary Shelley, as she became, was born 13 years after General

[4] Sir Godfrey Kneller, 1st Baronet (1646-1723) was the leading portrait painter in England during the late 17th and early 18th centuries.

[5] Bingley, *Ibid*, p29.

[6] Mary Shelley (*née* Mary Wollstonecraft Godwin, 1797-1851) was a British novelist, short story writer, dramatist, essayist, biographer, and travel writer, best known for her Gothic novel *Frankenstein: or, The Modern Prometheus* (1818).

[7] William Godwin (1756-1836) was an English journalist, political philosopher, novelist and supposed 'free thinker'. He is considered one of the first exponents of utilitarianism, and one of the first modern proponents of anarchism.

Godwin; such wild inaccuracy suggests that the remainder of Harold's assertions should be treated sceptically. Alfred's comment to Arthur is more measured:

> Of course we knew that there was Royal Parentage, but the generally *(sic)* idea, was that General Godwins *(sic)* sire was either the Duke of York[8] or the Duke of Kent.[9]

Rather extraordinarily, however, Alfred continues, 'now we know that it was George IV', yet frustratingly fails to cite any evidence for this. For his knowledge of the general's early days, which were evidently spent at Kensington Palace, Alfred says he is indebted to his godmother and Queen Victoria's confidante, Harriet Phipps.[10] If so, he has taken a long time to reveal her 'secrets', as Phipps died 15 years before his letter was written. Phipps's papers were allegedly burnt at the time of her death. She is, however, known to have shared many confidences with Queen Victoria. The Phipps link is therefore apparently quite a strong one, but leads nowhere.

Alfred goes on to say (in the letter cited above) that General Godwin was brought up by a Colonel Baker or Barker. The Royal Archives registrar, Miss Pamela Clark, failed to find any trace of a Colonel Baker or Barker in the Royal Archives.[11] I have, however, discovered a very likely candidate:

> Sir Robert Barker, 1st and last Bt. was born circa 1732 at St. Anne's, Soho, London, England. He was the son of Robert Barker and Hannah Whitehead. He married Anne Hallowes, daughter of Brabazon Hallowes and Anne Jackson, on 4 November 1780 at Bolsover, Derbyshire, England. He died on 14 September 1789 at Bushbridge,[12] Surrey, England, without issue. He was buried on 28 September 1789 at Hammersmith, London, England. His will was probated in October 1789.
>
> Sir Robert Barker, 1st and last Bt, gained the rank of officer in 1749 in the service of the Honourable East India Company. He gained the rank of Captain in 1758 in the service of the Artillery, at Chandernagore and Plassey. He gained the rank of Major. He fought in the capture of Manila in October 1762. He gained the rank of Colonel in 1764 in the service of the British Army. He was invested as a Knight on 16 April 1764. He held the office of Provincial Commander in Chief at Bengal in 1770, signing the treaty of Pyzabad with the Rohillas against the Mahrattas. He gained the rank of Brigadier-General in 1770 in the service of the Honourable East India Company Service. He held the office of Member of Parliament (MP) for Wallingford between 1774 and 1780. He was invested as a Fellow, Royal Society (FRS). He was created 1st Baronet Barker, of Bushbridge [near Godalming], co. Surrey on 24 March 1781. On his death, his baronetcy became extinct.[13]

---

[8] Frederick Augustus, Duke of York and Albany (1763-1827) was the second son of King George III. From the death of his father in 1820 until his own death in 1827, he was the heir presumptive to his elder brother, King George IV, both to the United Kingdom and the Kingdom of Hanover.

[9] Edward Augustus, Duke of Kent and Strathearn (1767-1820) was the fourth son of King George III and the father of Queen Victoria. He was created Duke of Kent and Strathearn and Earl of Dublin on 23 April 1799, the same year he became commander-in-chief in North America. He was, however, only 16 or 17 at the time of General Godwin's birth; the Duke of York was four years older.

[10] Harriet Lepel Phipps, VA (1841-1922) was a confidential attendant of Queen Victoria. She was the younger daughter of Sir Charles Beaumont Phipps, Keeper of the Privy Purse and also a courtier and confidant of the Queen. Harriet was appointed Maid of Honour in Ordinary to the Queen on 3 March 1862 (giving her the courtesy rank of a baron's daughter), and later served as a Woman of the Bedchamber from 1889 until Victoria's death in 1901.

[11] Letter to the author, 11 June 2010.

[12] The older spelling of modern Busbridge.

[13] Source: www.thepeerage.com, personpage 13217.

Sir Robert would have been the Austens' Surrey neighbour, since his Busbridge estate lay next to Nore (and he would also have been a colonel at the time of General Godwin's birth). The question remains: why would this childless couple (according to their peerage citation) have been given the care of Henry Thomas Godwin? In 1785, Sir Robert and Lady Barker were living at 11 Savile Row: their move there might have been motivated by acquiring an extra family member. Their suitability as guardians, being childless and having a most distinguished career behind them, might have been suggested by George Legge, Lord Lewisham, the third Earl of Dartmouth, from whom they bought the house. George Legge was Lord of the Bedchamber to the Prince of Wales (i.e. the future George IV) in 1782-3, when Henry Thomas Godwin's mother would have been known to be pregnant. It would not have been difficult for this Earl of Dartmouth to suggest to the Prince of Wales – Lord of the Bedchamber being a usefully private post – that the Barkers would make sensible guardians for the child on its way. The connection is extended by Sir Robert having been Commander-in-Chief in India (1769), a post which was subsequently filled by Sir William Gomm (see next paragraph and Chapter 7), General Godwin's lifelong friend, whom he certainly knew well from the days when they played together as infants in Kensington Palace.

The other evidence in favour of the Prince Regent being the general's father is the suggestion of a quite calculated cover-up of the birth. In February 1957, an attempt was made to resolve the matter with the help of the College of Arms. RP Graham-Vivian, MC, wrote that Sir William's father having died young, he was brought up by Miss Jane Gomm and Miss MC Goldsworthy, both of whom were governesses to the daughters of King George III. This information[14] therefore suggests another link between General Godwin, around the time of his birth, and the then royal family.

General Godwin was himself a Gentleman Usher to the Prince Consort (Prince Albert) from 1827 to 'about 1851'. (It cannot have been later than 1851, as by that time Godwin was in Burma: in his role as commander-in-chief in the Second Anglo-Burmese War, see Chapter 6.) Mr Graham-Vivian discovered General Godwin's marriage date but failed, despite it 'normally being an easy matter', to locate the licence which, he speculates, 'might have been destroyed deliberately'. The investigation continues in the 1851 Census where the General claims that he was born at St Martha's.[15] This seems very unlikely, as St Martha's is merely a hill with only what was then a recently renovated chapel on top; it overlooks Chilworth Manor, which Robert Alfred Cloyne was renting at the time. Godwin also claimed to be 60, i.e. born in 1791, which sounds, when combined with the likely false birthplace, like obfuscating tactics. No records of a Godwin family have been found in the St Martha's area. Finally in support of this cover-up theory, we are told that Godwin's commission in the army was purchased when he was 15, thereby suggesting a wealthy but anonymous benefactor, someone so concerned about secrecy that even when Godwin rose to high rank through his merits as a soldier, the benefactor still refused to acknowledge his very able protégé.

In recent correspondence with the Royal Archives, the registrar has tended to pour cold water over this royal connection. Godwin, I was told, is not mentioned in Anthony Camp's *Royal Mistresses and Bastards*.[16] George IV did acknowledge two illegitimate sons and provided them with financial support; the registrar concludes that 'there is no good reason why he should not have done the same for any other such children'. Of course, there *is* a perfectly good reason: the child might well not have been publicly acknowledged if the mother were of a relatively 'low' social standing, such as a governess. The archives' registrar suggests that we are on less firm ground in the not so well-documented lives of 'Prinny's' brothers, the Dukes of York and Kent. The most accurate (one presumes) painting of the Prince Regent does show little resemblance to the general: of the

[14] Letter to RB Godwin-Austen from RP Graham-Vivian, MC, College of Arms, 27 February 1957.

[15] In *Seizing Opportunities*, Dr Richard Godwin-Austen is clear that the St Martha's registers for 1783, '84 and '85 make no mention of him.

[16] A mistress, however, suggests a semi-permanent position whereas we know nothing about General Godwin's mother.

three brothers, he most resembles the Duke of Kent. Weighing the evidence above, I am inclined to believe the story of a royal connection: the circumstantial evidence of the Barker-Gomm connection and the birth obfuscation are too telling[17] to be ignored. But which of the three royal brothers was General Godwin's father is impossible to determine.

General Godwin, whatever his background, enjoyed a most distinguished military career. The Napoleonic Wars had begun and it was an exciting time for a young officer ready to make his mark. After being commissioned ensign on 30 October 1799,[18] he joined the 9th Regiment of Foot (later the Royal Norfolk Regiment, later still part of the Royal Anglians). He was rapidly promoted, by all accounts because of his daring soldiering rather than any supposed influential parentage. He was promoted lieutenant on 9 August 1803 and captain on 28 March 1808. He began his service at Ferrol, where a force landed by a British fleet was repulsed by a strong Spanish defence. In this engagement, we can perhaps see the germs of his idea of combining marine and land army tactics which was General Godwin's major contribution to the development of tactical thought. After a period in Hanover, as adjutant of his battalion, he found himself once more in the Peninsular War (1808-14). He fought in operations leading up to and including the battle of the Douro (1809) in the British victory under Wellesley (as he then was) against Marshal Soult. From there, he was sent to Gibraltar and in October 1810 he volunteered under Lord Blayney to be part of the mixed Anglo-Spanish relief force which embarked on a disastrous landing at Fuengirola, near Malaga. A second relief attempt (of which Godwin was also a member) was made at Tarifa in 1811 but despite the defeat of a detached French force of 15-20,000 men under Marshal Victor, the siege was not lifted. In the following year, he was back in the Portuguese part of the peninsula, where he was severely wounded at the Battle of Barossa. The aim of the raid, which was to destroy the French siege works, was not achieved, but heavy casualties were inflicted on the two French divisions and the British, Portuguese and Spanish force was able to return to Cadiz unimpeded. The British considered the battle a victory. In this engagement, Godwin commanded two flank companies. For his services in the Peninsular War, Godwin was made a Brevet Major and CB.[19]

Evidently, it took time to recover from his wound as no further actions are recorded for him during the Napoleonic Wars. He was, however, on 26 May 1814, appointed major in the 5th West India Regiment. We next hear of him in 1821 where, on 26 July, he is made lieutenant-colonel of the 41st Foot (Welch Regiment), which he takes out to India in 1822 then on to Burma in the first Anglo-Burmese War from 1824 to 1827. In this conflict, he was present at every action, notably the capture of Rangoon, except for a period commanding a detached force where he conquered the province of Martaban in October 1824. (It was the British victory in this war, consolidated by two later ones, which allowed Haversham to be among the first Europeans to penetrate the subsequently conquered northern territories of Manipur and Assam: see Chapters 15 and 16.) After this war, Godwin exchanged to half pay, became colonel of his regiment on 10 January 1837 and was promoted to major-general on 9 November 1846.

Godwin was now an important figure in the Indian Army: he transferred to the General Staff in Bombay in 1850 and on 1 November 1851 was given command of the Sirhind Division. The second Anglo-Burmese War was not long in coming. The politics of that war, and Haversham's

---

[17] Haversham tried to solve the riddle of General Godwin's parentage. Letters to him covering a period of at least four years in the 1890s come from a Staffordshire branch of the family. The suggestion is that General Godwin was the (probable) great-grandson of a General John Godwin and his first wife, a Miss Heywood. That such people existed is indubitable, as the Staffordshire archives make clear: a Captain-Lieutenant John Godwin appears to have fought in the Royal Artillery at Culloden (1746) and in the same regiment at the siege of Gibraltar (1779-83). The connection is not, however, at all clear. It is well-intentioned, but muddled and lacking in definiteness, and cannot be considered reliable. Its main curiosity is that the Staffordshire Godwins of the time also appear to have a Barker among their names.

[18] Source: *Dictionary of National Biography*

[19] The Most Honourable Order of the Bath (formerly The Most Honourable Military Order of the Bath) is an order of chivalry founded by George I on 18 May 1725. The name derives from the medieval ceremony for creating a knight, which involved bathing (as a symbol of purification).

rather belated role in it as his grandfather's ADC, are dealt with in Chapter 6. Martaban was bombarded in 1852; Godwin swiftly captured Rangoon (again) in April 1852 and in June, Pegu, a victory so decisive that 'Pegu' was included on the General's coat of arms. Burma was annexed by Dalhousie's proclamation in December 1852. River operations up the Irrawaddy and some of its tributaries as far as Prome then followed. By July of 1853, the Burmese had staged their own *coup d'état* after which the new rulers came to an agreement with the British. Sir John Cheape replaced Godwin; grandfather and grandson then left for India. Criticism of Godwin's rather aggressive tactics was made by some of his enemies, notably Lord Halifax (see Chapter 7), but it appears in hindsight that these tactics were mostly successful.

Back in Simla after the war, Godwin awaited his KCB, but had been six weeks dead before it arrived. To make amends, Queen Victoria asked his only daughter, Maria, who had been married to Robert Alfred Cloyne by this time for 21 years, how the general's service could be recognised. Maria requested that, as she was the general's only child, the name of Godwin might be joined to that of her husband. Victoria complied with this request by letters patent on 24 October 1854, after which the family was known as Godwin-Austen. The general's portrait (see plate 3) and his medals survive. A retrospective assessment of his career, in the light of the enigmatic inscription on his tomb, is given in Chapter 7.

HAVERSHAM THE SOLDIER

Haversham's own military career lasted for 26 years, from being commissioned as a subaltern on leaving Sandhurst in 1851, until his retirement as lieutenant-colonel in 1877. Much of this time was of course taken up on survey and exploration work. His diaries say little about military action in Burma, largely because General Godwin had achieved victory by the time of Haversham's delayed arrival following his thigh-breaking accident in Calcutta en route from England to Burma. His services were, however, sufficient for him to be awarded the Pegu Medal with Clasp. He seems to have seen more definite military action a little later in his career when his regiment was required to guard rebellious native troops in Amritsar during the first winter of the Indian Mutiny. The assault he suffered (see Chapter 9) the following year (1858) was probably a consequence of the mutiny. Thereafter, in 1864, he was appointed ADC to Brigadier-General Dunsford at the sieges of the Bhutanese-occupied settlements of Dalingkot and Chamurchi. There was also at least one skirmish (1873) with the Nagas in the Burma and Assam years during which shots were exchanged. Other than that, Haversham's considerable energies were engaged elsewhere.

BROTHER OFFICERS AND DESCENDANTS

This period of imperial expansion marked the heyday of Godwin-Austen military endeavour. Two of Haversham's brothers, Godwin (known as 'Old Godwin Godwin' because he had been christened 'Godwin' before the family surname change) and Frederick ('Freddie') were both in the 2nd Battalion, 24th Regiment and fought in South Africa, chiefly in the Zulu War (1878-79). Godwin, by then 34 and a captain, commanded 'B' Company. He was severely wounded during operations near East London, Eastern Cape Colony (as it was then known) and was therefore not present when his company, commanded by Lt Gonville Bromhead, fought the battle of Rorke's Drift from late on 22 January and throughout the night. There, the 24th Foot won seven out of eleven Victoria Crosses, the largest number ever awarded to any one regiment in a single action. Though Godwin was absent because still recovering from his wound, his dog, Pip, had been adopted by the Company and is depicted taking an active part in the battle in Alphonse de Neuville's painting.

Earlier that day, and about ten miles away, companies from the 1st and 2nd Battalions, 24th Regiment, and other units, fought the Battle of Isandhlwana, where they were all but destroyed in that famous military humiliation, generally regarded as the first time a technologically inferior army had defeated a modern British force. Nearly 1300 men were killed by the Zulus, including about 800 Europeans, all within two-and-a-half hours of fighting in the afternoon. Godwin's younger brother Freddie was a 25-year-old second lieutenant in 'G' Company. He and his friend, Lieutenant Charlie Pope (commanding 'G' company), were among the last to be killed. A Zulu report recalled their heroic end – one officer had an empty revolver, and was shot, the other was killed, while continuing to fire, by a Zulu assegai. The most chilling detail in the Zulu report was that both officers were said to have had 'pebbles in their eyes' – both wore monocles, the only officers so to do. Freddie's valiant death is commemorated in a stained-glass window at St Mary's Church, Shalford.[20] This was by no means the last the Godwin-Austen soldiers would see of Africa.

In the 20th century, General Sir Alfred Reade Godwin-Austen, KCSI, CB, OBE, MC ('Uncle Reade', 1889-1963), one of the most likeable of all the Godwin-Austen soldiers, having brilliantly masterminded the British Contingent's withdrawal in 1940 through British Somaliland in the face of overwhelming Italian opposition, thereby went on to gain the remarkable accolade of Churchill's 'most detested general'. (He was also a brilliant mimic of his brother officers and of the 'insufferable' Montgomery, under whom he had served in Palestine.) Six months later, he had the satisfaction of returning to Somaliland, then Abyssinia, in command of the 12th (African) Division, to clear the invading Italians out of these countries. In August 1941, Godwin-Austen went on to command the XIII Corps of the Eighth Army in North Africa and was thus able to deliver the memorable despatch, 'Tobruk and I both relieved'. He resigned shortly after, following a dispute over command. A divisional commander, Francis Tuker, who was caught between Godwin-Austen and General Ritchie[21] later wrote:

> His *(i.e. Godwin-Austen's)* going was the latest of many misjudgements which had started to shake confidence in the leadership. We lost the wrong man.

This high opinion was corroborated by Corelli Barnett in *The Desert Generals*:

> *(the)* Eighth Army lost an able, strong general and a much-loved man.

General Reade (whose career accomplished a great deal more than this very brief summary mentions) brings us up to date with this line of family soldiers (other than with my equally brief reference to Lt Col Robin Godwin-Austen in Chapter 1), Haversham himself being one of the most distinguished.

Looking back over these various characters who have played their roles in the centuries-long Godwin-Austen drama, we are reminded of the ancestral forces which shaped Haversham's extraordinary life: the enthusiasm for foreign adventures, the daring of the explorer, the courage and spirit needed to live an isolated life in distant and largely unknown country, the ability to get on with men and women of lesser rank, indeed of all ranks; the sense of social responsibility; the interest in nature and natural history; the preoccupation with collecting; the dedication to systematic, academic thought and scholarship and the wish to 'improve the estate', very often spread out over lives of great longevity. It is therefore hardly surprising that Haversham's story is as varied and as colourful as the following chapters show.

---

[20] For some of the detail about this battle I am indebted to Guildford local historian Margaret Dierden and her article, 'A Fearful Conflict'.

[21] Lieutenant General Neil Ritchie (1897-1983) fought in France and Mesopotamia in the First World War. He won the Military Cross in 1918. After Dunkirk, Ritchie served as chief of staff under General Auchinleck, who became commander-in-chief of British troops in the Middle East in July 1941. Four months later he appointed Ritchie as head of the Eighth Army, although Ritchie's last command had been as leader of a battalion in the First World War. Ritchie was replaced as head of the Eighth Army by Montgomery on 25th June 1942.

CHAPTER THREE

GRANDPARENTS, PARENTS, BROTHERS, SISTERS

Sir Henry Edmund

The Austens may have founded their fortunes on sheep-stapling, but by the time of Haversham's grandfather, the genial but headstrong Sir Henry Edmund Austen (1785-1871, the second Austen knight in the post-medieval family), their social rise had been assured. Sir Henry was appointed deputy, then lieutenant, then magistrate of Surrey, and became a sheriff in January, 1810. At court, he obtained the rather exciting privilege on 9 January 1832 of being named one of the 'Gentlemen of the King's Most Honourable Privy Chamber in Ordinary'. Although this was an honorary post, it must have been worthwhile during the year of the Great Reform Bill, which Sir Henry strongly supported, to have had William IV's ear at this time of constitutional crisis.

This dedicated estate-improver, in one portrait (see plate 2), resembles several Austens, including Haversham, being good-looking and bright-eyed. His smile shows a comfortable, almost Pickwickian complacency. He clutches his cane like a field-marshal's baton, with a lordly self-confidence, while something approaching playfulness crinkles round his mouth and eyes. Thanks to the family fortunes increasing during his long life, mainly because of his profitable interest in forestry, Sir Henry enjoyed a reputation as a local humanitarian and benefactor. As the tombstone inscription in the family vault states:

> Sir Henry E. AUSTEN, interested in forestry, and planted largely on his estate;
> he also knew the value of maps, and had excellent ones of his property.

... an obvious source of his grandson's unusual talent for map-making.

Sir Henry Edmund was a genuine Liberal, one of several such Austens: as well as the Great Reform Bill, he is known to have supported Catholic emancipation and the abolition of slavery – and in 1839 he objected to the proposal that unmarried women in the workhouse be compelled to wear yellow stockings![1]

This expansive knight gained substantial properties when he was only 12: on Robert Austen's death in 1797, Sir Henry inherited his estates of several thousand acres. The estate crossed the parishes of St Martha's, Bramley, St Nicholas, Holy Trinity, St Mary Godalming, Dunsfold,

[1] I am much indebted to Margaret Dierden for these interesting snippets about Henry Edmund.

Hascomb(e), Chiddingfold, Cranleigh and Ewhurst, all prime Surrey land, and nowadays enclosing one of the wealthiest definable areas in Europe. The exact amount of land is hard to calculate; my best estimate is 3-4,000 acres. In 1797, the Shalford Manor Farm estate alone comprised 434 acres. A *Times* obituary for Arthur Godwin-Austen in 1948 mentions 1,800 acres. However, this territory had, frustratingly, been entailed on the instructions of his father and was run by trustees because Sir Henry had married just short of his 21st birthday.

His main residence, Shalford Park – no longer just 'House', having been expensively extended and modernised in the Adam style[2] – represented the Austens' rocketing social rise: the house's Tudor brickwork had been plastered over (see plate section) and an extra storey added. Pevsner tells us that it became, 'effectively all heavy stuccoed Classical of 1797. The main (north) front of nine by three bays has an elegant porch of four Ionic columns shielding a fanlight, up to the best Dublin standard.'[3] Other notable features were a large, semi-circular bow-fronted drawing-room containing a fine 1609 (possibly 1601) fireplace, with a music room above it. Stables, entrance lodges and a summerhouse on the Monks' Walk were added later.

The garden was laid out and an avenue of limes planted. It perfectly reflected the taste of the time for the picturesque, for nature rearranged to look more intensely 'natural', since it enjoyed fine views across parkland to the water meadows of the River Wey, and beyond there to the Gothic ruined Augustinian abbey of St Catherine's.

Sir Henry, however, did not live in Shalford Park for much of his life. By 1820, he was spending his time at Versailles, which was much cheaper (and probably a good deal more fun) than London or indeed Paris. Shalford Park was let for £300 a year during much of Sir Henry's voluntary exile after the restoration of the French monarchy in 1814.

Sir Henry's French sympathies were obvious from an earlier part of his life. One curious story attached to Shalford Park in Sir Henry's pre-Versailles days is that he harboured French exiles, perhaps even a French king-in-waiting, from the Terror: such an action would at least be in keeping with his Liberal principles. The story is told by a local paper-mill worker:

> On a spring day in the year 1793 or 1794, a stranger presented himself at the little mill and, showing my grandsire a blank note with certain water-marks, inquired if he could undertake a quantity of exactly similar paper. The bargain was concluded on the spot and time for delivery agreed upon which the stranger was asked to leave his address but he replied as he was continually travelling he would return at the period agreed upon to receive his paper, and he departed, leaving a heavy sum as deposit. A few days after the expiration of the stipulated time he returned in a post-chaise and, being satisfied with the perfect imitation of the water-mark, he had the paper placed in his carriage, with which he departed, leaving another for a similar order. These transactions were several times renewed, until on one occasion the unknown required for a new order various changes to be made in the water-marks. The 'former' or 'form-maker' of the mill was sent for. He was a young man named Longhelt, a native of Germany, with whom I remember to have talked in my youth. The stranger explained to him the alteration he wanted to be made, and sat down by his side for the purpose of seeing him begin his labour. Longhelt, who had been drinking, resented the intrusion, and getting impatient at the stranger's interventions, he waxed furious and threw the form at the visitor's head. The latter went to complain to my grandfather, who advised his customer not to attempt the supervision till the morrow, when the modifications were successfully made. Further alterations to the form were continually made, but after

---

[2] I am indebted for a good deal of the anecdotal family information in this chapter to Lt Col Robin Godwin-Austen, in personal conversation, and to Dr Richard Godwin-Austen, in personal conversation and from his book, *Seizing Opportunities*, The Memoir Club, Durham, 2008.

[3] Ian Nairn & Nikolaus Pevsner: *Surrey*, p452; Penguin Books, Second Edition, 1971.

> some time the stranger departed to be seen no more.
>
> Some time after this my grandfather learned that this mysterious patron was no other than the Comte d'Artois, afterwards Charles X, king of France and Navarre, and that the note paper he had manufactured was for the purpose of being converted into false assignats,[4] with which France was at that time inundated.[5]

Sir Henry is not mentioned as such in the story, and there is no independent evidence, although the Comte d'Artois is most unlikely to have been resident at Shalford Rectory without Sir Henry knowing and approving of it. The truth about the Austens and the 'French exiles' will probably now never be known, although at least one rather sycophantic letter of gratitude by a French priest for providing some sort of safe haven confirms that this part of the Austens' long history has some basis in reality.

In 1805, aged 20, eight years after having succeeded to the *Austen of Shalford* title but not having quite attained the necessary age of independence from the estate's trustees, Sir Henry married Anne-Amelia, daughter of Robert Spearman Bate, one of the directors of the Honourable East India Company (HEIC). This Austen-Bate marriage has been generally seen as 'imprudent': it led to the entailment of the Austen estates in a way that, 90 years later, made it difficult for Haversham to stave off bankruptcy by selling land.

Not much is known about Haversham's grandmother, Anne-Amelia, other than that she died at Dieppe on 5 September 1837, aged 53, while returning from the Pyrenees. The inscription above her tombstone in the family vault in Shalford church tells us that she had been visiting 'the baths which she had been using for the recovery of her health'.

This HEIC connection probably marks the first link with India in Haversham's background. It was hardly surprising as, thanks to Robert Clive's victories at Calcutta and Plassey in 1757 – after which he acquired fabulous wealth – the company was expanding at a spectacular rate. (At this time, another director of the HEIC was Richard Chicheley, or Chichele, Plowden, a descendant of whom became Haversham's first English wife.)

Sir Henry died in Cheltenham, aged 76. His funeral at Shalford was a grand one:

> The hearse, drawn by plumed black horses, brought the coffin covered with his personal hatchment to the central front door of the entrance façade. His son was, 'so pleased with the effect that that he commanded the funeral procession to "go round again" so that he could enjoy the scene proceeding round the gravel path in front of the house ... "and round again".'

Sir Henry was, however, still very much alive when Haversham was born on 6 July 1834 at Ogwell House in Devon. Although he spent his first four years there, then at Chilworth, a rented manor next to the Austen Surrey estates, Haversham would, given his grandfather's wealth and ancestry, have been brought up as a country gentleman, and a Surrey one at that, i.e. not quite a provincial. In their *Introduction*, pp22-3, to their *Surrey* volume (1971) of *The Buildings of England*, Ian Nairn and Nikolaus Pevsner argue a strong case for Guildford (Shalford is one of its satellite villages) deriving its style – not just architecturally – from London. Haversham would have been well aware that he was the eldest child of his generation in a remarkable old family, and therefore an heir who could enjoy the privileges of wealth, influence and social distinction with even royal connections. With such an awareness, however, he would also have learnt of the responsibilities which went with managing land, property, money, servants and tenants.

Haversham's other grandfather was, of course, General Godwin. His influence, until his

---

[4] *Assignats* were paper money issued by the National Assembly in France during the French Revolution. They were issued after the confiscation of church properties in 1790 because the government was bankrupt.
[5] Source: http://www.weyriver.co.uk/theriver/industry_8_papermaking.htm.

sudden death in India, over his grandson's military, social and moral conduct was enormous. It is fitting that Godwin's distinguished military career, and the way he looked after Haversham while still quite a vulnerable young cadet at Sandhurst and his ADC in Burma, is therefore described in the chapters which enclose this one.

If these general influences are clear enough, then the detail of Haversham's early life and formal education are correspondingly uncertain. We do know, and in some detail, that his father, Robert Alfred Cloyne Austen (1808-84), had become interested in the new sciences of geology and palaeontology while at Oxford, largely thanks to the influence of his mentor, William Buckland.

GEOLOGY AND PALAEONTOLOGY

These new sciences were generally not much more than fashionable activities in the late 18th and early 19th centuries. At first, they were enjoyed only by amateurs, but gradually became scientific disciplines as the importance of fossils to geology and biology became understood. Interest was highly stimulated by the discoveries of Mary Anning (1799-1847) in Dorset in 1811, when she found the first complete skeleton of an ichthyosaur, her brother having discovered the skull of what appeared to be a large crocodile a year earlier. It was an important find, and was soon described in the *Transactions of the Royal Society*. Mary Anning's discoveries became key evidence for the new theory of 'extinction', an idea which contributed to early thoughts on evolution. It was Buckland's proposal in 1829, based on observations by Mary Anning, that oddly shaped fossils then known as 'bezoar stones', which were sometimes found in the abdomen of ichthyosaur skeletons, and which when broken open often contained fossilized fish bones and scales as well as bones from small ichthyosaurs, were fossilized faeces; Buckland named them 'coprolites'. Such a discovery helped to reconstruct primitive ecosystems; it also suggested that creatures such as ichthyosaurs were long extinct because they did not much resemble any living species – rather than simply continuing to live on parts of the planet as yet undiscovered – and therefore became one of the earliest significant steps in the development of palaeontology in itself and as a prop to the Theory of Evolution.

HAVERSHAM'S PARENTS

Both of Haversham's parents were striking to look at. Robert Alfred Cloyne's slightly pinched Victorian severity, rather resembling a tightened-up version of Thomas Hardy (including the moustache), could not hide the Austenian twinkle in his eyes. Maria enjoyed a Winterhalter style of classical beauty, as a vast family portrait at Papplewick makes clear (see plate section).

Robert Alfred Cloyne was an original thinker, a considerable scholar, a distinguished scientist and collector and a responsible manager of the Austen estates. He was educated at Midhurst Grammar School, Sussex (founded 1672, closed 2008. Sir Charles Lyell (1797-1875), another great geologist of the period, was also educated there, as were HG Wells and Ian Serraillier). He was then sent to France, to a 'semi-military' school, or to the 'military school in Paris', probably the St-Cyr military academy. In 1826, he entered Oriel College Oxford, of which, shortly after graduating BA in 1830, he became a Fellow. He then spent some time as a student at Lincoln's Inn. Among his early friends were Charles Lyell and fellow geologists Leonard Horner and Roderick Murchison, who sponsored, also in 1830, his election to the Geological Society of London.

Haversham's father is mainly remembered among scientists for his contributions to our understanding of the geological character of southern England, and as one of the first European palaeo-geographers, as well as for his work in defining the structure and marine biology of the English Channel and for his prediction that a coalfield would be discovered beneath the younger rocks of Kent. His first original work was devoted to the limestones and slatey rocks of south-east Devon. Henry de la Beche, who in 1835 founded the Geological Survey of Great Britain, encouraged the young man by relying upon him for the geological lines on the map covering the district between Dartmouth and Chudleigh, an approximate reflection of the relationship between Captain TG Montgomerie and Haversham in the Kashmir Survey 30 years later.

After 1840, when he moved to Chilworth Manor, near the Austen estates in Surrey, Robert Alfred Cloyne began to devote his energies to the local geology (although he also evidently knew enough about Anglo-Saxon history to give a public lecture on it at the Guildford Institute on 30 October 1851).

He also worked on the geological structure of the Weald, the chalk-land strip running across south-east England between the North and South Downs. An original idea began to form, of ancient seas advancing and retreating over western Europe. To follow this up, Austen visited the coalfields of northern France and studied the geological structure of the Ardennes. In 1856 he produced what remains his best-known paper, suggesting a possible cross-Channel extension of the continental coal measures to beneath south-east England. He maintained that the coal-bearing strata of England, France, and Belgium were once part of one continuous formation. Austen's views attracted high-level interest. In a letter from JD Hooker to Charles Darwin on 4 August 1856, we see that he is making a name for himself:

> Have you read Austen on the possible extension of the coal fields? that *(sic)* does appear to me outgeologizing Geology, with a vengeance!

(Hooker is referring to the pages printed and corrected by Edward Forbes before his death in 1854 for his book on the natural history of the European seas. This text was subsequently edited and continued by his friend, Robert Alfred Cloyne, and was published in 1859.)

In his last paper, published in 1879, Robert Alfred Cloyne was still advocating these views. They were not, however, proved (and on this occasion Hooker was proven wrong) until six years after his death, when a borehole drilled at the foot of Shakespeare Cliff near Dover showed the existence of the predicted coal measures beneath the chalk and led to the development of the Kent coalfields; these turned out to be sufficiently extensive to be worked until the 1960s.

Austen was, moreover, a pioneer of the history of the English Channel and was among the first marine geologists. He was elected a fellow of the Royal Society in 1849. In awarding him the Wollaston Medal of the Geological Society of London, the highest honour for a geologist, Sir Roderick Murchison declared in 1862 that he was, 'pre-eminently the physical geographer of bygone periods', a description amply justified by his essay on the European seas. Haversham, therefore, if his later scientific work is anything to go by, must from an early age have known about scientific professional methods, standards and achievements and realised that he could count some of the finest scientists in England among his acquaintances.

Given this background and early rapport between father and son, it is not surprising that relations between Robert Alfred Cloyne and Haversham remained generally affectionate and only occasionally strained, probably because the two men possessed so many similarities. They were, however, politically opposed – Robert Alfred Cloyne, like his father Sir Henry, being a Liberal, while Haversham was an active Conservative.

But we can say with confidence that Robert Alfred Cloyne's work, interests and support played a powerful formative role in shaping Haversham's character and career. He supported his son, for example, by travelling up to London on 11 January 1864, despite not being in good health, to hear Haversham's reputation-making paper *On the Glaciers of the Mustakh Range* being read at the Royal Geographical Society. His active exploitation of powerful social connections in developing Haversham's army career and in guiding him through the turbulent waters of his marriage to Pauline in the 1860s show just what a strong influence this father exerted over his extraordinary son. There were, however, some bumpy moments in the relationship, mostly connected with Haversham's uncertainties about life in the army at quite an early stage in his service in India and his hasty decision to marry Pauline.

It was mainly William Buckland's teaching which had inspired the young Robert Alfred Cloyne. To further his studies, he had moved to Devon, where most of the major discoveries in these new sciences were being made, very shortly after having made a socially very suitable marriage to Maria Elizabeth, the only daughter and heiress of Lieutenant Colonel (as he then was) Sir Henry Thomas Godwin, KCB, on 23 July 1833, at the fashionable church of All Souls, Langham Place, St Marylebone. (Despite the interesting vagueness about who Godwin's father was - see Chapter 2 - he was clearly not a man without a fortune.)

We know regrettably little about Haversham's mother, Maria Elizabeth Godwin, the general's daughter and heiress. She was born on 15 July 1813 and died aged 90 on 12 January 1904. While she undoubtedly possessed a long-life gene, evidence of royal descent[6] through the longevity idea is inconclusive: her father was 69 when he died in India of influenza, or perhaps typhoid fever with complications, and George IV, perhaps her grandfather, was only 67. Neither parent of George III (who lived until he was 81) was long-lived, although his grandfather, George II, reached 77. (It is, however, curious that the English wife of Haversham's son by Kudikji,[7] Emma Theresa Smith, lived until she was 96, from 3 September 1861 to 22 February 1958. She had, of course, no previous blood connection with the Godwin-Austen family. Haversham's younger brother, Chads, moreover, married Elizabeth Ann Warnes, again without any blood connection, and she lived until 1955, having attained the age of 100 years and one month!)

RECORD-BREAKING SIBLINGS

Easily the most remarkable aspect of Maria Godwin's gene surfaces in the long lives of Haversham and his 17 siblings. Their ages make astonishing reading:

Haversham (89), 1834-1923
Giuliana[8] *Maria (92), 1835-1927*
Sophia (85), 1836-1921
Helena (101), 1837-1939
Charlotte (104), 1839-1943
Victoria (102), 1840-1942
Beatrice (92), 1841-1933
Caroline (80), 1842-1922
Alfred (90), 1844-1934
Alice Mary (88), 1845-1937
Elfreda (78), 1846-1924
Robert (22), 1847-69
Emily (98), 1849-1947
Edward George (84), 1850-1934
Arthur (died one day after birth), 1852
Frederick (25), 1853-79 (killed at Isandhlwana)
Edmund Chads (77), 1854-1933
Harold (87), 1856-1943

This list contains three centenarians, four nonagenarians and six octogenarians, with the total age of the siblings being 1394 years, almost certainly a British, perhaps even a world, record. (A written family source, apparently correct in most other respects, states that Caroline died in 1933. I have not been able to verify this, or the 1922 (family tree) death date from any other source. The total Godwin-Austen sibling age total could therefore be increased by 11 years. Also, some birth months

---

[6] See Chapter 2.
[7] See Chapter 9. Haversham says in a late letter that the name could be written 'Kudji' or 'Kudikji'. In family correspondence, the spelling 'Kudidji' surfaces and appears not to be correct. My preferred spelling is as above.
[8] The wrong spelling of 'Guiliana' survives in at least two family documents.

are not recorded and could therefore make the total be incorrect by up to six years.) Maria must have been pregnant for 162 months out of a total of 269, from November 1833 (nine months before Haversham's birth in July 1834) to March 1856 (Harold's birth month), well over half of these 22 years and five months being spent bearing children.

Apart from bringing this large brood into the world, Maria was responsible for the family's distinctive name change to Godwin-Austen. General Godwin died suddenly in 1853, and was buried in the fashionable Raj hill-station of Simla, only four months after his final victory over the Burmese. Sadly, the notice of his honours of KCB and being declared colonel of his regiment only arrived some time after his death (see Chapter 7); the honours were therefore received at court on her father's behalf by his daughter Maria. On meeting Queen Victoria, she was asked if the Godwin name could be honoured in any other suitable way. Maria replied that, as she was the last to bear the Godwin name, it might be kept on by being united with the name of her husband. Victoria granted this request by royal letters patent on 17 October 1854. It has been asserted[9] that Haversham did not use his new double-barrelled name until 'well into the 1860s'. In fact, a letter of his as early as 1855 is signed 'Godwin-Austen', so there can be no argument about any resistance to his change of name.

With so many younger brothers and sisters to look after, Haversham would therefore have added to his early experiences a sense of family responsibility and perhaps of knowing where he stood in crowds. His dutiful letters home, his looking after his unmarried sisters in the 1880s and 1890s when he was approaching bankruptcy, and his securing employment and advancement for his youngest brother Harold, all testify to a man who took his 'eldest son's' responsibilities very seriously.

Haversham's eldest brother, Alfred (because he had been christened Alfred Godwin before the name change, came to be known in the family as 'Old Godwin Godwin') began life in conventional circumstances before going out to fight in the Zulu War as captain of a company in the second battalion of the 24th regiment. He was wounded in Natal and was therefore invalided out of action when the shocking Zulu attack came, on 22 January 1879, at Isandhlwana. Only a few tens of British troops survived out of several hundred men; Lieutenant Freddie (from 'Frederick', not to be confused with Alfred) Godwin-Austen was, however, not one of them. Quite what Alfred thought about the loss of his younger brother, when but for a chance wound he would probably have died too, is not recorded but quite easily imagined. (The story of Alfred and Freddie is told in more detail in Chapter 2.)

When Alfred married after his retirement, however, in 1885, his wife became known in the family for her 'inability to take a firm view or decision about anything'.[10] Probably as a form of domestic escape, Alfred adopted a racy about-town style, ('he dressed flamboyantly and carried a silver-topped cane')[11] and visited entertainments judged 'vulgar', such as the music hall. As this lively character – there is something of Edward VII's raffishness about him – felt himself more and more alienated from his feckless wife, he began to take up with the children's governess, Annie Bentley. Unfortunately, Annesley, his eldest son, discovered the relationship when he was only 18, in 1903, and confronted his father head-on about it. By all accounts, a spectacular row ensued and Annesley was thrown out of the house. The two did not speak again until Alfred was induced to attend Annesley's wedding in 1931, when he was 87.

Haversham's sisters were also strong characters. Charlotte, for example, when aged about 90 (depending on which family story you go for) broke her hip when she fell over then down some steps while walking in the orchard at the Manor House in Pirbright. No-one was around to answer her cries for help; she, with the kind of ferocious determination which Haversham was later to exercise along the world's longest glaciers – and in a remarkable reflection of the aftermath of his

---

[9] Mason: *Abode of Snow.*
[10] Richard Godwin-Austen, *Seizing Opportunities*, p17.
[11] Richard Godwin-Austen, *ibid*, p18.

femur-breaking accident in Calcutta – crawled back into the great house, but took the [illegible] round in order to avoid the kitchens! (She remained upstairs, bedridden, for the ensuing 14 or so years of her life, the longest-lived of any of Haversham's siblings, and is said to have died cursing Hitler.)

The happier occasion of Charlotte's wedding in 1866 describes a stage in the ever-changing relationship between the Austens and their locality. The *Surrey Advertiser* reports the village bunting showing 'the widespread respect for the young lady's family'. A local builder built a nuptial archway over the south-west end of Shalford's main street; it was decorated by Shalford House's own gardener and a local farmer. On the side of the street which the bride would first see, banners were erected which amiably but uninspiringly declared 'Shalford Greets You', then on the other side 'We give you joy'. Beyond the arch, the village houses were dressed with banners, flags, wreaths and garlands.[12] The path from the church gate to the porch was lined with schoolchildren holding flowers. By the 1890s, as financial disaster glared at the Godwin-Austens, such a jolly country scene would become much harder to create.

Among the more junior brothers and sisters, we come to Chads, the founder of the American part of the Godwin-Austen dynasty. His unusual name seems to have been inspired by his godfather, Admiral Sir Henry Ducie Chads, GCB (1788-1868), who saw action from the Napoleonic Wars to the Crimean War. Chads's wanderlust began in 1873, with an attempt in New Zealand to take up the ancient Austen work of sheep farming. He met his future wife, Elizabeth Ann Warnes, on the ship out: they were married shortly after arriving. After three less-than-successful years in New Zealand, they boarded the steamer for San Francisco, then took the railroad via Santa Fe to Las Vegas. Along the way, they probably made a diversion to Montana then briefly to Canada. By 1880, they had reached Liberty, Texas. From there, they travelled by covered wagon to the Texas panhandle and reached Tascosa, at one time a rival to the notorious Dodge City. They set up home on the LX Ranch, thereby completing an odyssey since leaving England of almost 30,000 miles. This free-spirited couple then began to live an authentic 'Wild West' life, virtually a cliché of innumerable Hollywood cowboy films. One incident in 1886, described by Chads in a letter home, sums up this kind of 'B' movie life:

> Feuds occurred, and there was not much to quarrel about but cows. Unfortunately, four good men lost their lives, one of them a disinterested party, Mr Jesse Sheets. The other participants were Ed King, Frank Valley, Fred Chitton, Len Woodruff, Chas Emery, Louis Bausman and the Catfish Kid (John Gough). The object of the affair was Ed, who had been an inspector, the others having rushed into their death through being his friends. I had been recently appointed Justice of the Peace, but was not qualified. It was not a desirable office in those days; I had been forced into it as there was to be an inquest. (That reminds me of a former Justice of the Peace when there was a celebration going on in one of the saloons. He ran in, crying out, "Kill 'em, boys! Kill 'em! I'm the coroner and get $2 a head!"). All this occurred on a Saturday night and Sunday morning. HH Wallace came to my house at 2am, calling, "Get up quick! There are four dead and I don't know how many wounded!" I got up and went down town; things looked threatening. There were two factions, and the sheriff had enough to do to prevent further trouble. He appointed lots of deputies, and no further trouble occurred. All went to jail bar Len Woodruff who had been wounded most of all, but he managed to get as far as Mr Briggs's ranch after stealing a bottle of brandy from a boy who was taking it to Mrs Briggs, supposedly sick.
>
> Sunday was the same as any other day; I was sworn in as Justice of the Peace, a jury summoned, and off we went. There were forty-two witnesses. It was the most

[12] *Surrey Advertiser*, 7 July 1866. I am most grateful to Margaret Dierden for this helpful illustrative reference.

> wonderful court ever held. Deputy Sheriffs all round, armed to the teeth. The trial ended with a hung jury. No sooner had the case ended than the Catfish Kid went back to Tascosa and deliberately murdered some inoffensive old man, for what, no one knows. For this he went to the pen, for some sixteen years. This is not a bad record for an outlying place like Tascosa: many have worse.[13]

Evidently, all this excitement proved a little too interesting for Chads and Elizabeth. A later part of his memoir tells us that they retired to Channing, Texas, where Chads died in 1932, the reasonably contented owner of an orange farm.

Of Haversham's other brothers and sisters, perhaps the strongest remaining mark was made by the youngest, Harold. Haversham secured a position for him in the Indian civil service. He was promoted to Assistant Commissioner of the Andaman Islands, a long way down the Bay of Bengal, and spent 13 years there, probably from early 1881 to about 1893. He was certainly there on 31 December 1881 when a massive earthquake struck these islands. It is not known what exactly its force was on the Richter Scale, only invented in 1935, but it was evidently enough to cause widespread structural damage. Harold reported the details to Haversham, who published them in *Nature* magazine 25, (23 February 1882):

> I see among the Notes in your last issue (p. 325) that there has been widespread seismic disturbance in Asia, including Ceylon, but unfortunately in no instance is the date given, which would have added very greatly to the value of the record. It may be interesting to give an extract from a letter I have just received from my brother, Mr. Harold Godwin-Austen, from Port Blair, Andaman Islands, which very probably was connected with the disturbance in Ceylon, and if so, it covered a very considerable area of the earth's surface, the distance being about 750 miles between the two places:
>
> 'Port Blair, January 2. We had a very bad earthquake here on December 31, 1881, at 7.52 a.m. I thought the place was going to pieces. There has been a good deal of damage done to work and pucca *(brick)* buildings, and we had five high and low tides in three hours after the shock, and the sea did not quiet down all day. Since then we have had two or three slight shocks.'

Not only would the earthquake have been intriguing to a geologist to whom plate tectonic theory was unknown, as Alfred Wegener's work of 1912 was still some 30 years away, but Haversham could also not have failed to make comparisons with his own frightening yet scientifically interesting experience of a strong earthquake in Cachar, northern Burma on 10 January 1869 (see Chapter 16).

While Harold's posting to the Andaman archipelago might conjure up images of idyllic, palm-fringed tropical islands, the reality was actually very unpleasant. Many of the natives were still resisting western visitors by as late as 1975, where one attempt at friendliness by a visiting film director was met with an arrow in his thigh. Much worse was the islands' history as a British penal colony (founded in 1789, although it fell into disuse following high mortality rates from disease after 1796) and, eventually, the world's second concentration camp. Although used as a fleet rendezvous for the Burmese War, reconstruction, using the prisoners as slave labourers, did not begin until 1857. The colony became notorious in Britain following the murder in 1872 of the Viceroy of India, Richard Southwell Bourke, while on a visit to the settlement, by a Muslim convict, an Afghani Pathan called Sher Ali.

From the time of its development after 1858 and following the Indian Mutiny which had begun in the previous year, this settlement, known as 'Kala Pani' (black water), became primarily a

---

[13] Letter to Mr J Evetts Haley from Edmund Chads Godwin-Austen, Channing, Texas, 4 August 1925.

dumping ground for political prisoners: Harold would have confronted some dreadful human suffering. And the criminal character of the islands may link to the writing of Sir Arthur Conan Doyle: the Sherlock Holmes novel, *The Sign of the Four* (sic), published in 1890, features a villainous Andaman Islander; it is one of several of Conan Doyle's stories to introduce unsavoury native characters from the Andamans. The link comes about through Conan Doyle having lived not far from the Godwin-Austens, from 1891 when he was in London, then as their neighbour after moving into Undershaw, at Hindhead, about eight miles from Shalford, in 1897. This link is, however, highly circumstantial.

More exciting as far as I am concerned was Harold's invitation from the great but enigmatic explorer, Ney Elias (1844-97), to join him on an expedition to Yarkand which took place in 1880. (Yarkand, coincidentally, was the fabled Chinese Central Asian trading city which Haversham had tried to find the way to when attempting the Mustagh Pass in the Karakoram in 1861; it was on the intended route – subsequently wrecked by rebellious Kyrgyz camels – of my first expedition to Central Asia in 1989.) Harold had probably been introduced to Elias through Haversham at the Royal Geographical Society. Haversham was elected FRGS in 1859, Elias in 1865; Elias received the Founder's Gold Medal in 1873, largely for his work in Central Asia and in mapping the course of the Yellow River. Elias and Haversham had a common friend in the genial Sir Henry Rawlinson, one of the great 19th-century diplomats and antiquarians, who was President of the Royal Geographical Survey from 1874-75 and of the Asiatic Society from 1878-81.

Haversham certainly played a protective and promotional role towards his youngest brother:

> *You mention Mr Godwin-Austen, Deputy Commissioner in Berar, Retired. He is my own brother. I took him to India and got him into the uncovenanted Civil Service.*[14]

On Rawlinson's recommendation, Elias was engaged by the Government of India to fill minor posts in Calcutta and Mandalay, and took part in a failed expedition from Burma to China in 1874, at the same time as Haversham was toiling through the northern Burmese and Assamese jungles. In 1877, Elias was attached to another failed expedition, this time to Kashgar (now more usually called 'Kashi' in Xinjiang, Western China, and the hub of the Silk Routes). He stayed on as a joint commissioner in Ladakh, one of the earliest explorers of which had been Haversham, in 1863.

In 1879, without government permission, Elias made a very dangerous high-level crossing of the Karakoram and reached Yarkand. Here, he carried out productive spying and trade work. In 1880, the Indian Government somehow approved this expedition and gave permission for Elias's next expedition to Kashgar and Yarkand, this time accompanied by Harold. The *Dictionary of National Biography* is wrong in stating that Elias was accompanied by 'HH Godwin-Austen' who is, of course, Haversham. Harold's initials ('HGA'), and date, were later discovered carved on a rock near Dras, north of the Zoji La: for the curious circumstances surrounding this carving, see Chapter 11. (Elias carried out one further great Central Asian expedition, across the River Oxus and over the Pamirs, in 1885, before acquiring an enduring liver complaint; he eventually died of blood poisoning.)

The main family inheritance (Haversham's dissolute but charming son, 'Bad Arthur' having been legally bypassed) descended from Alfred to Robert Annesley Godwin-Austen (1885-1977), who also enjoyed an unconventional life. After the violent quarrel with his father about the governess, he had to make his own way in the world and became, appropriately enough for one of Haversham's nephews, a surveyor. After the First World War, and using his surveyor's training, Annesley, with generosity and discretion, did what he could to help the aged Haversham out of his financial worries by trying to extract an increase in his Civil List pension, as well as obtain financial support from the Royal Geographical Society, as will be seen in Chapter 19. Then he joined the colonial service in Persia and East Africa (where he was paid in solid gold, sensibly converting it

[14] Letter from HH Godwin-Austen to Mrs H Barclay, 12 June 1923.

before leaving, bar one coin, into Persian carpets), before returning there after marriage at the interesting age of 46.

The remaining family character who is hard to avoid is Arthur, Haversham's only surviving son from his first English marriage. I will leave a detailed description of him until a later chapter, however, as his life represents a part of Haversham's legacy rather than what shaped it. Arthur misbehaved, mostly in the matter of debts, womanising and drink and conduct unsuited to an officer and a gentleman. Such misconduct grew worse as time went on: when Haversham died, the estate, even in its much reduced state, was put into trust in order that Arthur could not get his hands on it.

The themes running through the spectacular lives of this extraordinary family are not difficult to find: family solidarity; a liking for hard, adventurous travel and willingness to take considerable risks; strong willpower; unusual physical and mental toughness; an ability to get on with people at every social level (with the exception of the servant-avoiding Charlotte); the occasional tendency towards impetuous errors. Haversham showed these traits perfectly, as will be seen when we look at his great expeditions through wild mountains and deep jungles.

It can also be seen that within such a diverse family, with its quite strong sense of togetherness, Haversham needed to exercise considerable self-discipline in order to foster such a positive family character. It is likely that he gained much of this self-discipline at Sandhurst, although it can be readily imagined that Robert Alfred Cloyne did not, when it came to the upbringing of his son and heir, allow a very long leash.

PART TWO

APPRENTICE SQUIRE AND SOLDIER

CHAPTER FOUR

OFFICER CADET

Early Years

HAVERSHAM WAS BORN ON JULY 6 1834, almost certainly at West Ogwell House in Devon, described as, 'now the property of General Thos. Wm. Taylor, C.B ... a handsome seat, pleasantly situated in a small deer park, finely clothed with wood.'[1] The property had been leased to Robert Alfred Cloyne by General Taylor, the Commanding Officer of Sandhurst and a friend of General Godwin. Haversham's siblings next to him in age, Giuliana Maria, b.1835, and Sophia, b.1836, both have their births registered there. Existing accounts of Haversham's life have him born in Teignmouth. While it is true that Robert Alfred Cloyne appears to have lived there for a short time while preparing for the arrival of his new bride, there is no evidence that Haversham was born there. And since it is likely that Maria would not have journeyed while pregnant to Devon until her husband had found a suitable home for her, it seems equally likely that Ogwell House was where Haversham was born, not long after her arrival.

The stay in Devon did not last long. In 1838, when Haversham was four, the Austens, with the first three of their 18 children, moved back to Surrey, to take up residence at Shalford Park, then Gosden House (at Bramley, just outside Guildford; now a special school) and Merrow House (on the eastern fringes of Guildford, an unattractive 'boxy' building built in 1802, now converted to flats for elderly people). They then moved to the attractive, if architecturally eclectic, Chilworth Manor, below St Martha's Hill at the southern base of the North Downs, which Robert Alfred Cloyne rented for 33 years until inheriting the main Austen estate on the death of Sir Henry Edmund in 1871.

Haversham therefore spent his first four years in the Devon countryside and seaside; it is not difficult to see where his interest in natural history and geology began. A good part of that time must have been passed with his father among the extraordinary colours of the geological folds in the cliffs and the stony beaches of the 'Jurassic Coast', the inspiration for several of Haversham's later watercolours and still a chipping-ground for amateur fossil hunters.

Despite his living on the Austen family lands from 1838 onwards, what we know of Haversham's early education remains somewhat hazy until his arrival at Sandhurst in January, 1848. Before then, a 'tutor', a Dr Stanley, is referred to in passing, possibly hired to help with preparation for the Sandhurst entrance exam, although in one obituary he is also said to have inculcated a love of science in the young Haversham. In *The Royal Grammar School Guildford*,[2] however, DM Sturley

---

[1] Source: http://genuki.cs.ncl.ac.uk/DEV/WestOgwell. It is now called Gaia House.
[2] *The Royal Grammar School Guildford*, DM Sturley; Governors of the Royal Grammar School and DM Sturley, 1980, p136, 2nd Edition, 1991.

mentions 'HH Godwin-Austen' as one of the school's more distinguished 19th-century pupils. Sturley's source is not known and subsequent fires at the school have obliterated any relevant records. Sturley is nonetheless probably correct; he was known (he died in 2008) as a meticulous historian and archivist. Also, the list of distinguished 'old boys' in which Haversham's name appears is otherwise correct. (A good example of one of these would be Col Algernon Durand who, when British Resident in Gilgit, North-West Frontier Provinces, in 1891 led one of the most exciting expeditions in British imperial history, the daredevil night-time storming and capture of the Hunza clifftop during which three VCs were won. The story of the campaign is very readably told by *The Times* campaign correspondent, EF Knight, in *Where Three Empires Meet*, with a kind of swashbuckling breathlessness very well suited to the courage of the British and native Indian forces. (Durand also welcomed to Gilgit Haversham's greatest successor in Karakoram exploration, Sir Martin Conway, when Conway was embarking on his long 1892 expedition, much of the planning advice for which had come from Haversham.) Furthermore, since it was the 16th-century Approved Man of Guildford, John Austen, who co-founded this King's School, later known as the Free School then Royal Grammar School, something of a family tradition of sending the eldest son to the school had evolved. Two of Haversham's younger brothers, Frederick – 'Freddie' – and Chads, for example, are known to have been pupils there.[3] And the RGS was only about four miles from Shalford Park, thereby being a doubly logical choice for the (just) teenage apprentice squire.

Haversham himself was in no doubt about his educational fate. In a memoir, he claimed that:

> It had long been decided that my career was to be in the Army, and I believe at seven years of age, my grandfather, then a Colonel, in the 41st Regt ... obtained for me a nomination for the R.M. College ...[4]

Clearly determined that Haversham should be a soldier, and with his father's agreement (as a later letter between them proves), General Godwin ensured his grandson's entry into Sandhurst, only about a dozen miles from Shalford, in January 1848 when Haversham was only 13.

By the time that his formal education had been completed, Haversham had, apart from his subject-training as an officer, been taught about geology, palaeontology, marine biology and the law. And he would have seen his father hard at work in the Chilworth study as he corresponded with some of the greatest 'cutting-edge' scientists of the day, for example Charles Lyell, Charles Darwin and Henry de la Beche. Such correspondence would have made him aware of the great city institutions, like the Royal Society, with their scholarly papers, meetings and medals as well as their exciting new work in science and collecting. A later letter confirms his father's influence:

> *(I was)* working with my father from the age of 12, and when I went to Sandhurst at 14 ... I made a Survey of the country round Guildford for my father, to illustrate his papers on the Gravels of the Peasemarsh – which taught me a lot – ... [5]

But while being brought up in Devon and on the Austen Surrey estates, Haversham would also have lived close to nature, to life in the fields, hills, woods and coastline, with all the natural history stimuli that these could provide. And he would have become familiar with the seasonal duties and expectations of the farmer. As letters late in his long life prove, these early experiences made a profound impact and never left him.

---

[3] Another source suggests that Haversham was educated at 'Kensington Grammar School', which existed at this time – c1845 to the end of 1847 – but which seems to me to be more of a conflation error, compressing 'Kensington', where Haversham spent much of his later years in the great scientific research institutions, into '(Royal) Grammar School'.

[4] HH Godwin-Austen: unpublished *Memoir*, Nore, Godalming, 1901.

[5] Letter by HH Godwin-Austen to Buller Newton, 7 April 1923.

Before departing to complete his education, at Sandhurst, it seems that Haversham might have enjoyed the mid-19th century equivalent of a gap year, although in his case this was more of a sort of grand tour through central Europe which cannot have lasted more than a shortish summer. Clearly, going abroad during the 'year of revolutions' did not put him off. The itinerary is not clearly defined: the only source is a dozen untitled pencil drawings (dated 1848) in the Papplewick collection. The scenes are generally rustic, though they include a number of studies of houses with their architectural details very carefully drawn. The best clue about where his journey took him is a river scene, showing the impressive Hammerstein castle (now, however, a ruin), about half way between Koblenz and Bonn, on cliffs overlooking the Rhine (the only titled drawing in this collection). Drawings near the end are of coastal scenes, suggesting that if Haversham had indeed visited the Rhineland, he had returned via the Mediterranean. The drawings are meticulously exact, if rather flat and somewhat dispassionate. But the detail which has gone into the occasional ethnic figure's dress foresees the exciting realism of his work after settling in India. (It must be added that these drawings could simply be copies of those seen by Haversham in a book, though their careful dating perhaps suggests otherwise. If these are simply copies, then Haversham's grand tour occurred only in his imagination.)

SANDHURST

In the aforementioned, unpublished, 50-page personal memoir written in 1901, Haversham describes his formative experiences there with the kind of detail which shows that, for all its rigours, he remained deeply attached to the place. Sandhurst was quite like a relatively benign version of several of the public schools of the era. Many of these imposed tough living conditions, for example, Marlborough, where the boys complained of chronic hunger to the point where their misery led to a four-day rebellion in 1851; order was restored only after much physical violence.

Evidently quite a well-run place under General Taylor, Sandhurst was nonetheless not without its history of indiscipline. Haversham recalls a rumbustious event which took place about 15 years before his arrival, the memory of which was still fresh.

The Battle of Bagshot

> I recollect hearing a tale of a visit of William IV [6] to the college – then still fresh in the mind of people about there – the King came to inspect the cadets and present colours, & was taking his departure – the cadets being at their dinner – in a hall on the ground floor. There was a large gathering of loyal people from the neighbourhood and from Bagshot, near the King & there was little enthusiasm & cheering shown – at least so thought the cadets – for they all got out of the windows, over 300 of them, and as the King left began knocking off the bystanders' hats, and eventually left off with the promise to come & visit the offenders next day, at Bagshot. This they carried out, falling in and marching from the Parades ground, armed with sticks, and had a regular fight there with the Bagshot people but on their return to College – the ringleaders, who they had elected as officers from among themselves for the occasion were expelled & the others put under arrest, but I remember how proud the cadets were of that "Battle of Bagshot" even in my time.[7]

To develop this theme, Haversham tells another story of a cadet's farcical insubordination:

---

[6] Reigned 1830-7.
[7] HH Godwin-Austen: unpublished *Memoir*, Nore, Godalming, 1901.

> A Cadet in his last year, we will call him B—, was a prisoner in the cells, they were generally called 'the holes'. Some time in the evening he rang his bell and when the Sergeant answered it, he shewed him a pane of glass, saying he had not cracked it. The Sergeant unsuspiciously went up to the window to inspect the crack when B— who was a particularly strong young fellow, tripped him up, rushed out of the room and locked the Sergeant in. He then having the keys went to see who was in the other cells. There were two or three in for idleness, who B— at first released, but as they were fresh joined and not to be trusted he locked them up again. He next went to the Sergeant's room where he found the key of the outer door, let himself out, locked it on the outside, and broke the key in the lock and made off. Time went by until the chapel hour. As we marched in, many noticed the absence of the Sergeant and his batch of prisoners. We took our seats, but the Chaplain did not come in. Then the Captain on Duty despatched a Sergeant to see what the delay was caused by. Before long, we were in possession of the whole story when the Cadets in the cells came up to their rooms, quite the heroes of the hour. The Sergeant sent up could hear the locked-up Sergeant shouting from inside, but the door could not be opened without a blacksmith and a ladder. Search was then made for B— who was found at last in a public house at Blackwater. He offered a determined resistance and gave the Sergeant great trouble. He was sent to his friends and we none of us ever saw him again or knew when he left.[8]

There was, however, an accepted method of protest against the authorities:

> Disapproval of unpopular acts or orders took the curious form of a run on the emptying of the jugs of beer, every cadet drinking two glasses to his usual one, while shouts of 'swipes!' rang from one end of the dining-hall to the other, in order that a large number of *(beer)* cans might be brought up from the cellar. It ended in noise and insubordination, and the Responsible would have to call in the Captain of the Day for only his presence would stop the disorder. By him a good rowing was administered, and a few punishments were given; the object was gained, the Governor would have the affair reported to him, and on the other side feelings were relieved.[9]

Such events were the exception rather than the rule. Haversham recalls that there was little bullying, 'apart from some restrictions of dress placed on the newly joined'. Well-established routines and a firm disciplinary code seemed to regulate daily behaviour quite well. Punishments, compared to the kind of thrashings at, say, Marlborough or Eton at that time, were not severe.

> No corporal punishment could be inflicted on a cadet, the most severe punishment being dismissal for disgraceful conduct. Next came solitary confinement in the cells, close and open arrest, and punishment Guard Duty. For entering a public house a number of days in the cells was invariably awarded. For disorderly conduct in study, the Professor could ring his bell, the Sergeant on Duty would shortly appear and remove the Cadet to the cells. The whole proceeding was amusing at times: the Professor opening his desk and fumbling for the key of a little door about 4 inches square which protected the handle of the bell. Having rung it, then came the interval until the Sergeant's appearance, and if the culprit was hardened, he would amuse the rest of the class by a cheeky argument with the

[8] HH Godwin-Austen: unpublished *Memoir*, Nore, Godalming, 1901.
[9] *Ibid*, p43

> Professor – on one occasion, I saw a Cadet put on a Professor's hat and cloak and walk round the hall of study to say goodbye to the others.[10]

And it is clear from Haversham's ironic comment that other forms of 'military' punishment were largely ineffective:

> ... a book being read aloud to us, as a warning of an account of how cadets many years before had waylaid and stopped a coach somewhere on Bagshot Heath, and letting the horses loose. Whether it was thought by the authorities that it was a solemn warning to us cadets I don't know, but I should now think rather the reverse.[11]

Discipline was also maintained through a rigid daily routine. Reveille in winter was at 5.30 or 6am, 'a hateful sound to us'. Then there was a two-hour study period before breakfast at 9am. This meal consisted of, 'a pint of milk served in a large white bowl, either hot or cold'. A Cadet had to give a week's notice when he wanted a change. Tea could be had, with a Surgeon's Certificate, but it was 'very weak'. There was no limit to bread, 'which was good, and about 1oz of butter per Cadet'. The next meal, dinner, was taken at 3pm. The diet was monotonously uninventive: roast leg of mutton one day, roast shoulder of mutton the next, with roast beef on Sunday. Two large dishes of potatoes, always boiled in their skins, were provided for each table. Knives and forks were wiped on the table-cloth 'of the very coarsest material' and the fork cleansed by being driven through this coarse cloth several times. Dinner-time procedure was, like the punishments, quite peculiar and not a little like the workhouse world of *Oliver Twist*.

> The Responsible, while we were drawn up in line facing the tables said a rapid Grace, called out 'Break' and we rushed to our seats, a scramble then took place for the potatoes, each Cadet clutching as many as he could get hold of. In two seconds these dishes were empty and the peeling going on. The head of the table would begin to carve the joint. He first helped himself liberally, then cut for the Cadet at the bottom of the table, next for his four neighbours. The joint, by this time looking very small, was taken by a waiter to the bottom of the table and the four juniors got what was left. Sometimes this was very little indeed, nothing perhaps for the last but the bone.
>
> For second course, we had baked Plum Duff called 'Stick jaw' every other day and on Sunday, and rice pudding on intermediate days. The suet was certainly cut large and received the name, 'Bagshot Diamonds'. We had as much bread as we asked for. We never had cheese. To drink there was beer, a very watery drink, and it was never known to have got to any Cadet's head, no matter how much he might consume. Dinner was a most noisy, rough, affair.[12]

For this soul-destroying diet, there were occasional compensations of the kind every schoolboy loves: on Good Friday (a term Haversham continues to use in his diaries in place of the calendar date), after the order 'break off', every cadet in the College would rush to Luke's cake shop in York Town[13] to get hot cross buns. Another anticipated treat was 'saucer pies' – fruit of various kinds, usually rhubarb or damson, covered in pie crust, though the price was, 'out of all proportion to the amount of fruit which could be got into the saucer'. And in summer, cadets could enliven their existence from a 'Tuck Barrow' which had 'cooling fizz'.

---

[10] *Ibid*, pp17-18.
[11] HH Godwin-Austen: unpublished *Memoir*, Nore, Godalming, 1901.
[12] *Ibid*, pp29-30.
[13] A village (at that time) near Sandhurst.

Within this disciplinary framework, compulsory 'hygiene' enjoyed a rather understated place. Bathing was practised daily and replaced the more cursory 'tubbing'. A couple of years later, in his Burmese diary, Haversham resolutely records when and where he bathes – usually in some lagoon of the Irrawaddy Delta – virtually every day, with a disarming lack of concern for the diseases lurking there. (His Burmese diary is strewn with references to diarrhoea afflicting him and his fellow officers.) Beyond a jug and basin, no means was available for washing oneself all over and no warm water was supplied, even in the severest frost. Haversham proudly declares that during his time at Sandhurst there were no epidemics, which he attributes (with 50 years' hindsight) to the clean water supply and the fine laundry (underwear was changed twice a week).

If these aspects of Sandhurst life show how Haversham's legendary strength and stamina were fostered, then the detailed description of his studies tells us about the joys and limitations of his academic training.

The entrance exam seemed to trouble him little, perhaps because Robert Alfred Cloyne, Lieutenant-Colonel Godwin and General Taylor were all well acquainted:

> Some dictation and arithmetic I got through without any great strain, then came the Latin not so well. The Professor did not press it very hard and helped me out ...[14]

Family connections undoubtedly made things easier for Haversham in the earlier stages of his career, as will be seen in Chapters 7 and 8, but it is also true that he succeeded in advancing his career through his own great talents and unrelenting hard work.

The Sandhurst course lasted three years, so that there was 'ample time to learn every subject thoroughly'. The focus during Haversham's academic career remained on exams although, 'there was no cramming'. Passing six public examinations entitled a cadet to a Commission in the Line without having to buy one. There were three compulsory subjects: Euclid, First Fortification and Surveying. Optional subjects were Latin, French, German, History, Geography and Maths, which was subdivided into Trigonometry, Algebra, Mechanics, Differential Calculus, Astronomy and Second Fortifications. Cadets were also compelled to attend a confirmation class with the chaplain, an unsurprising requirement as the muscular Christianity which supposedly underpinned the 19th-century expansion of the Empire developed. Football and hockey were the main curricular sports. Haversham, like all other officer cadets, was of course also taught horsemanship. In the final exams, the history questions were all viva voce; some took 'many minutes' to answer. The exam on the 'Attack and Defence of Fortresses' was held around models in the college museum. If a cadet passed more than six subjects (which Haversham did), among which were Trigonometry and Second Fortifications, he received a Certificate of Superior Qualification.

In his *Memoir*, Haversham writes at some length and with some nostalgic appreciation about the different subjects he studied. Of these, the most important to him, and which answered his talents best, were Drawing, Triangulation and Surveying. Among his recollections of those who taught him, the strongest influence was Captain Petley, '*a most accomplished artist*' whose '*penwork representation of ground could not be surpassed at that time*'. Petley's influence shows in the architectural exactness and fine penwork of Haversham's watercolours and drawings, such as those of the pagodas in Burma. Contrary to the rather vague and uncertain 'old French pictorial school' mentioned in the *Dictionary of National Biography* and parroted elsewhere, Haversham tells us that Petley had as his instructor (when a cadet at Sandhurst), '*a French professor many of whose originally and beautifully executed works were still among the examples we had to copy*'. (Petley's artistic influence on Haversham is looked at in Chapter 22.) French influence, according to Haversham, also extended to a '*rigorous method of triangulation and the measurement of an arc of the Meridian*'. '*The French,*' he continues, '*were years in advance of this country in all appertaining to surveying and the delineation of ground.*'

---

[14] *Ibid*, p2.

It was a Mr Burr, who had been in the Ordnance Survey, whom we need to thank for first teaching Haversham how to use a plane-table. '*For sketching ground,*' he tells us, '*we had to do at least two sketches of a square mile in extent on a scale of eight miles to one inch,*[15] *starting from the cadet's own measured base – no system could have been better or more practical.*'[16] Mr Burr also taught triangulation, using a theodolite, '*on a line of road and plotting the same afterwards*'. '*Those cadets,*' he goes on – and I suspect at this point with characteristic modesty he means himself – '*were excellent draftsmen and capable of making a readable sketch of any bit of country.*' He acknowledges that he owes his career in the army service to this high-quality course of teaching: '*It gave me my appointment in the Topographical Survey of India.*'

That appointment was indeed the outcome, but Haversham, by his own admission, during his first 18 months or so at Sandhurst was rather idle. When he first joined, he preferred the rough and tumble of cadet life and the privileges of greater freedom of movement allowed him after he began to show a talent for drawing and artwork, rather than pursuing any determined attempt to work hard. He sorrowfully recounts what happened as an example of his irresponsible behaviour:

> I had the misfortune to once spend the night in the cells. I was there for idleness. Reported by the Junior French Professor. I had finished my imposition: several pages of Telémarque.[17] I had got restless, there was absolutely nothing to do. I hadn't my penknife, even, to cut my name on a stool or the window, for this article and the contents of pocket had been turned out by the Sergeant before being locked in. I could hear the voices of the Cadets at hockey on the green, and longed to see what they were at. With a good deal of trouble, I managed to climb up and sit on the top edge of the lower window sash, and look through the narrow opening of about four inches. My glimpse of the outer world was not long-lived, there was a sudden slip, my boot struck a pane of glass and split it right across. When the Sergeant came in with my bread and water for tea, he spotted the broken pane. He said, 'You will have to remain here all night and I will have to report you to the Lt Governor.' When time for tattoo arrived and I heard the Sergeant unlocking the cells to march the other obstinate idlers down to chapel, he opened my door and kicked my roll of bedding into the room. In the dark, I really did feel miserable then. I believe I wept. It was very still. Not a sound reached that out of the way part of the building. The next morning, I was marched to the Guard Room and thence to the Lt Governor's office, explained there what I had done, had a good wigging and was ordered back to duty.[18]

With self-critical hindsight Haversham develops his observations on his poor conduct:

> I was particularly idle and thoughtless in my first two years, the reports issued each half year of conduct and progress were not so pleasing to my parents as they might have been. The fact was up to 15 I had not begun to realise there was a future before me and I hardly thought I should or could get my Commission. There was always a certain number of well-to-do boys who will not work, and such would say, 'if I do not get my Commission here, my father will purchase one. I shall get into the Army all right.'[19]

---

[15] A slip by Haversham. He could only have meant 'eight inches to one mile'.
[16] *Ibid*, p37
[17] Perhaps Fénelon's *Télémaque* (1700) which has as its theme the existence of kings for their subjects, not vice-versa.
[18] *Ibid*, pp20-1.
[19] Despite the inconsistency in the tenses, the transcription is correct.

It is at this point that General Godwin returns to the story. The general, with or without the help of a royal father, had risen rapidly through the ranks by merit and he had clearly decided that Haversham was not pulling his weight. A scene ensued not unlike those between King Henry and Prince Hal in Shakespeare's *Henry IV, I* and *II*, where a serious confrontation has a nervously amusing undertow of irony.

> One day, a Sergeant came into the Hall of Study I was in, and said to the Professor, 'Gentleman Cadet A 30 is to attend at the Lieut. Governor's Office'. That always meant a hamper from home to be opened by the Clerk and its contents inspected and passed, or a rowing and subsequent punishment, so it was with mixed feelings I proceeded to the Office and presented myself ... I was soon passing through the green baize door that divided the Clerk's office from the Lieut. Governor's own. There to my surprise I saw my grandfather General Godwin, just appointed to a Command in India, his manner towards me I soon noticed was not quite so affectionate as usual. He said he was very busy but had come down especially to see me all the way from Town, and was sorry not to hear a good report from General Taylor. He asked leave for me for a short time, as he said he would like to have a talk with me before he returned by the next train; he bid *(sic)* goodbye to the Lt Governor and we went out to the front of the College, where we walked up and down the Parade. I had a 'mauvais quart d'heure' or so, my Grandfather's language being rather strong when he was angry. One sentence long rang in my ears, 'Damn it, Sir, if you imagine for one moment if you don't get your Commission at this College, I shall purchase one for you, you are very much mistaken and if you want to enter the Army as you say you do, it will have to be as a private for I will have nothing more to say to you.' He extracted a promise that I would work hard and then cooled down into his usual kind way and I walked with him to Blackwater Station to see him off and returned to the College. He had given me many and many a present, and been the most affectionate of grandfathers but that visit was the best and kindest thing he ever did for me. I began from that time to think, and began to work hard, and found I could get on just as well as others, and in some subjects better. The next time I saw my Grandfather was at Prome, in Burmah[20] *(sic)* ... I saw him through his last illness and laid him to rest in the pretty cemetery of an Indian hill station. He was a very dear, good friend to me.[21]

The way in which the story is teased out, and the way it quickly turns from being in deadly earnest back to familiar friendliness suggests that here are two family members who know that they fundamentally get along but see no harm in a bit of self-dramatisation to take the sting out of what could otherwise have been an emotionally ugly occasion.

This confrontation with General Godwin was one of the turning points of Haversham's life. From then on, he evidently worked hard and matured so that on graduation day his commission stated:

> Gentleman Cadet Henry H Austen, being qualified ... and having accordingly, passed six Public Examinations, and having over and above the required Qualifications made such further Progress in his Studies as ... entitles him to a Certificate recording the Special Approbation of the Commissioners; We are of the opinion that Gentleman Cadet Austen, whose conduct has been exemplary, is worthy of this Honorable Distinction, and do grant him our Certificate accordingly.

---

[20] See Chapter 6. Haversham arrives in Burma in 1852 as General Godwin's ADC.
[21] HH Godwin-Austen: unpublished *Memoir*, Nore, Godalming, 1901.

It seems, therefore, as though Haversham's education had reached a happy ending:

> Never in the whole course of my life have I ever felt so supremely proud and happy as on that evening. My father and mother and a sister had come over and been present at the Examinations and Riding School ... I had the world before me ...

Perhaps Haversham had read his *Paradise Lost* by this time, also.

Haversham's talents and privileges, however, could not protect him from the inexorable movements of the wheel of fortune. Some apparently innocent observations made on the day he entered the college and met his commanding officer, Sandhurst's lieutenant-governor, Major-General Thomas William Taylor CB, give us an early clue about the future difficulties which Haversham, who undoubtedly had an eye for female beauty, was going to land himself in. A good-looking man himself, his comments, sketches and diary observations over many years testify to his lifelong interest in women. His surviving diaries are peppered with observations on what women look like, particularly their hair and dress. This interest, within a few years, was to lead him into a passionately indiscreet Islamic liaison followed by a painful second marriage, both of which, despite a much happier third marriage, were to haunt him for the rest of his long life.

> At the time of our visit *(the day of the Entrance Exam)* there was a fine handsome son in the Artillery at Home *(General Taylor's)*, and two unmarried daughters, both pretty girls *(the Taylor family were noted for their good looks)*. The youngest took me under her especial charge so that I went to bed that night in a more happy frame of mind as regarded the ordeal of the next day.[22]

In the meantime, the young ensign prepared to set out for Burma, as aide-de-camp to his grandfather, General Godwin, the commander-in-chief of the British forces in the Second Anglo-Burmese War. He was about to embark on a voyage which, like the rest of his eventful life, brought one surprising adventure tumbling after another.

[22] *Ibid*, pp1-2.

CHAPTER FIVE

A PASSAGE TO ASIA

Into Adulthood

ON 15 JUNE 1852, GENTLEMAN Cadet Austen began his accident-prone journey into adulthood. The sea voyage to India was a banal, if dangerous, step for a young officer of that period on his way to join the Indian Army, generally recognized as the most professional part of the British Army then operating. The first changes in his life were geographical: a five-month sea voyage via Cape Town to Calcutta with the exciting sights and sounds and smells of South Africa then Bengal; the relatively short voyage across the Bay of Bengal to Rangoon with its surrounding swamps, lagoons, watercourses and jungles of the Irrawaddy delta. Then, a few months after the British victory in the Burma campaign, Haversham moved on to enjoy the comforts of a fashionable Raj hill station, Simla, the uplands near which were to exert a defining influence on his scientific life. The humid Burmese period and more comfortable interlude in Simla were soon followed by the dust and glaring dry heat of the Punjab. After only a short time there, on being posted to the Great Trigonometrical Survey in April 1857, he walked among high mountains for the first time in that most European and Alpine part of the sub-continent, Kashmir. The novelty of the ethnological and natural life of these wildly different areas stimulated his imagination and thoughts to shape his career as a surveyor, an explorer and a scientist.

Perhaps more important than new travel experiences, however, were Haversham's emotional discoveries. He learned what it was like to be in great pain when he broke his femur in Calcutta, or to be ill and in fear for his life when he caught smallpox while crossing the Bay of Bengal; he learned about hatred and violence, when he was beaten up and knocked unconscious in 1858, probably as a consequence of the Indian Mutiny, and he saw death close up, as the only member of his family present when his grandfather, General Godwin, died. Following the general's death, he must have learnt what it was like, for a time, to feel lonely and far from home, despite having a regiment of brother-officers around him. On the other hand, there were happier emotional experiences – Haversham made friendships which endured; he discovered the pleasures of sex, as well as all its awkward consequences. It was also a rich period for the stirring of aesthetic emotions as can be seen in his lively and beautiful watercolours of the Burma campaign, the Punjab, Kashmir and Zanskar.

Haversham's early career was also shaped by his powerful connections. In 1854, his sovereign[1] had changed his surname by personal decree. In the army, despite starting out as a lowly ensign, he was quickly made ADC to two generals (Godwin in Burma and Reed in the Punjab), and his

[1] For the circumstances of the name change, see Chapter 3.

posting to the Topographical Survey was overseen by Sir Andrew Waugh, the Surveyor-General of India himself. Haversham must have realised that his family was his fate. (His father's assiduous role in helping Haversham to obtain these positions is looked at in Chapters 7 and 8.) He never exploited his fortunate circumstances in any obvious way – he was too well-bred for that – but it must have been a source of comfort and confidence to him, particularly when aligned with his natural talents and training, to know that powerful men and women were aware of, and approved of his work.

LEAVING ENGLAND

Haversham was posted to Burma very soon after obtaining his commission. Before departure, a painful family leave-taking had to be endured. Haversham therefore took the train home to Chilworth, the manorial village curled up at the southern foot of St Martha's Hill, just south-east of Guildford. Pevsner is reasonably complimentary about the manor, which Robert Alfred Cloyne had been renting since 1840, but adds with some waspishness that the best thing about the place is its walled garden, laid out in terraces up the side of the hill in a 17th-century design. With a playful irony, Haversham describes how he was met by Mama (Maria Godwin) who, having received his letter with the '*bad news*', was said to be '*very much cut up about it.*'[2] Papa was perhaps rather sensibly out at an election meeting in Guildford. On his return, however, father and son talked over the matter during dinner and '*discussed a bottle of good claret.*' The following day, a Sunday, was at first little better. Sisy (sic), (Sophia) perhaps Haversham's favourite sister (if mentions of her in his surviving letters are anything to go by), having been crying all day, arrived at tea-time. The other sisters had been obliged to walk over from Guildford; '*Sisy's crying set half the girls off crying also,*' as he mischievously remarks, '*for it seemed catching.*' During the afternoon, however, Haversham made his escape from the tearful women by making the short but steep ascent, past the 17th-century gunpowder mills and the Bronze Age henge monuments, to the recently restored (15 May 1850) St Martha's Chapel, standing at about 175m above sea level on the summit of the hill of the same name. The views from it south, east and west are extensive; in 1852 much of the foreground view would have been over Godwin-Austen land. This was a private leave-taking of his homeland, '*I saw the beautiful view that I shall ever remember … hills I have seen so often and never looked so hard at before.*' Having frequently walked to the top of St Martha's, I can vouch for its commanding view, said to be over parts of eight counties, as the ground falls away suddenly in three directions. One looks over the little, well-ordered, grass-green dips and wooded folds of 'leafy Surrey', across the North and South Downs; it is as pleasant a prospect as one could hope to find in southern England.

The whole leave-taking scene in some respects is worthy of Jane Austen, one of whose brothers, Rear-Admiral Charles (1779-1852), died of cholera while in command of the British fleet off Burma just as General Godwin was taking over the land forces.

CHATHAM TO CAPE TOWN

Haversham's journey to the sub-continent was to begin at that Conradian port,[3] Gravesend. Robert Alfred Cloyne began by accompanying Haversham to Chatham. On the way, the young man observed 'Druidical structures' near Maidstone, actually a neolithic chambered cairn now known

---

[2] Most of the material in this chapter relating to the voyage to Burma and his period of duty there is taken from two of Haversham's surviving private diaries, *Journal of My Voyage to India, (JVI)* and *Journal from Burmah to India, (JBI)*.

[3] See Joseph Conrad: *Heart of Darkness*, the expressionist story of a commercial adventurer who penetrates the evils of European colonialism in Africa.

as Coty House (or Kit's Cotty House) – his archaeological interests were evidently already aroused. Rather more mysteriously, on arrival in Chatham he went to a jeweller, bought a locket and headed for Rochester, where he gave it to 'Susana'(?).[4] If she was his girlfriend, it must have marked the (unwilling) end of their relationship as she is not mentioned again in the diaries. Its being a locket, and necessitating a return journey to Rochester for its handing over in person, both suggest special effort and expense for a woman important in his life. In return, he was given, 'a housewife and a bunch of flowers'[5] (the domestic nature of which suggests either a former governess or a girl dropping hints about marriage). Whoever 'Susana' was, she, too, 'cried immensely'.

The embarkation on 15 June followed a military farce. Haversham describes the scene with the judgementalism of the virtuous young:

> It was on the morning that we left Chatham that I witnessed, what was, I think, the most disgraceful scene I have ever witnessed. Before departing the troops were on parade, and Major Kelly addressed the men, saying that there was to be no cheering as they marched off. The men naturally resented this. He told them that if there was any cheering the Band would have to go back to Barracks & they would march through the streets without any music. Of course, when the order, 'right turn, quick march' was given the men at once burst into cheers. Major Kelly shouted, 'band, you know your orders' and they marched back to Barracks. Then followed a scene. The men got absolutely out of hand, & many of them broke ranks. Some had fiddles and tin whistle pipes, & these were put in front. They were joined by men and women from the streets, who brought boards which they beat with sticks, & with this music we marched through the streets of Chatham.[6]

For this largely monotonous voyage, Haversham was attached to the 16th Royal Irish; he was to join his regiment, the 24th Foot (later the South Wales Borderers), in Burma. In its long record of cheerful courage, whatever the odds, the Royal Regiment of Wales and its predecessors have won 29 Victoria Crosses, and over one hundred Battle Honours. (See the regimental website.) Family connections with the regiment are durable: Lt Col Robin Godwin-Austen, Haversham's great-nephew, retired some years ago as the regiment's commanding officer.

The ship to transport Haversham to India was the *Lord George Bentinck*, named after Lord (William) George Frederick Cavendish-Scott-Bentinck (1802-48); it was 'a Bombay built teak barque of 590 tons and about as safe a vessel as you could go to sea in, though slow'. Life on board resembled a mobile college: several of his fellow-travellers would, much later, make their own mark on history. One companion was Ensign Esmond(e), the grandfather, though Haversham mistakenly calls him 'father', of the Irish Nationalist politician Sir John Lymbrick Esmonde (1893-1958) and father of the Nationalist MP Dr John Joseph Esmonde (1862-1915). And, although he does not mention him, also on board was the future Lord Roberts of Kandahar,[7] one of the stars of British imperial history. Much of Haversham's time was spent with a young fellow-officer, Pocklington, who went on to marry Haversham's eldest sister, Giuliana Maria. When they weighed anchor, there was 'not one (officer) with dry eyes'; Haversham was, after all, not quite 18.

Haversham's mental expansion on the first leg as far as Cape Town, which they reached 80 days later on 3 September, can be measured in several ways. First, for the young surveyor were the

---

[4] Haversham's handwriting is not clear. The word could be 'Sunna', which seems unlikely. And the spelling 'Susana' is quite unusual.

[5] Housewife: a pocket-case for needles, pins, threads and scissors *(OED)*. Such a gift might also suggest either a woman hoping to be domesticated or, more probably, a former nanny or governess.

[6] HH Godwin-Austen, *JVI*.

[7] Field Marshal Frederick Sleigh Roberts, 1st Earl Roberts, Bt, VC, KG, KP, GCB, OM, GCSI, GCIE, PC (1832-1914) was one of the most successful commanders of the Victorian era.

various geographical or topographical sightings, such as the Nore[8] (an evocative name, being the same as the Godwin-Austen farm, the family's oldest Surrey property, where Haversham spent his final two decades), the Bay of Biscay, the Straits of Gibraltar and Madeira. The ship then seems rather worryingly to have been affected by adverse winds, as Haversham's next sighting of land is off the coast of South America, near Pernambuco, close to the easternmost extremity of Brazil.

A more important form of learning is shown in Haversham's natural history observations. In late June, he notes his first grampus.[9] About the same time, Esmonde shoots a porpoise, though Haversham is 'unsure whether mortally or not'; in mid-July, south of Madeira, he records his first sight of a flying fish; sharks, pilot fish and more grampuses are seen later the same month; in August, somewhere off South America, he spies his first whales; as they approach Africa, a Cape pigeon is noted and, on the 'Glorious Twelfth', the beginning of the grouse-shooting season, he observes an albatross, another one of which is shot a fortnight later by the captain, with complete disregard for the warnings of the Ancient Mariner – the sailor's belief that to shoot an albatross brought catastrophic bad luck appears to have been an invention by Coleridge. Finally, on rounding the Cape in sight of the Lion's Head, 669m, between Table Mountain and Signal Hill, he spots his first penguins. Haversham was no doubt thrilled to see these exotic creatures, and would also have enjoyed the sense of having passed through several different climate zones. (I once unforgettably experienced something similar in 2008, on land, in the four-hour descent from Bhutan's second-highest pass, the Thrumsing La, which had deep snow and Alpine conditions on top, but bananas and pineapples growing at the bottom, a passage through three major climate zones in the space of a few miles and 3000 vertical metres.) The sightings also confirm that Haversham already possessed a confident knowledge of natural history, and it is likely that what he saw provoked thoughts on the adaptability to their different environments of the creatures he identified.

The voyage was also measured by the changing emotions running through the crew. (Having spent my Sixth Form in what up until then had been a boys' boarding school where only one weekend per term was permitted as an exeat, I can understand with painful recollection the ways in which adolescent males turn on each other and react to relentless and inescapably close contact.) On only the second day, two of the crew were caught smoking below decks – a fire rather than a health hazard – and were 'made prisoners'. In late July, Haversham rather nonchalantly describes a verbal outburst then fight among some of the men, followed next day by their court-martial and sentencing to 50 lashes, a callousness reflected in the indifference shown towards the burial at sea in the following month of a soldier called Murphy.

Being an officer and above all this brouhaha below decks, Haversham, to his credit spends his time in what he considers constructive pursuits rather than just idling on deck. He reminisces about Sandhurst with the Commanding Officer, Colonel Grattan ('of the famous Irish family'); he learns navigation under the guidance of Captain (of the ship) Edgell, and the chief mate; he begins to teach himself Hindi, the first of many Asiatic languages which he comes to speak fluently; and in a more manual activity, he mends sails on deck with the sailors and learns the art of splicing and knot-tying. While these skills would have practical applications for Haversham in later life, what is also striking is his ability to get on with all ranks, a theme which runs through his expeditions with his unusually fair and considerate treatment of his 'coolies', to his sympathetic attitudes much later in life towards his tenants and labourers as his estate was moving towards bankruptcy. Fortunately, he also spent much time in sketching. The endpapers of the diaries are covered with elegant and precisely detailed drawings, or doodles, of the ship, fellow soldiers, faces, dolphins, even a tropical island. Many of these were later turned into consistently detailed and attractive formal sketches and watercolours.

Some monotony cannot be avoided, however. To relieve it, shortly before arriving in Cape

---

[8] A sandbank at the mouth of the Thames where it meets the North Sea.

[9] Also known as Risso's dolphin.

Town, the *Lord George Bentinck* passes through a violent tropical storm. Haversham's description has a nicely Victorian ring to it:

> ... we were destined to have a blow before we got *(to the Cape)*, the Captain predicted it at dinner-time, as the glass had been going down fast although there were no outward signs. But about 6 o'clock, as the sun set, a long streak of black clouds could be seen rising in the NW and the men were busy taking in the sails so we were ready for it. In a very few minutes the clouds were overhead and the wind began to whistle through the ropes, which very soon became a regular gale of wind, thick leader clouds flying through the air, but very clear between and every now and then the moon shining through them, which had a beautiful effect adding to the wildness. About 9 o'clock we hove to being rather too near the shore to be comfortable and in the morning we put out from it by going S by W. The wind was very high and a heavy sea was running which rolled (us) about dreadfully but towards the afternoon the breezes moderated and we were making for Table Bay.[10]

The almost complete lack of interest in any punctuation other than commas gives his description a kind of racy breathiness which no doubt reflects his evolving emotions of apprehension, fear and relief. His expressions are strikingly similar to those used by Apsley Cherry-Garrard in describing the horrors of the Antarctic winter, in that most riveting of travel stories, *The Worst Journey in the World*.

After the storm, better times lay ahead: as the ship rounded the Lion's Head, near Cape Town, Haversham was sentimentally reminded of 'the chalk hills near Dorking'. (Comparisons with home always risk anticlimax: when staggering over the Shingo La, 5064m, into Zanskar in 1992, I was moved to exclaim from the top of the pass how much the countryside appeared to resemble the Ben Alder area, a remark unanimously treated as laughable by all the other members of the expedition.) On arrival in port, on 3 September, he is brought papers which tell of General Godwin's capture of Pegu, the Burmese stronghold – Haversham was to have an 'easy' war. (In fact, the only fighting action Haversham experienced in his 26 years as a soldier occurred when guarding captives in Amritsar in 1857 and at the siege and recapture of some small towns on the Assam-Bhutan border in 1865.)

It is not recorded how often Haversham landed at Cape Town, or any other parts of Africa, on his various furloughs during the following 26 years. As a continent, however, he seems to have left Africa to other Godwin-Austens: the luckless Freddie and Alfred in the Zulu War; the highly successful General Reade Godwin-Austen in Abyssinia and North Africa during the Second World War; Annesley in East Africa; Robin Godwin-Austen, having been born in Tanganyika, being with his regiment in Botswana at the time that Rhodesian UDI was declared, followed by guard duty on the BBC broadcasting post at Francistown, Botswana's second city, and Richard Godwin-Austen, who, with Robin, was evacuated as a child from Cyprus, not long before the German invasion of Crete, then travelled by taxi via Port Said to Suez and from there by tug to a ship, *The Empress of India* which took them on to Durban, Natal and thence, thanks to kindly people from whom they rented a house, to a secure, nearby village called Kloof.

Nonetheless, Haversham's first meeting with the Dark Continent was an enthusiastic one. After three days in port without shore-leave, his youthful impatience surfaces:

> Nothing must be more provoking than being in a strange port and not being able to get on shore.[11]

---

[10] *JVI*, entry for 2 and 3 September 1852.
[11] *JVI*, entry for 6 September 1852.

Soon after this complaint, however, the Boer authorities allow him to hire a horse 'superior to the hired horses to be found in England', and off he trots on a ride round the Lion's Head. On the way, ominously enough, he notices 'some reasonably good-looking ladies' – it is of course not so much the quality of their looks which in itself is important, but that they provoke Haversham's appreciative thoughts, which is our signal of trouble to come.

Haversham's exploration of the Cape extends to an easygoing 18-mile carriage trip to the coastal town of Constantia. His description of his day out, with the colonel and a fellow junior officer, Stackpoole, tells just what a sharp observer Haversham had already become:

> The drive was delightful and novel, the road no different to what one was accustomed to, the long teams of oxen and mules, some wagons having as many as 14 harnessed to them, trotting along making a tremendous rattle, which no English horse would have stood for a moment, the drivers with their Malay hats and long bamboo whips which they used beautifully, generally several black women in them getting a lift to Cape Town, with their curious dresses, mostly white and red handkerchiefs bound round their heads, some of them who had white blood in them were pretty good looking. But the regular natives are a very ugly race, they seem to be a very happy people, talking a curious language which they called African *(i.e. Afrikaans)* which is very corrupted with Dutch and English words. They laugh most immoderately at everything, especially when you chaff them a little. Their houses and themselves are exceedingly clean, and would shame many in England, and I never saw a beggar while I was there. The place swarms with children. Along the road were some very nice houses well wooded all round with fir and oak and other trees which I never saw before. The flowers are beautiful, many that one sees in England and taken immense care of, grow wild, especially geraniums and a sort of rododendron *(sic)* ... There is a beautiful white fir also which they call a Silver Tree. In about an hour and a half we arrived at Constantia, having passed through ... two small villages. The view here was lovely of the blue mountains which are about twenty miles off across some flat marshy country, and rise up ... immensely steep with the tops covered with snow. Table Mountain and the range from it is a regular island which seems as if it had been once separated from the mainland being only joined to it by the marshy ground mentioned before. Arrived at Constantia, we went to see the store and taste the wines. We were very nicely treated by the owner, a Dutchman, after asking us into the house we went to the store where were immense butts of wine all in rows. There are four kinds made from different vines, viz Poutac which is the dearest, very little being made, ... Red and White Constantia, but they are very sweet wines, only just to be used as a liqueur ... like the juice of raisins. Some they had was very old which improves it very much. There we saw a wonderful oak, an English one which had been trained into an arbour, the trunk of it was about 12 feet high, the branches from this were trained up into a cup shape all round and then brought over at the top again making a sort of room, to which one ascended by steps ... We went down into the garden where the orange trees were very large and quite yellow with fruit. From this we went on to the other place after thanking him for his civility with a hint that perhaps the Captain would be coming to taste his wines. At the other place ... we saw a collection of Cape birds which are very beautiful and of very bright colours. There are not very many wild animals. There used to be a great number of hyenas but one summer there being a disease among the horses, these animals ate them and then most of them were destroyed. Seemingly, they shoot panther, but very seldom. On Table Mountain there are baboons, bucks and rock rabbits but in small quantities ... We arrived back at about 4 o'clock and then walked about the town, went to the hairdresser so that the Col had his hair cut. At 5 ... had dinner which

> was pretty good, the Col and myself then went on board and the others stayed on shore as I was on duty next day.[12]

The diversity and minuteness of Haversham's observations are impressive. He undoubtedly had not just a sharp eye for people, animals, birds, trees, plants, places, language and objects but had already (he had turned 18 on board) acquired a degree of knowledge which could make sense of what he observed and organise it systematically: for example, his comments on Table Mountain are early versions of his geological observations of the Karakoram and Himalayas on which so many of his later scholarly papers were based. And his remark about the 'pretty girls' anticipates his momentous affair with Kudikji only five years later.

Four days remained in Cape Town after the Constantia trip. Most of these were spent on enjoyable, if routine, rides and sketching, including finishing some fine sketches of Table Bay, visiting hotels (for meals and drinks) with Stackpoole and Pocklington, or walking and gossiping in the public gardens.

FROM THE CAPE TO CALCUTTA

Having drawn in their breath, on 12 September 1852, after nine days in South Africa, the little barque and its military crew set sail for the second city of the Empire, Calcutta. This section of Haversham's diary produces many weather observations (and a daily record of latitude changes perhaps because the ship made better time despite a few becalmed days after crossing the Tropic of Capricorn): like so many Victorians, Haversham loved tables and lists. He gleefully records making 'a 100 miles per day' with 'a best running of 230 miles'. He again faithfully records the birds and fish observed, such as (more) penguins, 'an immense whale', dolphins, a linnet and more flying fish.

Ships without engine power are a frustrating mode of travel for a young man, however: although only 180 miles from Calcutta by 2 November, the ship was obliged to tack in order to reach the mouth of the Hoogly, the mighty river which opens into the sea at Calcutta. This delay was doubly irksome because on first sight of land at False Point – a famously good harbour in Orissa state, India, but so called because it is often mistaken for a headland one degree further north and therefore nearer journey's end – a gale blew up and drove the ship back right down the Bay of Bengal, so that False Point, true to its name, was not seen again for three weeks.

By 6 November, however, the *Lord George Bentinck* had reached a point 40 miles from the Indian coast. Various things are spotted in the water: one object is mistaken for a small boat, and turns out to be a dead bullock; another object is then thought to be a dead animal, but turns out to be a species of aloe, with swollen, round roots perhaps suggesting legs. More pleasantly, Haversham observes many species of butterfly and moth fluttering round the ship. As they head north through the Bay of Bengal, Haversham is 'impressed by the enormous distance the muddy water of the Brahmapootra, (now 'Brahmaputra', or in Tibet, the Yarlung-Tsangpo) and the Ganges extends south'. By 10 November, being so near the coast, messages and letters from home ('thank God they contained good news, especially of Mama') are brought on board by post-boat. Haversham hears about the death of the Duke of Wellington on 14 September 1852; unlike Garnet Wolseley's famously fulsome reaction, Haversham rather boringly offers no personal response. He also hears that General Godwin's campaign is going well; Prome, the Burmese capital, has surrendered. Again, there is no personal reaction – we do not know if Haversham was glad or sad to have missed out on the action.

On 12 November 1852, after a voyage of 61 days from Cape Town, and a total of 141 days at sea from England, Haversham set foot on Asiatic soil for the first time. This important moment was preceded, however, by an event both shocking and vile:

---

[12] *JVI*, entry for 7 September 1852.

> We did not take a tug at the entrance to the Hoogly, our captain deciding to work up to Calcutta by sail. The native boats soon came alongside with fruit and vegetables, & the chief amusement of the sailors was throwing bits of pork at them. This alighted on their bodies, and they immediately rushed with fearful grimaces to the water to wash themselves as the pork destroyed their caste. This, of course, amused the soldiers immensely.[13]

Small wonder that the atrocities of the Indian Mutiny were only five years away. Like any English gentleman, Haversham has two things to do on landing. First, he makes himself comfortable by visiting Mr McKinlay, about whom we know next to nothing other than that he was the co-Managing Director of Gillanders and Arbuthnot & Co Ltd, the person who would look after him in the right style. Gillanders & Arbuthnot, already a prosperous company, first started supplying banking, financial and trading interests in India in 1819, and is still thriving (having become a plc in 1947) in Kolkata.[14] Haversham then goes shooting: within 24 hours of landing, he has bagged snipe, parrots and doves. A little surprisingly, in this apparently destructive rampage lies the foundation of one of the greatest collections of bird skins amassed in the UK. (For the sale and re-housing of Haversham's collection of 3,750 bird skins, the bulk of which were collected in the Kolkata area, see Chapter 19.) Otherwise, he entertains himself by recording strange sights and experiences: the cotton factories, the dead bodies floating down the Hoogly, being shaved by a native barber and looking round the bazaar (probably the Lal Bazaar, as the Hogg or 'new' market was not opened until 1903). 'Calcutta astonished me: everything is so different.'

Just as reassuring to this somewhat open-mouthed young traveller was the sudden re-appearance of his brother-officer, Pocklington. They went out for drives in a buggy and breakfasted or stayed over at Spence's, an imposing, if rather dully-designed, building opened in 1830 as one of the first hotels in Calcutta for visitors from Britain. This easy-going existence was interrupted by the exciting news that Haversham had been told he was to proceed almost immediately to Burma on the SS *Mozuffer*, a 500-hp naval steamship of 1,500 tons which would shortly play its own part in the Burmese War.

But Haversham's excitement is short-lived and suffers a complete reversal: only a week after arriving, and with considerable understatement, he describes 19 November as 'an unfortunate day for me'. In handwriting markedly fainter and with less accurate self-expression than usual, he describes with amazing coolness, considering the excruciating pain he was in, what had happened:

> ... at about 5 o'clock I found Pocklington on the course.[15] We had a canter round the Circus.[16] We rode together towards Allipore,[17] and when returning broke into a canter. I soon found that the excitement of this put my mount out of control & finally it bolted. I was able to steer clear of many riders but at last I approached a Fort Point[18] bridge crossing the outlet from the Fort moat into the Hoogly. Just as I approached the bridge, a buggy left the line of carriages to pass another conveyance, &, there being no room between it and the side of the bridge, I dashed into it. The collision sent the buggy clean over with the two occupants – a lady and gentleman – and broke both the shafts. I was thrown right over the near wheel. I soon found that something was wrong with my left leg which was moving about in an extraordinary way, and I held it with my hand. Some Sepoys of a native guard

---

[13] *JVI*, pp4-5.
[14] See www.gillandersindia.com
[15] Here, the usual routes for horse-riding taken by the young men-about-town.
[16] i.e. the centre of the European part of Kolkata.
[17] Alipore is a district of south central Kolkata, near the fashionable Tollygunge area and the great bridge over the Hoogly.
[18] A venue for the Kolkata Rowing Club.

> close by ran out, & one of them brought a brass vessel of water & gave me some to drink. I was soon picked up, and put into a Dooley,[19] to be conveyed to Clive Street. The pain was so great, however, through the jolting of the Dooley, that when passing Spence's Hotel I told the bearers to take me in there. I was taken into a bedroom on the ground floor and in due course Dr Jackson,[20] a very old Calcutta medical man, came and examined me. I remember so well the first thing he said was: 'Lucky you are so thin; we'll soon make a good mend of it.'[21]

The grisly setting of the break follows:

> ... then began the operation of pulling my leg down and feeling for the bone which being done they put on the splints and bound it tight up – the thigh bone was broken and rather high up, but only a simple fracture.[22]

While one has every sympathy for Haversham's pain – having once shorn off a slice of my knee-cap 'like the top of a boiled egg' in a motorbike accident, I can readily feel the agony suffered in the *dhooly* and during the bone investigation – the event smacks of the modern phenomenon of the over-enthusiastic young driver trashing his car not long after passing his driving test. The femur, however, mended soundly enough for Haversham to move with astonishing speed and sturdiness over the mountains of Kashmir and the Karakoram no more than five years later.

When an active but accident-prone young man is confined to bed for seven weeks, without the benefit of modern electronic diversions, what does he do? Boredom was, at times, a problem: Haversham declares that he 'never had such a dull Christmas', perhaps a euphemism for feeling homesick. For the first few weeks of his recovery, however, he joyfully records receiving a stream of visitors 'among the young officers of the various detachments who were arriving from home and proceeding to their regiments up country'. Pocklington brought news of the war. On 30 December, Pegu, the principal Burmese province, was formally annexed; Haversham had missed the war and would be required for nothing much more than assisting General Godwin while he commanded the army of occupation until a treaty was signed. Many of Haversham's other visitors are listed; their presence further helps to refute the assertion that Haversham was a naturally tetchy character. Every day in his diary he describes some form of sociable activity. (The tetchiness seems to be more usually associated with the exacting nature of his scholarship at a much later stage of his life.)

He did make one new friend, Baddeley, a hotel guest who had assisted on the day of the accident:

> I remember I was wearing long Wellington boots at the time (of the accident), and when it came to cutting the left off I said to the man with the knife, 'Now look out and don't go cutting my foot!' This man, quite a stranger to me, staying at the hotel afterwards turned out to be a thoroughly good fellow & I saw quite a lot of him. His name was Baddeley & he came and saw me every day & would sit reading to me for many an hour. He also advised my getting a *(illegible: probably 'text book')* and learning Hindustani. If it had not been for him, my six weeks would have been very dull ...

The advice to learn Hindi was sound: not long after reaching Peshawar, in 1855, Haversham obtained his Army 'Hindustani' certificate. (His diary also contains six A5 sides of Anglo-Burmese transliterated vocabulary.)

---

[19] Doolie or dhooly: a covered litter consisting of a cot or frame suspended from four bamboo poles. (Hobson-Jobson).
[20] See: *Who Was Dr.Jackson: Two Calcutta Families, 1830-1855* by Mary Bennett
[21] *JVI*, p8.
[22] *JVI*, entry for 19 November 1852.

Haversham spent 46 days in bed. He then (on 11 January 1853) started to move about on crutches. As soon as he became mobile, he rather bravely headed for Fort Point and sketched the scene of the accident, then applied to be posted to Burma. He was told that the 80th Regiment was going to Rangoon, and was attached to them. The Commanding Officer was Major Holdich, the father of Sir Thomas Holdich, KCMG, KCIE, CB (1843-1929), later president of the Royal Geographical Society. He is best known as Superintendent of Frontier Surveys in British India and author of numerous books, including *The Gates of India*. He became a lifelong friend of Haversham and wrote the best of Haversham's several obituaries. On this voyage, Haversham adds:

> One of the Ensigns going with the draft was Wolseley of the 80th. That was my first meeting with the afterwards famous general ...

Field Marshal Garnet Joseph Wolseley, 1st Viscount Wolseley KP, GCB, OM, GCMG, VD, PC (1833-1913) served in Burma, the Crimean War, the Indian Mutiny, China, Canada, and widely throughout Africa, including his brilliantly executed Ashanti campaign (1873-74) and the Nile Expedition in Sudan against the Mahdi in 1884-85. His reputation for efficiency led to the late 19th-century slang expression, 'everything's all Sir Garnet', meaning 'all is in order'.

The *Mozuffer*, however, was not due to sail until February, 1853. The remainder of Haversham's time in Calcutta continued to be occupied in routine social engagements, such as going to a ball, in dress uniform, while irritably observing that it was '*very provoking not being able to dance*',[23] and having his hair cut, which the barber '*certainly did with a vengeance*'. This new note of awareness of appearance chimes with two references to visits to people identified only as the 'S---s', coyly suggesting the possibility of some love interest.

Perhaps the most important formative experience, however, remained. On 27 January, for the first time, Haversham visits the Indian Museum in Park Street. It claims on its website to be 'the ninth oldest regular museum in the world', (whatever 'regular' means) and was housed at the meeting place of the Asiatic Society at 1 Park Street Calcutta from 1814 to 1878, after which it transferred to its present venue. Haversham's first observation reveals that this visit was a pivotal event in his personal development as a scientist:

> They have some very fine skeletons there of everything from man downwards. One is struck with the similarity of ourselves and the monkey *(illegible: 'link'?)* in the formation of the bones ...

This is a clear reference to Haversham's sympathies with the Theory of Evolution, despite the publication of *The Origin of Species* still being six years away. (It is well known, however, that Darwin's theory had been in wide circulation for perhaps 20 years before the publication of his book.) It was a significant visit: just over 20 years later, the Indian Government asked Haversham, by then perhaps the finest natural historian working in India, to arrange the museum's extensive ornithological collection. Work of fundamental importance was also done by him at this time on the Indian malacological and conchological collections.

Returning to the *Diary* account, we are given a brisk series of events leading up to departure for Burma. Haversham relates that his preparations consist of seeing the pay officer for his 'dibs' (pay in rupees) and making his leave-taking calls. On 5 February, he boards the *Mozuffer* for its departure for Rangoon, this event being marred for him by being kept awake as the ship got up steam and by the many mosquitoes. It was these insects which were about to make his voyage much more irksome and dangerous than he had expected.

---

[23] *JVI*, entry for 24 January 1853.

CHAPTER SIX

THE ANGLO-BURMESE WAR

To Burma

HAVERSHAM'S TROUBLES ON THE NINE-DAY voyage to Rangoon began with his observation that, 'As an instance showing how careless sanitary arrangements then were, & how few precautions were taken, ... there were cases of smallpox among the crew.' This terrifying disease, now effectively eradicated, came in terrible epidemics and was usually fatal. In 1849 nearly 13% of all Calcutta deaths were due to smallpox.[1] On the day after departure, this cool young officer recorded that:

> I was seized with a great drowsiness and I found I could hardly keep my eyes open ... In the morning I found a rash on my face and body ... The doctor was sent for and I was told that I had got smallpox. The intolerable discomfort of such a thing, bad enough to myself, was still greater to my brother officers, for the only place in which I could lie was on a sofa in the gun room where they assembled from time to time for meals. The irritation of the smallpox was terribly increased by the swarms of mosquitoes in the place. The only way I could keep it down was by dabbing my face with a wet towel. Extraordinary to say, none of the occupants of the gun room went down with the disease, but a few of the sailors did. Very soon after our arrival in Rangoon I recovered from the attack, & was able to go on shore and then ride about and do my duty, although I was rather weak.[2]

It is a remarkably casual description of a frightening illness but is at least consistent with the restrained account of the acute pain which Haversham must have suffered when he broke his femur. On arrival at Rangoon, his 'weakness' delayed Haversham from going up river straight away to see General Godwin at Prome. The war against the Burmese, meanwhile, had reached its endgame. While Haversham no doubt felt that he had reached his intended destination, his stay in Burma turned out to be not just quite short (seven months) but also the springboard for postings to at least five other parts of the sub-continent before his return to Burma in 1866 as the first explorer and surveyor of its northern border with Assam.

---

[1] *Proc R Soc Med.* 38 (3): 135-40, London 1945.
[2] *JBI*, pp10–11.

THE BACKGROUND TO THE SECOND ANGLO-BURMESE WAR

The *casus belli* for the Second Burmese War was the alleged 'oppression' of British traders by the Burmese. The Burmese court at Ava was bound by the Treaty of Yandaboo, 1826, to protect British ships in Burmese waters. But a dispute, clearly manufactured by the British, arose between the Governor of Rangoon and the British ships, the *Monarch* and the *Champion*.

The Governor-General of India, Dalhousie,[3] quoted Lord Wellesley's pompous assertion that 'any insult offered to the British flag at the mouth of the Ganges should be treated as an insult offered at the mouth of the Thames'. Dalhousie, however, recorded in his personal diary his aversion to resolving the matter by war, although given that Lambert, the 'combustible commodore', went out to Burma with Dalhousie's full connivance, one wonders if the diary entry was more intended for self-justification in the eyes of history than recording any personal scruple. The Burmese eventually removed the Governor of Rangoon but this action was not considered sufficient. Commodore Lambert deliberately provoked an incident and then announced a war.

The Burmese kingdom offered little resistance. The first real blow was struck by the British on 5 April 1852, when the port of Martaban was taken. Rangoon was occupied on April 12 and the Shwedagon Pagoda, the Burmese equivalent of St Paul's Cathedral, was entered two days later after heavy fighting, after which the Burmese army retired north. Bassein was seized on 19 May. General Godwin led this engagement in person, with 800 men (a mixture of British army regulars and native Indian troops): having started early, the attack was concluded by 6pm.[4] Pegu was taken on 3 June after some sharp fighting round the Shwemawdaw Pagoda. When the fighting stopped during the rains the politics then began: the approval of the East India Company's court of directors and the British Government was obtained for the annexation of the lower portion of the Irrawaddy River valley, including Prome.

Lord Dalhousie visited Rangoon in July and August; he decided that to dictate terms to the Court of Ava by marching to the capital was not how the war should be conducted. Complete annexation of the kingdom was judged unachievable, both militarily and economically, for the time being. As a consequence General Godwin, who bitterly resented having to deal with the Royal Navy under the command of Lambert, a mere commodore, after the death earlier of Rear-Admiral Charles Austen, occupied Prome on 9 October, having met only slight resistance from the Burmese forces under the command of Lord Dabayin, son of General Maha Bandula (who had been killed in the First Anglo-Burmese War). Early in December Lord Dalhousie informed King Pagan that the province of Pegu would henceforth form part of the British dominions. After that declaration, General Godwin's heraldic crest was emblazoned with a collar inscribed 'PEGU' on the Godwin lion rampant.

The proclamation of annexation was issued on 20 January 1853, and thus the Second Burmese War was brought to an end without any treaty being signed. The war resulted in a revolution in February 1853 at the Court of Ava, with King Pagan Min (1846-52) being overthrown by his half brother Mindon Min (1853-78). Mindon immediately sued for peace. No treaty was ever signed although trade resumed between British Burma and the Kingdom of Ava until fresh hostilities broke out in 1885.

By what his supporters considered wise policy, Dalhousie attempted to pacify the new province. But in language and culture it was utterly different from Dalhousie's India, and could never have been successfully integrated into the Indian system. The result of the war was to add to the Indian empire an expensive new military and political dependency which did not even generate sufficient

---

[3] James Andrew Broun-Ramsay, 1st Marquess of Dalhousie KT, PC (1812-60) was an expansionist Governor-General of India.

[4] General Godwin's despatch.

taxes to pay for itself. Why Dalhousie was so obsessed with increasing the size of a territory that could not pay its way has never been explained.

General Godwin played a highly successful military part in achieving Dalhousie's political aims. Although he resented having to deal with Dalhousie's 'combustible commodore', it gave him the opportunity to co-ordinate land and sea forces, particularly in the spectacularly successful amphibious assault on Rangoon. (Haversham's diary is peppered with references to troopships moving forces and supplies between Pegu, Rangoon and Prome.) This strategy of amphibious warfare, a development of the 16th-century concept of Naval Brigades, now the Royal Marines, was probably General Godwin's greatest contribution to the history of military ideas.

GROWING UP IN BURMA

When Haversham arrived in Rangoon in February 1853, still aged only 18, the war itself was over, although the overthrow of King Pagan was still going on at this time and, according to rather dubious Burmese sources, resistance continued in British-occupied Pegu for three years. All that remained was the formal surrender of the Burmese forces and the declaration of peace which took place at Prome on 20 June 1853. Haversham witnessed this 'cast of thousands' spectacle, and produced a magnificent, beautifully detailed watercolour of it.

Haversham's seven months in Burma were a powerful formative experience. Old activities were revived, such as the observation of wildlife, particularly of new creatures such as crocodiles (which he wrongly calls '*alligators*') and flying foxes; monkeys were much in evidence, and he enjoyed his first ride (of many) on an elephant, with the general when they visited the White Pagoda.[5] There was a farcical incident when he met a Captain Lattock (? the name is not clearly written) who was hunting for his mongoose which had escaped from its box.[6] More seriously, on a rather tetchy day, he wrote, '*I am getting tired of writing ... I think a Journal is a stupid thing, and not worth the trouble of writing*'.[7] He went on to contradict himself by immediately stating:

> An elephant died at the Beach, going into the water. Dropped down of apoplexy. It was immediately cut up and carried off by the Burmese, who will eat anything – lizards, snakes and any animal that dies.

As descriptive writing, this comment hardly compares with Orwell's *Shooting An Elephant* (1936), but the abrupt, short, first two sentences nicely echo the elephant's sudden collapse, and the wider conclusion helps our understanding of how at least part of Haversham's mind was working at this time. Inevitably, much of Haversham's interest in local nature is taken up in hunting and shooting it. With 21st-century hindsight it is rather uncomfortable to record that Haversham, generally at dawn or dusk and almost daily, shoots flying foxes and crocodiles, as well as pigeons, doves and plover.

Rather happier is his interest in plants and fruit. He enthusiastically eats mangoes for the first time, '*delicious, the best fruit I have had since leaving England*'[8] or '*I bought some very fine cucumbers of a market woman: 6 for 2 annas. 3d.*'[9]

Like many other naturalists and gardeners of this imperial era, he wants the new species he sees to be reproduced in England, '*Saw a very pretty flower growing in bunches on a sort of shrub, also got some convolvulus seeds for ... Chilworth.*'[10] Chilworth Manor still has gardens full of fine exotic trees

---

[5] *JBI*, entry for 15 March 1853. Since many Burmese pagodas are painted white, it is difficult to tell which one this is, although the best candidate appears to be the pagoda at Mingun, not far from Rangoon.
[6] *JBI*, entry for 16 March 1853. The mongoose was presumably at the captain's as a deterrent to snakes.
[7] *JBI*, entry for 4 April 1853.
[8] *JBI*, entry for 29 April 1853.
[9] Anna: one-sixteenth of an Indian rupee.
[10] *JBI*, entry for 28 April 1853.

and shrubs, though it is not possible to know which specimens originated with Haversham. There is usually an Open Day around May – gardeners will not be disappointed. At least one green-fingered friend of mine is critical of Haversham for bringing the convolvulus, with its enthusiastic ability to dig itself in then spread, into this country, although he was no doubt not the only offender. Quite early in his stay he admired the jungle, which he considered 'grand in the extreme'.[11] It is to become one of Haversham's most extraordinary characteristics that he seems equally at home in both the densest jungle and the highest mountains on Earth.

An even happier set of observations, although loaded with ironic consequences, is of the local women. He often comments, with some relish, on the native ladies' appearance and dress: one week after arriving, the assembly of thousands of Burmese for a major religious festival provoked the following appreciative observations:

> The dress of the women is very picturesque, generally of a yellow colour, striped with blue and red, reaching from the middle to the feet, going once and a half round and fastened in front. Above this they wear a red silk thing, passed several times round, but not reaching above the armpits ... over this again is an open jacket like ours but made of white linen and sometimes of black velvet. Their heads are bare and their hair is combed off the forehead and twisted up in a knot at the back of the head with a sort of curl hanging down and some flowers stuck in it. They have very beautiful black hair, put up with great care. The young girls are often very pretty and are fairer than the natives of India.[12]

It is not just the sensuality of this description which is striking but the comparison at the end: his description is artistically as well as sexually appreciative. Within a fortnight of arriving, he had 'painted the figures of Burmese women'.[13] The diary's endpapers contain six female sketches, although one looks to be of a European woman.

One thought-provoking entry, towards the end of this diary (by which time Haversham had returned to India) is perhaps linked to the above:

> Dull hot day. Very hot. Went over to Granville (?) in evening and had talk with him about ~ [14]

The final symbol resembles an 'S' on its side. It might be a cipher for 'sex' (although it could be the name of a third person whom Haversham might not want others to know about; there is only one other example of this in both diaries). If the final symbol is indeed a cypher for 'sex', then that would hardly be surprising: the affair with Kudikji began, at the latest, two years after this bit of code-writing. His other descriptions of Burmese women are similar: they vary between artistic appreciation of their tastefulness and enjoyment of their elegance in dress and appearance ('*these dresses are very picturesque and graceful. I envied some of the shawls the Burmese women had on, of red silk and gold – they are very tasteful in doing up their hair*')[15] and a kind of just-repressed, scarcely self-knowing sexual hunger.

Another theme much developed in his diary is Haversham's understandable worries about sanitary conditions and disease; '*jungle fever*', he claimed, '*broke his health*' during his explorations in the northern Burmese and Assamese jungles in the 1860s and '70s. He goes bathing, without really realising the risks he is running, in some of the Irrawaddy Delta lagoons virtually every day. Illness is never far away – diarrhoea (which even Garnet Wolseley suffers from) occurs often and is

---

[11] *JBI*, entry for 13 March 1853.
[12] *JBI*, entry for 21 February 1853.
[13] *JBI*, entry for 28 February 1853.
[14] *JBI*, entry for 24 June 1854.
[15] *JBI*, entry for 13 March 1853.

treated resolutely if ineffectively with castor oil; infections are treated with leeches, as many as 150 being applied at any one time (including, on one unpleasant occasion, six to each eyelid). Other unspecified illnesses, generally given the vague designation 'fever', are treated with saline solution and, more desperately, despite the heat (which on at least one occasion he records at above 100°F, with high humidity), by wrapping himself in a blanket until '*perspiration took the headache and fever away*'. Much worse, however, is the sudden death of a brother-officer, Harper, the assistant-surgeon on the *Lord George Bentinck*, from cholera, '*he was taken ill in the morning and died next morning*'. Haversham says he was, '*a strong young fellow, that one would think would never die, but cholera takes weak and strong alike.*'[16] Survival expectations are low: at one point, Haversham observes that the 18th Regiment '*only have 35 in hospital*' from disease.

To offset these illnesses, Haversham enjoys much friendly gadding about with brother-officers, particularly the early morning ride, which happens almost daily. One of his most frequent riding companions was Chads, presumably the son of the admiral who was the godfather of Haversham's youngest brother, Edmund Chads (b.1854). When he arrived in Prome, Haversham again bumped into Pocklington, with whom he shared a bottle of Moselle at tiffin; on another occasion, they 'demolish a couple of bottles of champagne' but this was an expeditionary field force, so the social events lacked the glamour or frequency of those enjoyed in India.

Set against this sociability is of course the duty of being part of the invasion force. As soon as he had recovered from his bout of smallpox, it being perhaps a fairly mild form as no subsequent pictures of Haversham show skin damage, Haversham sailed up the mighty Irrawaddy to meet General Godwin, rather like the uncertain narrator in Conrad's *Heart of Darkness*, although the General makes a rather unlikely Kurtz. The fog Haversham encountered on the way, however, resembled the blurring of everyday reality which makes such demands on readers of *Heart of Darkness*. On the *Desmond*, despite the unwelcome presence on board of cholera and dysentry, a lift was helpfully offered to a group of Jewish merchants from Rangoon; on a later trip, he transported some Burmese who are afraid of being attacked by dacoits (groups of armed bandits; the word is derived from the Hindi). They passed, rather languidly, other troopships on the way.

The meeting between general and grandson, rather like a minor-key version of Livingstone meeting Stanley, occurred at the General's house in Prome:

> I walked ... up to the general's house – at last I had got to him. I walked into the verandah where he was sitting up at the end. He watched me coming up and I declare he did not know me until I told him. I must have changed very much ... He was paler and thinner, but the same as ever. I was happy to get him at last, being stopped so many times ... [17]

Next day, Haversham and Chads reviewed the Burmese fortifications. (Although the war was over, there was a kind of siege or stand-off at Prome. The Burmese, convulsed in their own political upheavals, were refusing to sign a peace treaty.) As ADC to the general, Haversham did not otherwise need to do very much. Apart from accompanying him in his various activities (and that by no means all of the time), little military work is mentioned apart from sorting out the general's letters ten days after he arrives. His undemanding job meant that he was therefore able to put his Sandhurst surveying studies, and the geological knowledge he had acquired through his father and his own reading, to good use for the first time by mapping the almost impossibly complex channels of the Irrawaddy Delta. His talents in conducting this survey were to stand him in excellent stead before long, and in a much more surprising way than he could have realised: the quality of his work was noticed by Sir Andrew Waugh, the Surveyor-General of India, who was behind appointing

[15] *JBI*, entry for 13 March 1853.
[16] *JBI*, entry for 3 May 1853.
[17] *JBI*, entry for 15 March 1853.

Haversham to the newly-formed (1855) Kashmir section of the Great Trigonometrical Survey in 1857. Much later in life, Haversham recalled what an important moment this was:

> When I got to Pegu in 52/53 I devoted my time to its geology which afterwards was of great help to me in the Khasi Hills of Assam.[18]

Otherwise, he had time on his hands to describe a variety of military and political events. For example, on 4 April, he mentions, with a mixture of interest and superciliousness:

> An historical day ... the meeting of the English Commissariat with those of Ava – and they were treated with all possible honours. Major Mayhew, Captain Latter and Lieutenant Pearson waited on them on the Beach when they landed, and they also had a guard of honour and a band. The procession was very pretty with their golden umbrellas and their curious dresses in which they looked now like old women than anything else. The Mengee *(the Burmese military commander)* had a sort of pagoda thing on his head and his dress was one mass of embroidery in gold and silver. He had a large body of attendants and guards ... Each of the Commissioners had a guard of Sepoys and Irregular Cavalry, and the General's staff turned out very smart in epaulettes, cocked hats and all. The horse battery was drawn up by the side of the *(illegible: probably the name of the nearby river)* ... and fired a salute of 19 guns.

The Burmese, however, continued to refuse to sign the proposed peace treaty, and asked for an extension of a week. A successful outcome depended on how much territory the British wanted to acquire. Godwin and Phayre[19] are left to argue it out. As negotiations continued without any definite result, even celebrating the Queen's Birthday with a '*feu de joie*'[20] was insufficient to stop the drop in morale once the monsoon had set in. By June, the Burmese king was asking for a lifting of the British blockade. The immediate British response was to hold a troop inspection at Myaday (now Baday Island): two companies of the 61st, two of the Madras Artillery and four companies of the 35th infantry were assembled, no doubt in order to intimidate the Burmese.

Haversham moved to Rangoon, and then, once more, up river to Prome. For no apparent reason, and of course all the more intriguing because of it, the diary has no entries between 15 and 27 July. Just before that gap, however, Haversham asserted that this was, ' ... *a very happy time indeed. I should not have cared how long it lasted. I don't at all like the idea of going away* ... '[21] A happy remark which contrasts with his grumpy assertion a month before that, '*I regret having to stay much longer in Burmah.*'[22]

Nothing indicates what caused this complete change of heart. From this point in late July onwards, Haversham was taken up with leave-taking as the military campaign came to an end. He left Prome on 29 July. At Rangoon, celebrations took place on 2 August as General Godwin handed over command to General Steel,[23] then, in an impressive ceremony, the subject of a picture in *The Illustrated London News*, General Godwin departed on 3 August for India. This departure, however, produced an unexpected but long-term effect on the accident-prone Haversham:

> They fired a salute ... the first gun as the General put his foot on deck. I was in

[18] Letter to Sir Aurel Stein, 9 April 1923.

[19] Lieutenant-General Sir Arthur Purves Phayre, GCMG, KCSI, CB (1812-85) was a career British Indian Army officer who was the first Commissioner of British Burma, 1862-67.

[20] Firework display.

[21] *JBI*, entry for 14 July 1853.

[22] *JBI*, entry for 8 June 1853.

[23] Brigadier-General SW Steel KCB, commander of the Madras Division (b. and d. dates not known).

a boat which came up just as the General was on the deck and as I came up the side the gun over the companion ladder was discharged a few feet over my head, a 68 pounder. The noise was terrific and made my ears ring for the next two days.

It seems probable that here was the source of the deafness which, alleviated only by an ear-trumpet, afflicted Haversham throughout his old age. His compensation was his war service Pegu Medal and Clasp after the peace agreement was concluded. And he was promoted to ensign, also in 1853, although the exact date is uncertain.

Another incident is acknowledged elsewhere. On Good Friday (25 March 1853), for example, he was kicked by a pony and declared, '*I am always getting something.*'[24] Accidents did often occur in the early part of Haversham's career, although much of what went wrong can be put down to inexperience through having to adjust to an alien world very suddenly, with only the kind of limited preparation provided by Sandhurst. Sometimes, however, he witnessed accidents which occurred to others; Haversham just happens to come across them:

> ... I had to pass through some Burmese houses. I heard a great row and ran to see what it was but I found a couple of Bengalis beating some Burmese with sticks. I immediately fell upon the Bengalis who of course as usual had begun the row by snatching some pice[25] from a Burman, and made them move off sharp.[26]

Haversham was not a tall man – he could not have been more than 5ft 4in – but he clearly had confidence in his own strength, his sense of justice and no doubt his British officer status. This was not the only incident of its kind:

> ... a Bengali was brought up by some Burmese for beating one of them with an immense stick, for which he caught it himself, a European soldier taking his hands and holding them high round a post while another accustomed to these sort *(sic)* of things gave it him over the back with a cane. The fellow made a tremendous row about it ... and rolled on the ground as if he was half killed.[27]

One other unexpected incident shows his interest in local customs or practices:

> ... as we *(Chads and Godwin-Austen)* were coming home, we passed a house where something was going on so we stopped and went in, and it was a young girl who had been knocked down by a stone striking her in the eye thrown by some boy ... We recommended vinegar and water but they would have nothing to say to it, and did a most curious thing rubbing their hands together till it *(sic)* is very hot, and putting it on the place. They then got some tobacco and water ground on a stone to a sort of paste and rubbed it into the sore, which brought her to very soon and made her cry out about it.[28]

Apart from these rather painful sorts of accidents (at one point, a soldier is mentioned as having drowned while bathing, in a way which suggested such events were a matter of course), the remaining unpredictable events were the arrival of the post:

---

[24] *JBI*, entry for 25 March 1853.
[25] Subdivision (one sixty-fourth) of the Indian rupee, the currency imposed by the British after the conquest of Burma.
[26] *JBI*, entry for 2 June 1853.
[27] *JBI*, entry for 26 May 1853.
[28] *JBI*, entry for 9 April 1853.

> There was a great opening of letters after dinner. Heard the news of another Prince | | | |.[29]

The four vertical strokes presumably refer to the birth of Prince Leopold on 7 April 1853, he being the fourth male child of Victoria and Albert.

Haversham observed many other aspects of Burmese daily life. For example, '*We watched some Burmese playing a sort of foot ball standing in a ring and kicking in from one side to the other – the clever way they kick it is very good.*'[30]

When it came to pastimes requiring a little more mental strenuousness, he played draughts from time to time and read *Eothen* by AW Kinglake.[31] And his facility for language-learning progressed, as he proudly remarks in his journal: '*I am getting on famously with the languages*'.[32] Just what Haversham was learning is a puzzle – in the back of *JVI*, he lists six sides of Anglo-Burmese vocabulary; but this language is not standard modern Burmese. (Nor is it Bengali, which he might have been expected to take an interest in while in Kolkata.) It may be a variant of Mon, spoken in southern Burma, and parts of Thailand, where Haversham was mostly stationed. Given that he refers to more than one language, one assumes he is keeping up with his Hindi at this time also.

He was at the same time making remarkable artistic progress; the watercolours of this period show a maturing sense of realism in the precision of his architectural and topographical drawing and in the way in which ethnic dress and physical features are so accurately depicted, as well as a growing ability to handle colour. The artistic stimuli were the pagodas, *pujas* (prayer ceremonies) and the festivals of Burmese Buddhism. Haversham was fascinated by them. They form a major theme of his finest watercolours; these, and others, are analysed in Chapter 22. This interest in Buddhism remained lifelong. In Haversham's garden at Nore, where he spent his last couple of decades, stood a beehive-sized stone model of a Burmese temple (which is still there). There are many diary entries mentioning his going to church, sometimes twice in a day, throughout his stay in Burma; how this regular Anglican churchgoer's interest in Buddhism evolved is dealt with in Chapter 24.

Where the Burmese gathered, whether for religion or holiday amusement, Haversham often made his way there also. One of the most obvious introductions to Burmese religious culture lay in observing the festivals. The water-festival ceremony, for example, he found particularly impressive. On St Patrick's Day:

> This was the time (9pm; 17 March) of the Great Water Festival. It was an extraordinary scene. The girls danced round a circle with large chatties[33] of sand and oil burning in the centre ... the most abominably noisy music of tomtoms etc. They move their feet about and bend their bodies into all sorts of attitudes, twisting their fingers about. Afterwards they had some acting which from the roars of laughter from the Burmese must have been very amusing and witty but rather vague to us. Got home at 1/2 past 11.[34]

It was an event which could occur anywhere throughout history; drama festivals and religious

---

[29] *JBI*, entry on 29 May 1853.
[30] *JBI*,entry for 28 February 1853.
[31] Alexander William Kinglake (1809-91) was an English travel writer and historian. He was born near Taunton and educated at Eton and Trinity College Cambridge. In 1856 he abandoned the Bar in order to devote himself to literature and public life. His first literary venture was *Eothen; or Traces of travel brought home from the East*, (London: J.Ollivier, 1844), a very popular work of Eastern travel, published anonymously, in which he described a journey he made about ten years earlier in Syria, Palestine and Egypt, with his Eton contemporary Lord Pollington. It was instantly successful. *JBI*, entry for 29 April 1853.
[32] *JBI*, entry for 14 July 1853.
[33] Burmese: a large boiler or cauldron.
[34] *JBI*, entry for 17 March 1853.

services have, after all, enjoyed an uneasy co-existence since the days of Ancient Greece. The Burmese have developed a sophisticated native drama; Haversham carefully observes their portrayal of the morality of everyday life:

> After dinner, went to a *(march?)* held in the street of the town ... It is very curious, but the man who acts the part of our clowns who always say when they come on the stage, 'Well, here we are',[35] the Burmese say, 'Here I've come again',[36] at which the Burmese laugh immoderately. At one part of the play, three men are trying to get the same woman for a wife and she chooses one of them. And the clown consoling one of them says, 'Never mind my good fellow, you buy *(?)* the cradle,' and it is a sort of low wit of that kind which goes down most.[37]

The festivals and the plays, however, are only the surface aspects of the country's religious and moral character. What grips Haversham, from the very beginning of his stay, are the pagodas. He had sketched one within three days of arrival. It was partly the curious architecture which attracted him, but the main motive seems to have been to observe – or take part in? – a *puja* ceremony. His visits to the pagodas, and pujas in particular, were recorded almost as often as his church visits. This enduring fascination for Buddhism might also help to contextualise his supposed ('*to all intents and purposes*') conversion to Islam before his marriage ceremony to Kudikji in 1858. It shows Haversham as a man of religious feeling, although evidently not a specifically Christian one, as the discussion on Haversham and Buddhism towards the end of Chapter 24 makes clear.

The diaries for this period record a life which could hardly be more alien from Surrey. Despite the monsoon drenchings, the terrible climate, the disease and the strains imposed by language barriers, Haversham had evidently taken to Asia. Even in old age he sometimes mentioned that he would rather be back in Asia – India specifically – than in England. When he set sail for India with General Godwin in August 1853, therefore, he had become used to a new way of life, and his journey into the beginnings of adulthood was almost over. Sooner than he realised, however, he was to encounter the great challenges of adult life – death, sex, marriage and a very unexpected career.

---

[35] The opening of *Twelfth Night*, for example? *('Well, this is Illyria, lady.')*
[36] This remark, and those following, give some idea of Haversham's considerable linguistic progress up to this point.
[37] *JBI*, entry for 28 April 1853.

CHAPTER SEVEN

TO THE NORTH-WEST FRONTIER

Death of a General

THE TRIUMPHANT GENERAL, ACCOMPANIED BY Haversham, left Burma on the steamer *Zenobia* in early August 1853. As they left the band played 'See the conquering hero comes'. Although it was described as a '*fair voyage*', many on board fell sick, '*not even venturing out of the longboat*'. On arrival in Calcutta, the general and his officers were invited by the Governor-General (Dalhousie) to lodge at the gloriously neo-classical Government House (later rebuilt by Curzon), although Haversham declares, perhaps a little perversely, that he would rather have stayed at Spence's Hotel. His time in Calcutta '*passed smoothly*' with dinners and '*rides on the course*'. After three weeks of such activities, and visiting old friends, Haversham and the general then left the imperial capital by river steamer up the Hoogly, on a voyage which, although they found it '*very comfortable*', would bring Haversham at a later stage to almost the opposite extremity of the sub-continent. Although with his usual care he noted the other passengers' names at the start of the voyage, he was more forcibly struck by the sight of a comet (only visible for one night), the scientist thus getting the better of the socialite.

The river voyage is described in the kind of close detail characteristic of someone enjoying the novelty of exotic travel. Grandfather and grandson steamed by the Nawab's palace at Morshedabad (Murshidabad). The Nawab at that time was Feradun Jah (1838-84). Haversham correctly noted *en passant* that this Bengali city is famous for its ivory and silks. And '*a great marriage feast*' was being prepared. Haversham and Chads walked to the bazaar, but complained that there was not much to entertain them. Haversham did, however, note a range of pretty wooded hills nearby, in which '*quite a difficult race of people*' was said to live, a chance remark which anticipated his arduous work in Burma 12 years later (see Chapters 15 and 16). The party then sailed past Purneah Nala, Dinapore and Patna (they are now of course sailing up the Ganges, in Bihar). Haversham, one regrets to record, showed that he had not yet outgrown adolescent immaturity when he confesses, '*I used to amuse myself with a pellet bow, pelting the natives on the banks.*'[1]

At Ghazepoor (Ghazipur), the leading city of Uttar Pradesh, General Godwin and Haversham undertook a rather Gothic excursion to Lord Cornwallis's tomb by moonlight.[2] Haversham also strolled by night into the '*native city*', which he judged '*a fine one*'. The steamer then ploughed on to Benares (Varanasi) where Haversham, showing that his passion for religious architecture still

---

[1] *JIB*, date unrecorded. Sometime in early October 1853.

[2] Lord Charles Cornwallis, 1st Marquess Cornwallis KG (1738-1805), was styled Viscount Brome between 1753 and 1762 and then known as the Earl Cornwallis between 1762 and 1792. He was infamous for the British surrender at Yorktown in the American War of Independence. He died in Ghazipur.

gripped him, sketched the minarets. His description of Varanasi's narrow streets as '*full of sacred bulls and fakeers*' suggests perhaps a possible hardening of the cultural arteries or just an inability to throw off the imperial sense of superiority which his education and upbringing would have inculcated in him. From Varanasi, grandfather and grandson progressed to Allahabad and so to the large military base at Umballa, now Ambala, a city at that time of about 50,000 people, in the Delhi district of the Punjab; they remained there for three days. The railway to Shimla was not to be opened until 1903, so they were obliged to reach it by road. (Originally called Simla, the capital of Himachal Pradesh, this popular tourist destination is often referred to as the 'Queen of the Hills' and was declared the summer capital of the British Raj in 1864.)

At Simla, the general was accommodated at the grand, if rather suburban-looking Barnes Court, the residence normally reserved for the commander-in-chief in India. Resembling an English country house – and even a Surrey one at that – Barnes Court was named after Sir Edward Barnes, commander-in-chief of the Indian army from 1832-3. This large rambling building with half-timbered gables boasted a Moorish ballroom and decorations by Kipling's father, John Lockwood Kipling. Barnes Court has enhanced its fame in recent times as the site of the historic Shimla Peace Accord (1972) between Pakistan's Prime Minister Zulfikar Ali Bhutto and India's Indira Gandhi.

The postwar process seemed to be running smoothly. But in a letter full of the irony of hindsight, the general complained, after his expected medals and honours failed to materialise in the mail of 8 October,

> Simla, October 15th 1853
>
> ... From the day when I embarked for Burmah till now, I never gave the subject of what advantages my position might bring with it, five minutes' thought. I may have been disappointed at finding myself the subject of falsehood and calumny, when I felt and knew that my exertions for the order, efficiency, and invariable success of the forces I commanded were creditable; but I cannot be disappointed as to rewards ...
>
> I am quite content to remain as I am. No title will ever be so dear to me as that of my rank as a soldier in the army I commanded, and from which army I received every possible mark of honourable distinction, confidence, and respect – this honour no one can take from me. I know what I am entitled to look to: taking the rank of the Generals of Division who served under me, and the extent and value of the acquisitions of this last war, I have a <u>right</u> to expect what my old Chief got for the war of 1824.
>
> I find the climate of this place invigorating and healthy, and can hardly believe in the change to warm clothing, after the constant perspiration in which I have lived for the last nineteen months. I am regaining flesh and firmness in my muscles, which used to hang about me. We shall all be much benefited by our visit here; in the meantime, my house at Umballa is being put into order for me. I propose to leave this about the 20th, when I shall have a good deal to do in my Division for the next five or six months.[3]

The 'old Chief' whom Godwin is referring to is General Sir Archibald Campbell, GCB, 1st Baronet (1769-1843). From 1824 to 1826, General Campbell commanded the British forces in the first Anglo-Burmese War, the longest and most expensive campaign in British Indian history up to that point.

Nine days later Godwin was dead. His KCB had in fact been despatched from London but never reached his residence until several months after his death. He and Haversham had spent

[3] *Burmah. Letters and Papers*, Major-General Henry Godwin, pp67-8. In a letter to Haversham dated 1 August 1854, Robert Alfred Cloyne refers to this book which he 'had caused to be printed'.

only about three weeks in Simla when the general caught influenza. Haversham recorded the general's decline remarkably dispassionately: '... *from the effects of that diarrhoea, and low fever, and natural decay of the system* (he) *died on Wednesday October 25th 1853.*'

The general died '*at the house of his friend*', the Commander-in-Chief in India, Field Marshal Sir William Gomm, GCB (1784-1875), who had become lieutenant-general in 1846, then was sent out to be the Commander-in-Chief of India; he arrived only to find that his appointment had been cancelled in favour of Sir Charles Napier, whom, however, he eventually succeeded (1850-5). One wonders if it is only a coincidence that Godwin's supposed mother, the governess Jane Gomm, and Sir William shared the same unusual name. Gomm's house was Barnes Court.

The general was buried in Simla. He rests in a fine tomb, which was sketched (see plate 15) by Lieutenant-General Fred Maisey in 1882. Maisey had served under Godwin in the Burma campaign. He notes that the site of the tomb under the largest fir tree in the cemetery had been chosen by Sir William Gomm, who also put up a memorial tablet in Simla church. The square tomb in the foreground is that of Colonel John MacKenzie, who had served in the Mahratta-Pindari war of 1817. His grave at Simla is inscribed, 'Sacred to the memory of Colonel John Mackenzie, Bengal Light Cavalry who departed this life on the 5th May 1856 aged 66.'

The general's death was marked in Shalford Church in the following way:

To the memory of MAJOR GENERAL HENRY THOMAS GODWIN KCB who joined the IXth Regiment in 1800. Served in the expedition to Perrol 1800. In Hanover 180_ . In the Peninsula. In the Campaign of the Douro at Jarifa Malaga. Fueriguerola Bar______ was severly wounded.

In India in 1821, as Lieutenant Colonel of the 41st Regiment. In Berma from 1824 - 1827, commanding the First Brigade of the Madras Division, in every action, and six separate commands, captured Martaran, Tantabain, and Sembike.

In India in 1851, commanding the Sirhind Division was Commander in Chief of the combined forces in Burmah, Arran, and Tenasserim, which achieved the Conquest of Pegu in 1853, and who died at Simla October XXVI MDCCCLIII. Fortunate in those he had commanded Unfortunate in those he had served

This record of fifty three years service is here placed by his only daughter MARIA ELIZABETH AUSTEN

The final comment about those he had commanded and those he had served has generated much debate. While the first part is clear enough, as a statement of admiration for his troops, borne out, one can well believe, by his various victories and letters, the second part is more ambiguous. If it stopped at being a purely political statement then it might refer to a single, powerful enemy on the Board of the East India Company. Because the war had provoked considerable political controversy, General Godwin's private thoughts were defensively published by Robert Alfred Cloyne, presumably in 1854 – no date is provided – as *Burmah. Letters and Papers* (London, Thomas Bosworth, Regent Street). Some of his remarks may clarify the inscription:

Rangoon, on board the 'Indus', June 21.

The war is now virtually over, and time alone will show the success of the line of policy we have pursued; the army will be broken up and some batta[4] distributed to it. There is no enemy near any of our stations ... You will have seen 'the Blue Book'[5] ere this, which shows clearly enough how unjustly and infamously I have been abused, and that for having carried out successfully all my orders. Lord

[4] Batta is an Anglo-Indian military term, probably derived from the Canarese *bhatta* (rice in the husk), meaning a special allowance of pay made to officers, soldiers, or other public servants in the field.

[5] A booklet of relevant diplomatic correspondence, usually pertaining to a war.

> Dalhousie's paper to the Secret Committee (November 3 1852) is one of the best, strongest, and well-reasoned documents I have ever read.

Further evidence of the treachery of the HEIC Board can be seen:

> Lord Dalhousie, in a note to me on my arrival here (Calcutta) says 'that he wishes to be the first to congratulate me on my return, and on the termination of a service which will be more justly appreciated when ignorance and malignity have exhausted their influence, and time has further developed the result.' This 'malignity' (a very proper word for what has followed me) puzzles me – knowing, as I now do, its source in that very Board by which India is supposed to be governed.

The villain of the matter was Lord Halifax, who was President of the Board of the East India Company. A *Memorandum to the General Commander in Chief, Horse Guards*, preserved in the archive, and written when the war's greatest victories had been won, crystallises his and Godwin's enmity:

> The Major-General has expressed himself in terms the most peremptory & has exercised his authority by advancing claims for the Army to an extent not only not justified by, but absolutely contrary to, the rules of the Service.[7]

The dispute centres on who commands the marines. The commander, declares Halifax, is in Portsmouth; General Godwin contravenes the rules by commanding the marines in Burma. It is an old story, good for our own times, of the commanders in the field being frustrated by the manoeuvrings of the politicians at home. The Shalford church inscription, however, might also allude more personally but in a veiled way to the general never having had his royal – or indeed any – parent publicly acknowledged. *The Illustrated London News*, which had reported Godwin's triumphal departure from Prome only 11 months before, commented:

> General Godwin, the veteran East Indian Commander, died at Simla on 24th October last. General Godwin, who had risen to his high rank through long and laborious service, was chosen to conduct the late warlike expedition against Burmah: his appointment and his subsequent dilatory mode of proceeding afforded subject for much public discussion. The General was, however, far from being unsuccessful: his progress proved slow but sure; to his great defeat of the Burmese on the 21st November 1852, and his subsequent conquest of Pegu, is, no doubt, owing to the victorious peace which our arms obtained last June from the Sovereign of Burmah. General Godwin has not long survived this prosperous ending of his campaign.

There then follows an unsigned drawing of the general, with the following caption:

> The above portrait has been engraved from a very clever and characterful pen-and-ink drawing, forwarded to us from Burmah, by a non-commissioned officer, who served under the general during the greater part of the Burmese War.

It could hardly have been Haversham's drawing, however, as he was not an NCO. It is in any case rather lifeless (though Haversham was not particularly good at drawing figures), and hardly shows

[6] *Burmah. Letters and Papers*, Major-General Henry Godwin, p57.
[7] Unpublished *Memorandum*, 25 November 1853.

the general as a distinctive military figure, which is how one imagines the young officer would have wanted to have portrayed his admired grandfather. Perhaps the last word on the general should go to one of his officers, whose comments remain on the back of the case in which his medals still hang:

> The General is very popular, being always kind in his deportment to all who address him. His disregard of his own personal safety, and of the bullets of the enemy ... the cool manner he rode about, when many about him were bobbling their heads, stamp him as a brave man. I heard him roundly rate, though in a kind way, some European soldiers for bobbing their heads. The cocked hat he wore rendered his vicinity anything but safe ... No man in the field was more exposed, or behaved with more coolness ... [8]

Frustratingly, no record remains of what Haversham feels about being the only member of the family present when the general died. This sudden disaster certainly produced negative consequences for this confident young man. Haversham had begun his career as the general's protégé; now, he would have to rise without the general's patronage. In one of his many letters to his son on this subject, Robert Alfred Cloyne carefully comments:[9]

> I must take the chance of this being in time for the Marseilles mail, to say, how truly happy your last, received yesterday, has made us all. I have always felt sure, since the dear General's death, what a loss you had sustained, and what a different position you would have held had his life been spared. I had hoped and even expected from the note I received from Sir William[10] that he had determined on doing something for you, after you had been some time with your Regiment. I firmly believe that it was the old man's intention to have so, and that it was My Lady[11] (who disposed of all his patronage) who chose Marwin *(?)* for reasons best known to herself. If any man's memory had a claim on Sir William, it was the General's – they began life as boys together, living in the same room, and they kept up their friendship through life. But Lady G never liked his old friends. She was civil – but she thought they might stand in the way, and from the time of his second marriage, all her influence was exerted to promote her friends & Scotch relatives[12] – I shall I suppose see them ere long, and I am glad that I shall be able to tell her Ladyship that you are on the Staff again and that you owe it to an old friend of your grandfather.[13] The only thing that makes me a little uneasy is the account you give of yourself. I do not like the bilious fever. Pray mind that – that no attack of this sort are *(sic)* dangerous, if you take them in time – and never, particularly in India neglect symptoms of coming ailments – get yourself put to right at once ...
>
> ... They now talk of a Staff Corps. I will see Genl. Witherall[14] as to what chance you might have of getting into it. Keep your Mathematics up ... [15]
>
> PS I went see the Review[16] at Spithead – but like somebody else was disappointed.

---

[8] Extract from William FH Laurie, *A Narrative of the Operations at Rangoon in 1852*.

[9] Dated Chilworth, 25 April, no year. Presumably 1854 since General Godwin had died in October 1853.

[10] Presumably Sir William Gomm, the Commander-in-Chief, (born in the same year as General Godwin).

[11] The General's wife had died in 1834; it is not known who 'Marwin'(?) is. The allusion is to Lady Gomm (1807-77), Sir William's second wife and a noted entertainer. (See further on in this letter.)

[12] Lady Gomm's maiden name was Kerr.

[13] Possibly Sir Andrew Waugh, although he was 26 years younger than General Godwin.

[14] I have been unable to trace General Witherall.

[15] Unpublished letter from RAC Godwin-Austen to Haversham, 15 April (1854: see n.17).

[16] Almost certainly the Spithead Review of 10 March 1854, before the fleet left for the Baltic under Sir Charles Napier, at the start of the Crimean War. Who was 'disappointed' is unclear – Queen Victoria?

In the following month, Robert Alfred Cloyne returns to his theme: he informs Haversham he has seen General Witherall again, though Haversham 'may have to wait'.[17] Having heard of Haversham's (unspecified) illness and being, I suspect, something of a hypochondriac, (although he did die of an unspecified and extended kidney disease), he adds with the kind of patronising tone more likely to produce a more unsympathetic reaction than the one hoped for:

> *(we were)* ... hoping to hear that you had got rid of the attack, for it is one which must not be allowed to remain upon you, and I was extremely sorry to hear that you had disregarded the just warnings you received respecting it. Why will young men always make a point of slighting the opinions of the Medical men?

Much in these letters anticipates events in Haversham's life: the influence of the powerful; illness - and the limitations of tropical and other medicine - even his successes as a surveyor.

After the general's death, Haversham's diary shows an understandable gap until 17 November 1853, when he found himself back in Umballa; he does not explain why, though looking after the general's house there would have been an obligation. No longer one of the general's dependants, Haversham might also have been required to find accommodation for himself in the large British army base. Perhaps there was nothing to explain: all he says is that he '*began again*' after '*a long season of idleness*'.

Haversham's first re-involvement in native life came on 18 November, when a great durbar was held for the marriage of the Rajah's son, '*a good looking lad about 14*'. Haversham was suitably amazed: all the local rajahs and their *jemadars* (the lowest rank for a Viceroy's Commissioned Officer), turned out to pay their respects to General Gomm. As Haversham fairly noted, '*They are a very fine race of people and beat us out and out in manners and coolness, not the slightest awkwardness*'.[18]

This durbar represented the high point of Haversham's stay at Ambala. On 23 November, he wearily recorded '*another day in Umballa*'. On 24 November, however, life was enlivened by the order for the regiment to be deployed to the distant but important British base at Sialkot, on the borders of the Punjab and Kashmir. (In 1849, the Punjab, including the present day North-West Frontier Province of Pakistan, had been annexed to the HEIC.) Haversham began by helping to load some camels. His diary then provides an excellent record of the daily life of a young British officer in the last years before the Mutiny.

TO THE PUNJAB

The regiment's slog across northern India, lasting 26 days, led Haversham first to the Gugger, a tributary of the Sutlej, itself one of the main tributaries of the Indus. To avoid the heat, marching took place early in the mornings, from about 5am to 7.30am. The regiment then shifted to Rajpoora, still in the Punjab, (about 36 km/22 miles south of the extraordinary city of Chandigarh which was later re-designed by Le Corbusier to become a kind of elegant Indian Milton Keynes). After the halt at Rajpoora, the convoy passed Sirhind:[19] they were therefore approaching the frontier provinces. By 1 December they had marched as far as the important garrison town of Loodiana (Ludhiana), the largest city in the Punjab, with an estimated population in 2008 of two million, standing on the Sutlej River's old bank. On 5 December they crossed the Sutlej; next day, Haversham bumped into 'Kirk' whom he knew at Chatham, this sort of chance meeting no doubt being a feature of Indian Army life. Although complaining of the cold by now, Haversham

---

[17] Letter from RAC Godwin-Austen to Haversham, 10 May 1854.
[18] *JIB*, entry for 18 November 1853.
[19] Meaning 'gateway to the Hindu lands (Hind)'.

nonetheless resolutely set out the following morning and reached Jelinder, (now Jalandhar, which in 1846 became the headquarters of the territory gained by the British after the First Anglo-Sikh War). As an officer, Haversham naturally proceeded on horseback. Finally, on 20 December the regiment reached journey's end in Sealkote, now Sialkot, situated in the north-east of what is now the Pakistani Punjab in the foothills of the snow-covered peaks of Kashmir near the Chenab river; it was, formerly, the winter capital of Kashmir. The city lies about 125 kilometres (78 miles) north-west of Lahore and only a few kilometres from Jammu. Haversham had arrived on the borders of Kashmir, where the greatest adventures of his life would begin.

As he travelled, Haversham recorded his daily routines in considerable detail. He would, for example, dismount to shoot geese or partridge and for breakfast one morning he ate heron, which he praised as '*excellent*'. And a tasty tiffin of roast duck and potatoes compensated for low morale one day. He further tried to relieve the monotonous routine of the march by going on a deer hunt; when stalking failed to work, he galloped towards the deer on his horse, but on firing, the horse shied and fell over on top of the thigh which the rider had broken in Calcutta. This clumsy fall prevented Haversham, who gave himself a good row for his thoughtlessness, from walking on the weakened limb for the rest of the day (28 November).

Illness or injury is indeed a banal and all too frequent occurrence, though usually of a minor kind. At Rajawalla, Haversham fell asleep on the ground as soon as he arrived, but woke with 'a most excruciating pain in my left shoulder so that sleep was out of the question' – the pain presumably induced by the way he slept as well as by the increasing cold. By morning (10 December), the pain had spread to his other shoulder, and as he plaintively remarked, he '*took two doses of Castor Oil with little effect*'. Thanks to a local (British) doctor and his pills, the pain then began to lessen, especially after bathing in hot water; by 15 December, Haversham felt well again. This illness, unimportant in itself, does nonetheless presage the sciatica and rheumatism which afflicted Haversham during the last 15 years of his life.

The pain, however, soon returned and was '*very bad*'. The doctor opined that Haversham had an inflamed liver, the cure for which was '*30 leeches on the side and other medicines*' (although it is not obvious how a liver disease and painful shoulders are connected). The pain diminished, rather amazingly, after that treatment, but Haversham mournfully remarked on 25 December, '*Not a jolly one again*'. (The previous Christmas he had spent, unhappy, bored and probably homesick, in bed in Calcutta as a result of his riding accident.) He thinks longingly and affectionately of his family, and how they will have received the news of the general's death just before Christmas. Next day – I wonder just how psychosomatic it was – the pain has almost gone.

The pain was closely associated with the weather. Although the days of non-stop perspiration in Burma were long gone, in several letters Haversham, perhaps with understandably low morale as his sense of isolation continued, declared that he would rather be back in Burma. He cannot get on with the cold: strange sentiments for the first serious explorer of the high Karakoram and the holder of a world high-altitude summiting record! The bad weather continued after the regiment's arrival in Sialkot. Lightning struck the barracks in February and knocked down a corporal and another man, the corporal being severely burnt. The pain returned in Haversham's shoulder, and an eye infection appeared: Dr Holloway's advice was to tell Haversham not to use the eye!

In the second half of March the weather improved, which eased Haversham's unhappiness and frustrations by allowing him to go out riding whenever he wanted. His eye infection, however, continued to give him problems: powders were taken and six leeches were applied. By mid-April, because he still felt very weak from the confinement and the various medicines, he feared he would not get to Kashmir but, remarkably enough for one of this wonderful region's first surveyors, dismissively calls that '*no great hardship*'. Towards the end of April, the weather was getting hotter: a temperature of 92°F was recorded, and Haversham thought of taking a month's leave in Simla, but the rigours of the journey decided him against it. Dust becomes the new irritant, although at night it was clear enough for him to observe a large meteor lighting up the sky to the north-west. He moved house, to share with a fellow officer called Head, where he thinks there will be much more room and he will be more comfortable. Haversham was, however, not the only one to feel wretched at living an arduous life so far from home: one of the soldiers shot himself on 5 May, the same day as the letters arrived from England.

This unhappy event is the first of several where Haversham records a suicide, on this occasion in greater detail than the May attempt:

> Last night another man of (the) 24th shot himself ... taking part of his nose off and then passing just clear of the eye taking a piece of his skull out of the size of a rupee – but not killing him, he being as calm and collected, answering every question put him. He said he had been thinking of doing this since the 11th and that he was quite prepared for death and did not fear it.[20]

On 20 May, Haversham had no fewer than 20 leeches applied to his ear, although it did him, he grumpily mentions, '*little good*'. By late May, all the '*gentlemen*' were complaining of the heat. For some, it inhibited their dancing at a '*reunion*' to celebrate Queen Victoria's 35th birthday (24 May). At the dance, Haversham listed six female partners, '*not one of which I knew, except the last*'; three were unmarried. His listing does show a clear and renewed interest in the opposite sex (who have not been mentioned since before leaving Burma), perhaps provoked by the heat. Not only did he '*lead off*' the ball, but did not get home until 2am. Unfortunately for him, he had to get up at 5am to go on parade, and he heard '*news which was rather disgusting*', although he fails to elucidate further. Another cryptic entry on the next day recounts that '*Holland and Hanahan* (fellow-officers who are frequently mentioned as riding and dining companions) *had had a Punchite on* ~ | |' Quite what the ciphers mean is anyone's guess but it certainly suggests that whatever the officers got up to with the 'Punchite', it was something to hide. (A Punchite was an inhabitant of Punch, or Poonch, an upland district of Kashmir. It lies in the beautiful Pir Panjal mountains, on the south side of the Vale of Kashmir about 80 miles north of Sialkot. Kudikji, though nor a Punchite herself, came from this area.)

Haversham's proneness to accident resurfaced on 7 June. While out riding at a canter his bay slipped and sent its rider tumbling over his head, to land on his face. He was '*stunned*' for a few minutes and found his mouth '*very much cut and full of sand*' because he '*had slid along the ground for some distance on my face*'. As he drily remarked, '*I shall get killed some day.*' (To compensate for this upset, on 17 June he was paid. His salary was 116 rupees, a respectable but hardly generous amount.) On Midsummer Day the whole day's diary entry is in Hindi; Haversham, commendably, never gave up practising his languages, an interest which even extended to the language of natural history classifications. He did however, late in life, observe:

> The 24th had a very clever interpreter at that time in Captain Greensill, a great chum of mine who poor fellow was killed at Delhi in 1857 ... I always dreaded being made interpreter of any regiment which in those days before the Mutiny meant never getting a Staff appointment.[21]

As the humidity built, conditions became more oppressive, necessitating moving his bed outside at 2am in order to snatch a few hours' sleep, although this disagreeable period came to an end on 25 June with a tremendous thunder and lightning storm. His tent was blown down and his servants (presumably having occupied the tent) were obliged to pass the night on his verandah.

On the march and during his stay at Sialkot, Haversham's daily life resembled that of many others: he would chat with fellow officers ('*Butler is as much a humbug as ever, but he is a nice fellow*'), play billiards, sketch, go for a bathe and ride, take tiffin and go to bed early, usually about 7pm. To break the monotony, the occasional VIP appears: at Jalandhar the officers dine with Lord

[20] *JIB*, entry for 5 June 1854.
[21] Letter from Haversham to Sir Aurel Stein, 14 December 1916.

Melville, famous only as the grandson of the first peer to be impeached by parliament and as the son of the Tory statesman. What interested Haversham rather more were the 'Hindoo' temples, although the mosques were also an artistic subject, this interest showing a new involvement with the world as his health improved. Another sure sign that he was fully restored was his stay at Dirle (?) for ten days where he played cricket, though scoring only a very modest 12 and 10 in each innings. While noting the change in the weather to greater cloudiness and rain, he also observed that the cost of bread has increased by a small amount.

In the middle of March the 'Hoalie' (now more usually Holi) festival took place. It was a joyous occasion, the time for renewal of friendships. Haversham enjoyed describing the strange action of '*all the bearers throwing a red water over themselves*'. (I experienced this festival in Kolkata recently: we were impressed by the street bonfires and derived a good running joke from some of the younger members of my expedition keeping up with tradition by covering their faces in red, purple and green vegetable dyes, without realising that such dyes took several weeks to wash off, much to the bemusement of the Bhutanese whom we were on our way to visit.)

An interaction with Muslim culture, on the other hand (Holi being a Hindu festival), occurred on 12 May, when a partial eclipse of the moon took place. Fireworks were let off. Haversham clearly enjoyed participating: '*This is a great night with the Musulmans – what they call a "shuherat"*.'[22] Although the pain in his ear returned the next day, he was still able to discuss a Cadet Parade with the colonel. Despite its being scarcely legible at this point, the diary also contains a sentence in transliterated Hindi (although all my attempts at translation have so far failed). Other sentences follow in later entries. Since he also mentions reading a history of the Sikhs at this time, Haversham is evidently absorbing the culture in which he finds himself, although the occasional word in Burmese also crops up. From a later observation when on his great expeditions, by the end of his time in India Haversham could be said to have enjoyed a reasonable grasp of Hindi, Bengali, Punjabi, Kashmiri, Balti, Burmese (Mon) and some Burmese dialects. He had also acquired a smattering of Pushtu, Gujarati, Zanskari, Tibetan and Dzongkha (Bhutanese). It is probably this facility for, and application to learning these culturally important languages which bound him to India with such endearment long after he had left it.

From about the end of February, the diary entries become much less frequent. Between 26 February and his last entry on 29 June, there are only 42 entries for 123 days. This change probably occurred because there were too many new experiences to record, and it was probably too exhausting to go back over them all. It is not known if Haversham continued to keep diaries such as these, although the occasional reference to a '*travel journal*' suggests that, in modified form, he did. He certainly kept notes during the big expeditions of his climbs, although these records have, sadly, been lost.

HAVERSHAM'S ARTISTIC DEVELOPMENT

During this period, Haversham's most important activity in his daily life was keeping up with his artwork. He visited the fort in Sialkot, said to be the oldest in what is now Pakistan, having been built by Rajah Salban in the second century AD, to sketch it, along with any number of military, religious and civil installations.

On 22 May, a most significant event occurred in terms of Haversham's artistic development. He called on '*William Carpenter the painter*':

> He showed me some drawings and gave me some hints on drawing figures in W. colour. Took a likeness of *(name illegible)* to try what I could do.

[22] Now more usually 'shuhrat'.

Next day he sketched his *kidmutgar*, a servant, possibly with Pathan connections ('Khidmatgar' being Pushtu for 'servant'), then rode back to barracks in order to write home and send back some deodar seeds.[23] (On my last visit to Chilworth Manor, I noted that an ancient deodar still stands in the grounds, although it is not possible to be sure if it grew from one of these seeds.)

This meeting with Carpenter is very important. Captain Petley at Sandhurst might have been the source of Haversham's technical artistic abilities, but it is the eccentric Carpenter who shaped Haversham's talent into what it became, at least as far as the sketching of figures within an urban or landscape context is concerned, a characteristic of almost all of Haversham's finished watercolours. We know that Carpenter and Haversham were in Peshawar at the same time in 1855. (The Regiment was stationed there from 28 November 1854 to December 1856.) Carpenter's *View of houses adjoining a stream, Peshawar, 1855* (see plate 12) with its architectural exactness and realistic depiction of locals in ethnic dress bears a strong resemblance to several of Haversham's watercolours of this period. (For a more detailed examination of Haversham's art, see Chapter 22.)

LIFE ON THE NORTH-WEST FRONTIER

As May moved into June, and the weather became very hot, Haversham spent his spare time (which seems to have been considerable) riding and sketching, paying particular attention to the 'coolies' as subject matter. Evidently, his artistic talent was becoming well-known among his fellow officers: Mangle and Clarke copy his sketches, and he finishes one for Parsons.

As the diary draws towards its close, the events described become more trivial: Haversham recorded a visit to the Colonel to view some '*Hindustani pictures*' and was offered white Kashmir cherries which were described as '*not bad*', this being explained by their being '*twenty-one days on the road and rather dry*', while at Halahan's (a fellow officer) one evening he enjoyed '*a fine tumbler of goat's milk*'.

And there, on 29 June 1854, the diary ends. It is the last detailed, day-by-day account of Haversham's life we get until we look at a surviving Burmese diary from the 1870s. We know, however, that Haversham was posted to Peshawar on 28 November 1854. Both he and his parents' anxieties were, apparently, over; he had obtained a secure post as ADC to General Reed, the commander of the Peshawar Division and an old Peninsular War soldier who is described as '*ageing but understanding*'. (General Reed, who seems to have been a fairly sick man when Haversham was appointed to his staff, died of cholera on his way to Delhi at the head of his troops, not long after the outbreak of the Mutiny in 1857.) The future Lord Roberts of Kandahar had been stationed there since 1851. Roberts was stationed at Peshawar during Haversham's time there, although they were in different regiments. He describes Peshawar in *Forty-One Years in India from Subaltern to Commander-in-Chief* as follows:

> Peshawar, too, notwithstanding its unhealthiness, was a favourite station with officers. To me it was particularly pleasant, for it had the largest force of Artillery of any station in India except Meerut; the mess was a good one, and was composed of as nice a set of fellows as were to be found in the army. In addition to the officers of the regiment, there were a certain number of honorary members; all the staff and civilians belonged to the Artillery mess, and on guest-nights we sat down as many as sixty to dinner. Another attraction was the 'coffee shop,' an institution which has now almost ceased to exist, at which we all congregated after morning parade and freely discussed the home and local news.

Peshawar is one of the most important Pakistani cities west of the Indus. It is one of the oldest cities in Asia, although now it is known as a frontier town and possibly the most heavily armed

[23] Himalayan cedar tree.

cities in that turbulent part of the world. What we know of Haversham's life there is defined by his sketches and paintings and some letters. He goes out of his way to describe, for example, a water-drawing method powered by bullocks in the countryside near Attock.[24] Haversham's fine picture of the *'bridge of boats'* over the Indus at Attock, formerly known as Campbellpur, now hangs in the Regimental Museum (the 24th Foot became the South Wales Borderers then changed into the Royal Regiment of Wales) in Brecon. The bridge was constructed from boats lashed together, aligned along the line of the stream, but unable to withstand each year's monsoon floods. Such a device for straddling the Indus was used by both Alexander the Great and Akhbar the Great; the British got the better of the flood problem by building a magnificent railway bridge in its place in 1883. There was also, however, a ferry. The Attock visit, where he observed the construction of the new bridge, was described in a letter home with his characteristic enthusiasm for technical detail:

> At Attock I went down the famous tunnel that is now being made under the Indus R. The shaft was about 100 ft deep from which I sank in a Bucket, the Gallery is 175ft and has now reached the centre of the deepest branch of the river but is very small) just room to get along. *(Illegible)*[25] may bore another shaft and Gallery all in course of construction on the other bank. The present work is only a trial. Should it succeed (which they have no doubt of now) it will be made large enough for trains – and will be a magnificent work, as the Bridge of boats is broken every year. The lock is a very hard heavy, black slate. – very little water comes in, and a pair of Bullocks *(i.e. water-buffalo)* can take it all out, it is drawn up in a large leather bag. The bullocks walk down an inclined plane – I went right to the end and did a little work with a pick and came back very dirty from the wet sides of the Gallery.[26]

Haversham produced several lively paintings of the bazaar in Peshawar – now as famous for its gunshops as for anything else – and there is a wonderful panorama of the Khyber Pass, (see plate 16) perhaps the most evocative spot in the whole of the sub-continent, the gateway to safety which the British army of the Indus, on its retreat from Kabul, had been attempting to reach when it was annihilated in Afghanistan only 13 years before Haversham arrived in Peshawar.

Haversham's artistic talents were not, as it turned out, just to be used as some sort of pastime. His associated abilities as a surveyor and draughtsman, already recognised for their quality in Burma, were about to reshape his career and bring to this young ADC a permanent distinction in the history of mountaineering.

---

[24] Attock is situated in the northern border of the Pakistani Punjab. It lies at the end of a natural corridor formed by the Kabul River where it flows into the Indus. It lies just 80 km/50 miles from Rawalpindi and 100 km/62 miles from Peshawar.

[25] Probably the name of the officer in charge of the drilling operations.

[26] A drawing and diagram of how the water-buffalo worked accompanied the written description.

PART THREE

MOUNTAINEER-EXPLORER

CHAPTER EIGHT

KASHMIR

'An Outstanding Talent'

Neither the Himalayas' snows, nor arid deserts, nor dense unhealthy jungle, nor hostilities … have been allowed to arrest the progress of surveys in our time.

– Sir Andrew Waugh

There is no work on which more distinction can be gained, or more fruitful in usefulness, from both a military and an economic point of view … It enlarges the mind … it gives an officer the opportunity of studying many other branches of science … it gives him the tact (sic) *also of dealing with all sorts and conditions of people.*

– HH Godwin-Austen

WHAT HAVERSHAM WAS UNAWARE OF when he arrived in the Punjab was that his work in 1853 on the Irrawaddy Delta had been noticed by the most influential topographer in India, Sir Andrew Waugh, the Surveyor-General. In 1852, Sir Andrew had accepted the calculation – made by several surveyors and not just, if at all, by Radhanath Sikdar – of the height of Peak XV as 29,002ft (8848m).[1] In 1856 Waugh named it Mount Everest, after his predecessor, Sir George Everest, (pronounced Eve-rest) during whose time as Surveyor-General the height of the mountain had been calculated. A local name, in line with Survey policy, was unavailable because Tibet and Nepal were closed to those who might have investigated the matter. Although Kashmir still remained outside British India when Haversham arrived in the Punjab, the British maintained a sufficiently good political friendship (just) with the Maharajahs of Kashmir, Gulab (1792-1857) then Ranbir Singh (1830-85), for them to be permitted free access to his extensive territories. Gulab Singh was the first maharajah of Jammu and Kashmir, the second largest territory of its kind in British India. At the time of the defeat of the Sikhs in the First Anglo-Sikh War, he was Chief Minister of the Sikhs. He 'suddenly perplexed the Governor-General by asking what he was to get for all he had done to bring about a speedy peace, and to render the (Sikh) army an easy prey.' As a result, the British transferred all the lands in Kashmir that were ceded to them by the Sikhs at the Treaty of Lahore to Gulab Singh for Rs.7,500,000.

(In his Presidential Address to the Malacological Society on 10 February 1899, Haversham recalls a mild joke about a conversation with Gulab Singh's son Huttoo; the prince had attributed a supernatural identity to what Haversham asserted was an elephant's tooth; that this conversation took place at Rajauri in Kashmir (the site of Huttoo's residence in the 'Vale of Lakes' and the garden of the old Delhi emperors) is known from Haversham's comment written on his watercolour of this village, dated 12 September 1854. See plate 20.)

---

[1] It is well known that the calculation was actually 29,000ft exactly, but that two feet were added in order to make this measurement more believable.

In 1855, Waugh had decided (perhaps with some prompting from Captain TG Montgomerie) that the Great Trigonometrical Survey of India could be extended to Kashmir, and appointed Montgomerie to organise this new section. Thomas George Montgomerie (1830-78) was born in Ayrshire and trained as a Bengal Sapper and Miner in Roorkee. Only 25 when appointed to take charge of the Kashmir Survey, he rose to become acting Surveyor-General of India from 1870-73. He then retired to Bath, after which he was promoted to lieutenant-colonel. He is said to be the model for Colonel Creighton in Rudyard Kipling's *Kim*.

Haversham's good work in Burma having been noticed, he was promoted and then gazetted Lieutenant Godwin-Austen on 27 May 1856; his appointment to the survey became official in March 1857. (Promotion thereafter progressed quite rapidly: he was gazetted captain on 29 October 1858, no doubt as a result of his highly successful first year's work.)

A darker interpretation, however, of Haversham's promotion has been offered. His affair with the native woman Kudikji (see Chapter 9), which must have begun before June 1856, would have generally been known about by the time he joined the Kashmir Survey, as Kudikji gave birth to their son in March 1857. Since an officer under 25 years of age was not allowed to marry, Haversham was being transferred to out-of-the-way Kashmir in order to let any likely 'marriage' take place out of sight. Either that or Haversham fell victim of a futile attempt by his commanding officer to separate him from his lover.

In a much later paper[2] Haversham claimed that he joined the Survey '*as a sort of accident*'. He says he was posted to General (then Lieutenant) Sir Peter Lumsden's department at Peshawar in 1856-7. On 5 April 1857, only a month before the outbreak of the Indian Mutiny, his posting brought him into the Kashmir Survey at Manganwar, in the southern suburbs of Srinagar, Kashmir's wonderful but isolated capital, still famous for its lily-covered lake with its houseboats and snow-capped mountains to the north, a kind of Indian Venice. On 5 May, he '*started work at once*' - one of Haversham's most likeable qualities is his briskness. He comes across almost invariably as a purposeful, systematic and thorough worker. (In Burma, a decade later, Haversham enthuses similarly about the '*opportunity for exploring interesting country*' rather than complaining about extreme physical hardship, a further sign of these highly positive attitudes which he brought to his work.)

His family was evidently delighted. Robert Alfred Cloyne congratulates his son:

> ... if only to say how much we were all agreed *(?)* at the news contained in your last respecting the appointment to the Survey.

He then returns to one of his rather grinding health themes in the same letter:

> ... but you must remember it is my special wish and request that you will not take work in the low, hot parts of India. I have no objection to such a country as you are now perhaps working in ... [3]

In reality, however, neither Haversham's blossoming surveying talents nor his passionate affair secured his appointment. Robert Alfred Cloyne, as his letter above makes clear, had been pulling strings on Haversham's behalf since the death of General Godwin and Lady Gomm's removal of her patronage. In the same letter (and in a few others) he recounts that he has seen, and requested help for Haversham from, Ross Donnelly Mangles (1801-77), the Whig (Robert Alfred Cloyne was

---

[2] *Exploration Survey Work, Professional Papers of the Royal Corps of Engineers*, Vol XX, Paper VI, 1894.

[3] Letter from RAC Godwin-Austen to Haversham, 19 March (1858, presumably, although the date seems a little late). While the telegraph had been operating since 1856, the reference to 'yours' near the beginning still suggests a letter. It suggests, if Haversham wrote in May 1857, that an exchange of letters between Kashmir and Surrey was taking eight months at that time. Given that Haversham's sea voyage from England to Calcutta took 141 days, that time seems about right.

a Liberal all his life) second Member of Parliament for Guildford who very conveniently was appointed Chairman of the East India Board in 1857. Other letters suggest that Robert Alfred Cloyne remained also on terms of some familiarity with Sir Andrew Waugh (hardly surprising, given Robert Alfred Cloyne's eminence as a geologist), but it is not clear exactly how well Robert Alfred Cloyne knew him; they would probably have met at the Royal Geographical Society, where Waugh received the Patron's Medal in 1857. Waugh would certainly have known the work of one of England's leading geologists of the time.

HAVERSHAM THE SURVEYOR

Haversham gained considerable praise for his work. As the great mountaineer-explorer Douglas Freshfield correctly remarked:

> The Survey was on the whole wonderfully accurate, if one considers the speed at which it was carried out ... the marvel is how even such accomplished surveyors as Colonel Godwin-Austen and those associated with him were able to get so great an amount of accurate detail in so short a time.[4]

Sir Andrew Waugh also described Haversham in the GTS Reports as 'exhibiting special talent and much zeal', his sections as being 'especially artistic', and referred to him as an 'indefatigable mountaineer'. As Haversham himself stated at a much later stage, when addressing young surveyors in England:

> There is no duty that is more attractive. What interest is so great or so absorbing as tracing out the run of a mountain range, or following to its many sources some great river system? What work so independent and so satisfactory when it is completed, and what work can be made so mathematically exact? I only wish I could begin again this pleasant life, and view once more the countries I have visited.

At the Royal Geographical Society meeting where the Kashmir surveyor William H Purdon read his paper, *On the Trigonometrical Survey and Physical Configurations of the Valley of Kashmir*[5] and once the new Kashmir map had been displayed, the chairman, Sir Roderick Murchison, (1st Baronet KCB, DCL, FRS, FRSE, FLS, PRGS, PBA, MRIA (1792-1871), from Muir of Ord, Ross-shire, a Scottish geologist who first described and investigated the Silurian system and was President of the Royal Geographical Society from 1851-3) added accurately, if rather tactlessly:

> The map was constructed under the direction of Colonel Waugh by Captain Montgomerie, and one of the most active persons in its compilation, besides Mr Purdon, was his *(i.e. Murchison's)* young friend Captain Godwin-Austen.

Haversham expanded at a later meeting on his own somewhat world-weary feelings about the vast size of his task when he read out his paper, *Notes on the Valley of Kashmir.*[6]

> It may be called ... immense work – that of weeks, months, and years which it has taken; the altitudes climbed, the long, wearing descents (for of the two, the last is most tiring and difficult when it is continued for any length of time and the

---

[4] Douglas Freshfield: *On the Glaciers of Kangchenjunga*, *RGS Journal*, Vol 20, p236.
[5] *Proceedings of the RGS*, Vol 4, 1859-60, pp31-3.
[6] *Journal of the RGS*, Vol 31 (1861), pp30-7.

> slopes great); the days and nights spent in taking observations from the many hundreds of trigonometrical stations; the pages of computation that this afterwards entails in working out the sides of the triangles; the drawing and working up of each plane-table section, and, the large number of men who share in the work, from the officers in charge to the native signallers.

Indeed, it seems likely that, at some point during the Kashmir Survey, Haversham became disenchanted with this very strenuous work, to the point of wishing to leave it. Robert Alfred Cloyne resisted this impulse very strongly, and proposed some more of the vigorous string-pulling which had originally helped to secure Haversham his position on the Survey (and the Staff Corps):

> I have followed you in your Summer work with the greatest interest and have recorded every scrap I could gather from all your letters, upon Vigne's map[7] – I trust that you have kept a full journal of all that you have seen, and if you have only noted the inclination of the great maps and marked the positions as well as relations after different *(two words illegible)* one another – you must have saved ample material for a good paper for the Geologists and Geographers. I hope that the little clinometer was constantly in use as well as the hammer – If you can find time during the winter I should advise you to put any observations into a systematic shape, and if you will send them together with any specimens over to me. I will let you have any points I may have to make, as to points which it will be advisable that you should work to.
>
> These words bring me to the point of this note. In one of your letters to one of your sisters, you talked of throwing up the Queen's Service, that the thought should have entered your head threw me into a fever. I can safely say that I have never met a man who had given up the Service who was heartily sorry for it. William M[8] told me the other day that he would give three times the price of the company he sold to get it back again. Do not for heaven's sake add yourself to the list of men who threw up just as the profession is beginning to hold out a prospect of military rank and position. Trust me, in public life, some military rank will be invaluable to you in your position.
>
> The Regt I see is to come home but if you wish to remain in to finish your special work, I think that a special case could be made of it if Waugh would apply, it could be backed here, and thus the position could be reserved. Let me know by return if this will do, and I will then set the machinery to work at once. Christmas is drawing in – not I think that *(illegible)* very cheerful, but I should like that you were with us and you may rely on it, we shall not forget you – and believe me your most affect. Father, R G-A.[9]

Haversham's 'good paper' turned out to be an astoundingly detailed and beautifully illustrated geological description, probably the first of its kind in India, of all the places he had visited in Kashmir, an astonishing achievement, given the youth of the surveyor and the complexities of the topography. Even for a non-specialist, the quality of the work fascinates, then – in its detail – astounds: cross-sections, stratification and composition notes, flow diagrams and so on. (This historic work remains in private hands at Papplewick Hall, Notts. See plate 19.) Haversham realised, however, that his work existed not just as an abstract scientific exercise. In the same paper, he examines the political implications of what he has accomplished.

---

[7] GT Vigne (pron. 'Vine', 1801-63) was the first European to penetrate the Karakoram to any depth. See also the beginning of Chapter 10.
[8] Unidentified.
[9] Letter from RAC Godwin-Austen to Haversham, dated Chilworth, 25 November, presumably 1860, as the Regiment was at sea on their way home by March 1861.

> The country *(Kashmir)* is not ours, and it is perhaps unfortunate for it is as well for us that it was handed over to the Maha Rajah Goolab Singh ... The whole of this district is admirably suited for European occupation, and no part of India assimilates more to a European climate. Had we kept the country, and had British colonies been formed in these hills ... they would have been the means of establishing with greater firmness our supremacy over the north-west of India and in the Punjab. At present, there are no roads, there not being such a thing as a wheeled conveyance ... in the Valley of Kashmir, all produce being carried on men's backs even to the Punjab. Were Anglo-Saxon experience, talent, and labour brought to bear on these lands, they would amply repay the settler ...

KASHMIR AND BEYOND

Haversham spent two 'seasons' (those weeks between May and June or July when surveying cannot be prevented by winter snow or the monsoon) in Kashmir, before succumbing to the '*enraged Sepoy*' assault which forced his return to England. Having left Dehra Dun, the Survey headquarters, on 1 March 1857, '*on a probationary footing*', (which suggests he was not present at his son's birth in Sialkot on 15 March), the survey party went on in April to enter the Poonch Pass as too much snow remained on the Pir Panjal mountains; in May, Haversham was despatched to the Kajnag Hills on the north-west side of the Vale of Kashmir.[10] These hills rise to around 4000m/c.13,000ft, with a few exceeding that height. He surveyed 510 square miles before descending to map a portion of the valley of the River Jhelum (another major tributary of the Indus). Phillimore declared that he helped to produce the 'first map of the Kuram Valley', on 'reconnaissance work by Lt Peter Lumsden and others'. As Haversham later explained, Lumsden was '*full of other work*' and '*he handed the completion of it over to me*'. He confirmed this point when he was presented with the Royal Geographical Society's Founder's Gold Medal in 1910: '*it was Peter Lumsden's request in Peshawar in 1856 to map the Kuram Valley which led to* (my) *appointment to the Survey*' when the Surveyor-General recognised the quality of his work. The Kuram Valley lies south-west of Peshawar, and provides access to the Khyber Pass and Afghanistan, a dangerous area: a relative of Lumsden's was murdered there in 1880.

Haversham's generally accurate *Abstract* (i.e. his own summary of his military career) tells us that he was surveying to the north-west of the Vale of Kashmir after visiting the Kuram Valley. Having completed that part of the Survey, Haversham was despatched in September to the wonderful alpine area of Kishtwar, probably only the second European to have reached there after Vigne, who had passed that way several times a little more than 20 years before.

(I am familiar with this beautiful area: my first sight of Kishtwar, in 1992, was seeing it burnt to ashes and its main street patrolled by machine-gun-toting Indian soldiers. The day before, my expedition, after an eventful 11-day trek over the crest of the Great Himalaya from Zanskar, had arrived at Gulabgarh, further up the River Chenab; there, our Liaison Officer quietly informed me that the local chief of police needed to see me because terrorists had shot four policemen nearby the week before. In his lookout post on an alluvial mound with a view over three intersecting valleys, the chief eventually declared that we were going to have to go out under military escort, as the bus from Delhi which I had arranged to meet us could not now pass through the terrorist-held area down-valley. The police chief reinforced this sudden, stomach-churning shock after our magnificent trek over the Umasi La (5342m/17,526ft) from Zanskar by instructing me to tell my young expedition members that we were to strike our tents at 2am, without using any lights, then

---

[10] Much of the information on Haversham's movements comes from the unpublished *Abstract* of his military service, although a few details have been added from the magpie military gossip-historian RH Phillimore.

board a local bus which would make its way out under armed guard along the dangerous cliff-face road above the Chenab valley until the Delhi bus could be reached. The tents were struck by moonlight; a kind of chaos then followed, as many locals (with their inevitable poultry) decided they wanted to escape by sharing our vehicle; we did not leave until 8am. I found the crumbling road, hacked out of the shivering cliffs high above the Chenab at least as frightening as any possible terrorist attack, though the combination of both kinds of anxiety made for an interesting inner cocktail. Kishtwar, the first town of any size we passed through, was reduced to not much more than smoking ruins, guarded forlornly by Indian troops. Fortunately, my terse instructions in the mess tent the previous night on how to conduct oneself in the event of capture by terrorists remained un-needed, and that evening we safely reached the wooded hill station of Batote, more or less out of the danger area; a welcome and long-lasting combination of undisguised relief and Kingfisher lager rapidly followed. Next day, we trundled down to Jammu, where two local people were killed when a bomb went off two streets away from our hotel. On the following morning, we passed the hole made by the detonation. Fortunately, risk assessments had yet to be invented at that time .)

At Kishtwar, Haversham surveyed about 800 square miles of the surrounding area, dominated by the striking peak Brahmah (6597m/21,644ft). (Haversham would not of course have climbed Brahma(h) as the season was wrong and the mountain too high and technically too difficult for someone of his inexperience.) It must have been a cold time. This part of the survey lasted until December; Dras, near the present Indo-Pak ceasefire line, and by repute the coldest place in India, lies only about 100km/62 miles to the north.

Although the survey work had been completed, Haversham was reminded that he was still a serving soldier, since he was posted for the winter, '*for a brief interval with his Regiment at Umritsur* (Amritsar) *guarding a disarmed native regiment*'. Apart from his assault in 1858, that was as near as Haversham got to the Mutiny. It does, however, suggest how stretched the British part of the Army of India had become at that point.

In February 1858, in the first of several steps into more difficult territory, Haversham was posted to the Jammu area, as his talent for combining artistic skill with precise surveying became known. He was required to survey 'that part of the Maharajah's territory adjoining the Sealkote district, together with part of the outer sandstone range', almost certainly meaning those hills which are a southern extension of the Pir Panjal, around Udhampur and south of the Chenab. These hills rise to over 4000m/c.13,000ft. Very regrettably, we do not know how many of them Haversham climbed. It was said, however, that he was responsible for putting Sonamarg – now a thriving ski resort – on the map. Again, he surveyed about 800 square miles. He describes this area as 'very intricate and difficult ground'. The nosey Phillimore tells us that, fortunately, his pay had gone up considerably – he earned an allowance of Rs250 per month (having been paid Rs116 a month when based in the Punjab); he was also given 'seniority above Melville and Murray', his fellow surveyors, this action supposedly having provoked some jealousy. But he never hesitated to recognise those of genuine talent, such as 'Johnstone (presumably Johnson), Beverley and Shelverton'.

Still, by May, Haversham had completed this portion of his work then proceeded to the Budrawar area (where he had 'married' Kudikji), just south of Dharamsala, the present seat of HH the Dalai Lama's government-in-exile. This area Haversham defines as south of his Kishtwar area of 1857, and lying between the Chenab (also known as the Chandra Bagha) and the Chata Dhar Range. ('Chata' appears to be a local word for a shepherd's hut. I do not know the exact meaning of 'dhar'.) To reach here, Haversham would have crossed the Indrahar Pass north of Dharamsala to reach the upper reaches of the Ravi, an important tributary of the Indus which it reaches via Lahore. The Ravi is one of the *Panj Nad*, the five rivers which give the Punjab its name. He describes the mountains as being about 10-12,000ft (c.3500m). Haversham again leaves no record of which summits he actually attained although some summits in that area are now designated trekking peaks. He had completed this section of the survey by the end of June.

From there, he returned to Kishtwar and struck up the Muro-Wardwan River, which has its source not far east of the Zoji La, the high pass which offers the most manageable of the three roads into Ladakh. Haversham walked as far as 'Ainshin' (now more usually 'Inshan'), about half way up that river, which joins the Chenab just above Kishtwar, and began the survey of that northern area. Indeed, we know from one of his watercolours that he penetrated almost as far as the then remote capital of Ladakh, Leh, since this painting is of Mulbekh Gompa (monastery), one of Ladakh's finest, and is dated 18 August 1858 (presumably the date of its completion as on that date he was working in Srinagar). The monsoon then set in – Haversham wisely returned to head office at Srinagar. He remained there until 23 August and took only eight days (we do not know by which route) to return to Ainshin, very efficient progress considering the muddy conditions underfoot.

Haversham's surveying was becoming ever more adventurous. He was now penetrating the edges of unexplored Zanskar, the 'land of white copper' (formerly known as 'Zaskar', the modern 'n' having been added to indicate a mild nasal sound). From Ainshin, probably by crossing eastwards over the Tramsar La, he reached a viewpoint at about 5300m/c.17,400ft on a plateau west of Faridabad, a hamlet or grazing ground just south of the Nun-Kun massif. A panoramic watercolour of these twins, full of mystery and of the wild desolation of the Nun-Kun area, survives (see plates 21 and 22). The remainder of this season was spent mapping what even Haversham concedes was '*high ground*'. His approach to those famous twin giants, Nun (7135m/23,409ft) and Kun (7077m/23,218ft)[11] marked his first close encounter with 7000m (c23,000ft) peaks – one can see the rapid progress he was making towards the following years' great challenges in the highest parts of the Karakoram. Another of his paintings, this time from the north of the Nun-Kun massif, suggests that he must have crossed the Nun-Kun area to its east, to come out on the Suru or Sangpoo River and follow its long valley north-west past the celebrated Ringdom Gompa all the way out to Kargil. He then continued to survey this watershed country between the Indus and the Chenab, a total of 868 square miles rising to 15,000 feet (c4500m), until the end of October. It was when '*commencing a fresh piece of country in the Low Hills*', i.e. on his way back to either Srinagar or Sialkot (the former is more likely) that he was attacked (see Chapter 9).

Haversham had proved that he was capable of doing great work in high country and, as we now know, on his return to Srinagar 18 months later (six months' convalescence in the Sialkot area; just over a year in England and several months on the return voyage to England), he was ready to tackle the highest and least-known mountains anywhere within the empire – the Karakoram.

---

[11] Also known as Ser and Mer.

CHAPTER NINE

A WIFE, A SON, AN ATTACK

Kudikji

HAVERSHAM HAD BEEN PROMOTED AND sent off to Kashmir almost certainly for reasons other than his pure talent and having the right social connections: by at least no later than June 1856, and probably since 1855, he had been enjoying an affair with Kudikji, the daughter of a landowner from Punch. Who was Kudikji? Little hard information about her exists other than that supplied by Godwin-Austen in the correspondence about the matter which suddenly arose very near the end of his life. He tells us candidly that:

> She was not a common woman but of good birth – land owners in the district of Poonch in the outer hills – not a Kashmiri.

Her likely ethnic origin, therefore, was Sudhun (sometimes written Sudhan), a tribe of the Sudhama *tahsil* (administrative area) around Punch, one of the eight administrative districts of the Indian state of Jammu and Kashmir. The Sudhun are recognised as the chief landowners and the principal ethnic group of this area. The 1838 census records about 25,300 males, with about 40,000 of a population overall (though the authority for this is disputed). According to Lt Col JM Wikeley in *Punjabi Musalmans* (sic), 'the Sudhans have pride of race and consider themselves superior to any of the other tribes of Poonch Kashmir'. However, they are not confined exclusively to Punch; a sizeable community existed on the lower ground in Rawalpindi, for example.

The Sudhuns' origins are unclear; most of what we do know derives from their oral history. There seems to be agreement, however, that they are high-caste Afghanis. They now even have a Facebook page, as follows:

> The Sadozai (Sudhun) are a lineage of the Popalzai clan of the Abdali tribe of the ethnic Pashtun. The lineage takes its name from its ancestor, Sado Khan. The Sadozai rose to power upon the death of Nadir Shah in 1747, when the Sadozai King Ahmad Shah Abdali united the Afghan tribes under his rule. He became the founder of the Afghan Kingdom that lasted until 1973. He was probably the only king in Afghan history who held the allegiances of a massive majority of the Pashtun tribes. Ahmad Shah assumed the title Durr-i-Durran, 'Pearl of pearls', and the early kingdom is known as the Durrani Empire. The Sadozai ruled Afghanistan

> until 1823, and briefly between 1839 and 1842. They also ruled Kashmir, Rawalakot from 1752 to 1819. The Sadozais originate from Kandahar province in Afghanistan.

As a religious sect, Sudhuns are mostly Sunni of the Hanafi branch of Islam. It has a reputation for putting greater emphasis on the role of reason and being slightly more liberal than the other three Sunni schools of law. The Hanafi school also has the most followers among the four major Sunni schools.

EDWARD

Haversham must have met Kudikji before June 1856 (assuming a normal pregnancy length), as she gave birth to their son, Edward or Eddy, also spelt 'Eddie' on occasion, at Sialkot on 15 March 1857, the very month of Haversham's appointment to the Survey. Haversham tells us that he and Kudikji were married in Budrawar in the border area of Indian Kashmir and Himachal Pradesh (just south of Dharamsala, the seat-in-exile of HH the Dalai Lama) in May 1858:

> We were married under Mahomedan rites by a Mullah from Budrawar, her mother was present and all my Survey Camp. To all intents and purposes I was then a follower of the Prophet.

In the opinion of Her Honour Judge Azmat Nisa,[1] this marriage was legal (under Sharia and Raj law) if we accept the following: first that Haversham fulfilled the condition of 'accepting one God and the Prophet Muhammad as his Messenger'. As a Christian by upbringing, Haversham would have accepted 'one God' as that God is the same for Christians and Muslims; the acceptance of Muhammad as his Messenger was never mentioned by him. Second, that Haversham's professing to be '*a follower of the Prophet*' would be sufficient in itself to make him a Muslim. (However, my personal inclination is to be rather suspicious of his phrase '*to all intents and purposes*'; there is something a bit slippery about it.) He also appears to have fulfilled the contractual requirement for two male witnesses, as there is no requirement, according to Judge Nisa, that such witnesses should be Muslim. And as the Muslim marriage could be considered legal, it follows that, under Raj law, the marriage would be recognised as it met the criterion of correctly fulfilling the Muslim conditions. Sadly, however, while the marriage itself might have been legal, Eddy was born outside it and remained consequently, in both Sharia and Raj law, illegitimate.

Eddy was adopted by a childless military couple, Mr and Mrs Thomas Milner, and was given the additional names Henry Hastings Godwin. This adoption Haversham describes as, '*with the consent of his mother*',[2] this phrase one suspects, despite its apparent neutrality, disguising quite a lot of pain. Precisely why Eddy was given to adoption is never made clear, but in November 1858 Haversham was badly injured (see page 86) and an important reason must have been his difficulty in being able to fulfil his fatherly role, as he had been invalided out to England in April 1859 when Edward was only 13 months old. We cannot define precisely how serious Haversham's injury was: any blow to the head must be dangerous; perhaps he thought he might lose his life, in which case a secure future for the boy would be being thought about. In any case, it might have been decided that an 'English' upbringing would give Eddy the best start in life; and it might also have protected Eddy from the strong anti-Indian feeling arising from the horrors of the Mutiny. The adoption would also have been carried out because the conventional, racist, social stigma attached to an

[1] Her Honour District Judge Azmat Nisa sits at Kingston upon Thames. She helps to provide judicial training on Sharia law and is Chairman of the Association of Muslim Lawyers.

[2] See Rita Ghey, *A Rich Inheritance*, p125. Published privately, printed by Imprimeries Quebecor, Mont-Royal, Québec, 1992.

Anglo-Indian (or in this case Anglo-Afghan) son would have made it impossible for Haversham to acknowledge his son in most of the circles he moved in.

In a letter to one of Edward's grand-daughters a few months before he died, Haversham says that the names given to Edward are new to him. Despite some disingenuousness in other parts of the same letter, this statement is probably correct, because of the confusion arising from associating 'Godwin' with Earl Godwin, the father of the loser of the Battle of Hastings, rather than Godwin the general.[3] In any case, Eddy took the surname 'Milner', which appears on his college diploma, marriage and death certificates.

Eddy did not enjoy a very happy life. He suffered from two widespread prejudices – against illegitimacy and against being Anglo-Indian. The law of the Raj at that time recognised marriages between Englishmen and women of other faiths as genuine if they were legal according to the beliefs of the relevant non-Christian faith. In the case of Fenton v Livingstone (1859), however, the House of Lords ruled that offspring of 'non-Christian marriages' as defined under English Law[4] were illegitimate; the judgement was 'conclusive against the recognition of the validity of a non-Christian marriage for making out a title to succeed to real property in England.' The ruling also stated that, 'members of all races where marriage is non-Christian are bastards, and incapable of taking settled property in England *ab intestato*' – 'from an intestate', i.e. one who dies without a will. So even if Eddy had been born within the marriage, he would still have counted as illegitimate and unable to inherit fixed chattels, with or without a will from Haversham, under English law. Eddy could therefore inherit nothing of the Godwin-Austen freehold lands in England, (although he could possibly have inherited 'chattels real', i.e. 'an interest in land less than a freehold, such as a lease' if it had been bequeathed to him. The phrase 'chattels real' is, however, not recognised everywhere and it is perfectly possible that 'chattels real' might not have been part of Eddy's non-inheritance either. Judge Nisa concurs with these legal opinions. She is clear that Eddy 'could not be a legitimate child and would therefore not automatically be entitled to a share of his parents' estate.'[5] From his subsequent family history, we know that Eddy evidently inherited (even as a gift) no landed property or even chattels of any kind from either of his parents.

Eddy was given at first a brief Church of England upbringing after adoption by the Milners. They then sent him to St Peter's College, an orphanage school in Agra, where he became Roman Catholic. (St Peter's College was founded in 1846, and is one of the oldest convent schools in India. It is now an Anglo-Indian Christian (Roman Catholic) Institution granted Minority Rights under Article 30 of the Indian Constitution, and still has a strong academic record.) In his letter to Eddy's granddaughter six months before he died, Haversham describes Eddy's career with less than convincing openness:

> When Mr and Mrs Milner adopted Edward they acted splendidly. I supplied the means (to his education) and when he went to Roorki they saw him through, he passed out well and got into the Public Works Department and did good work in it.

Eddy trained at the Thomason Engineering College in Roorkee (Roorki), generally regarded as one of the best in India. (Roorkee, a city in Uttarakhand, is also known for Roorkee Cantonment, one of India's oldest and the headquarters of the Bengal Engineer Group (Bengal Sappers) since 1853.)

---

[3] Not that we can be certain: the Milners might well have heard of General Godwin or recalled him being mentioned in conversation with Haversham. 'Hastings', however, does seem a very improbable introduction of a name by Haversham, as it has no family connection with him. There is an historical association, of course, between (Earl) Godwin and the Battle of Hastings.

[4] See Fitzpatrick, Dennis: *Non-Christian Marriage: Journal of the Society of Comparative Legislation, New Series, Vol. 2, No. 2* (1900), pp359-87.

[5] Judge Nisa's opinion, for which I am most grateful, was delivered to me in an email dated 3 June 2010 following an email correspondence giving the evidence and discussing it.

Some evidence[6] exists that Eddy remained very short of money during his education, although whether that was because Haversham's financial provision, as a married man on little more than a military salary, failed to meet his son's needs, or because the Milners used the money to cover school fees and little else, is not clear. At this time, he joined the Thomason Volunteer Rifles, one of many groups which were created after the Mutiny in order to prevent a similar surprise butchering of the civilian population. Later, Eddy achieved considerable respect as a public works engineer, after an early career as a civil engineer in Hyderabad State, though this profession failed to bring great financial rewards.

Although it seems that the Milners told Eddy who his real father was, it was treated as 'a dark family secret'.[7] It appears that after the adoption Haversham and Eddy never met, although when at his bungalow at Dehra Dun, where much of his surveying and scientific work would have been written up outside the 'field' season, Haversham could not have been staying very far from his son. It would of course have been something worse than undiplomatic to see his son during his marriage to Pauline (which lasted from 1861 until her early death in 1871: see Chapter 14), particularly after the birth of his legitimate son and heir, Arthur, in 1864; and from November 1865 until his retirement and return to England when Eddy was 20, he was very intensely occupied in the Burmese and Assamese jungles, as well as spending substantial periods (see Chapters 15 and 16) back in England on business and recovering from *'jungle fever'*. So not meeting Eddy could not just be dismissed as pure callousness. It is said in a privately published family history, although no direct evidence supports it, that, 'Eddy *(then)* rejected his father and all further financial help from him. Ever after, he looked with shame on his mixed blood origin.'[8]

The only indirect evidence for the first statement above is contained in the rather eccentric begging letter which Florence Barclay (one of Eddy's grandchildren) sent in 1934 to the President of the Royal Geographical Society. (For the letter's contents and interpretation, see Chapter 25.) Eddy could have done with any money he could find because, on 9 September 1879, he married Emma Theresa Smith (her second name implying a possible Roman Catholic upbringing, also). She seems to have been descended from a family called Hardinge, in Lincolnshire, although her grandchild Florence Barclay claimed, without evidence, a blood connection with Viscount Hardinge of Lahore, a Governor-General of India. Eddy and Emma produced 15 children (the initials of the first six names of whom spell 'MILNER'). It is said that once they began to have children, Eddy's true parentage was 'only whispered about'.[9] This large family then spread itself out through India, Australia, Canada and England.

Eddy, described on his death certificate as a 'Retired Divisional Engineer', died of a mastoid abscess in St George Hospital, Mumbai (this hospital having been originally designed to look after European patients only) on 19 February 1917, aged only 59, and was buried in the cemetery of Sewri, a suburb of Mumbai. Emma Theresa, on the other hand, managed to carry a long-life gene of her own and died as late as 22 February 1958, aged 97.

Only one other interesting point about the mysterious Kudikji comes from what Haversham tells us, '*She went with me into Baltistan when I was doing very hard work on the Great Glaciers there.*' This statement implies that she was certainly alive on their return from the first Baltistan expedition, which reached Srinagar in early September 1860. Haversham was absent from India, as a result of the Udhampur assault (see p86) from April 1859 until his return to Srinagar in the first half of June 1860. (Sources differ slightly, here: Haversham, in his unpublished abstract of his military service says the middle of June; Phillimore says 1 June.) Evidently, he immediately took up with Kudikji again, and she was enthusiastic enough to follow him into the wilds of the High Karakoram. After the expedition, he tells us that during the winter of 1860-1 he recessed at Dehra Dun, and by 5 April 1861 he had rather precipitately married Pauline Chichele Plowden.

---

[6] Ghey, *Ibid*, p129.
[7] *Ibid*, p122. The 'dark family secret' was on the Milner, rather than the Godwin-Austen side (I am told).
[8] *Ibid*, p126.
[9] *Ibid*, p126.

What happened to Kudikji between September 1860 and April 1861? The answer is pure speculation. Given that she and Haversham were reunited after Haversham's absence of 15 months i.e. that absence had not driven them emotionally apart, and that Kudikji then devotedly undertook the very testing experience of joining Haversham on his first Baltistan expedition, and given that it is highly unlikely that she would have joined Haversham in the winter recess to Dehra Dun, where a public liaison with a 'native' woman would have been frowned upon, the conclusion can hardly be avoided that, most probably in the autumn or early winter of 1860, the complications of a second pregnancy (conception having taken place during or just before the Baltistan expedition) led to Kudikji's (probable) death. (It is unlikely, although not impossible, to have been a death in childbirth, as that would have had to have occurred after February 1861 at the very earliest, by which time Haversham would have been making his wedding preparations with Pauline; a liaison with a pregnant Muslim wife while making arrangements to marry a Christian one would attribute uncharacteristically callous behaviour to the young officer.) The prevalence of illnesses (in themselves, and as a source of pregnancy complications) such as cholera cannot by any means be ruled out, of course.

Haversham commented finally on Kudikji, '*She was a good wife with me and if there is anything now left that I can look back on and love, it is her memory and the way she looked after me.*' While the liaison ('with' suggests more of a passion-based union than the more usual 'to' which might tend to describe all traditional wifely activity) was no doubt easily kept burning by youthful passion, Haversham's remark seems to me in keeping more with the rather disingenuous and placatory tone he uses in the letter to Florence Barclay in June 1923 than a genuine statement of feeling. The extensive and consistent evidence that Jessie Robinson remained the strongest and most enduring love of Haversham's life (see Chapter 17) suggests that Haversham's statement is more (self-?) deceptively sentimental than accurate.

'KNOCKED INSENSIBLE'

This affair, with its lifelong consequences, leads us, therefore, to yet another accident in young Haversham's life, this time one of the most serious. In the unpublished abstract of his military service, Haversham records that:

> In November (1858), when commencing a fresh piece of country in the Low Hills,[10] Lt *(he was actually captain by then)* Austen *(he had also been 'Godwin-Austen' for four years by then, and had signed himself thus from 1854 onwards)* was attacked near Udampur[11] by the villagers (incited by some rebel Sepoys who were in the village) was severely handled by them and had to be carried into Sealkote where he remained for the rest of the field season, finishing the plans of the preceding 7 months.)

It was a serious attack. Haversham has added in his own handwriting in the margin to this entry that he, '*receiv*(ed) *a blow on the head which rendered him insensible*'. Given that the attack occurred in 1858, Haversham's account of its motives seems perfectly believable: any vengeance directed towards a lone European where, owing to the remoteness of the attack, there would be little chance of revenge or legal retribution seems highly understandable. Udhampur, however, is only about 200 km/124 miles from Punch, and one cannot help thinking that, five months after their

---

[10] As opposed to the High Hills which at that time would have meant to Haversham the Pir Panjal and Kaj Nag Ranges, and the peaks around Kishtwar.

[11] Udhampur, population 60,000 (2001) lies in modern Indian Jammu and Kashmir. It serves as the district capital and the Northern Command headquarters of the Indian Army. Sialkot is about 80 km/50 miles from Udhampur.

marriage, Kudikji's offended relations (by her marriage to a perhaps less-than-convincing Muslim), and perhaps suggested by her father's non-appearance at the wedding, might have had a hand in the assault.

Whatever its origins, the attack sufficed for Haversham to be permitted leave, which he took in April 1859 when he transferred to the 24th Foot's Second Battalion and joined them in England. While there, he became a Fellow of the Royal Geographical Society, though he is named only as 'Austen' of Chilworth Manor; this organisation awarded him its highest honour far too late, 51 years later. And as the dress of the ladies in the foreground of his drawings makes clear, he made at least one trip to Anglesey, indicated by drawings of the Ruabon (Rhiwabon) Viaduct and the Menai Suspension Bridge.

That December, Haversham no doubt enjoyed a happier Christmas with his family than any since leaving home in 1852, although by 28 December we know from a letter that he had returned to Aldershot. (In this letter he signs himself 'Godwin-Austen'.) On 12 January 1860, the regiment had taken up quarters at Winchester Barracks. His return to India then progressed in stages. As he states in the *Abstract* of his military service, he managed to, '*Effect an exchange back into the first battalion*' then still in India. He left the Second Battalion in Cork. When he reached Bombay (now Mumbai), he wired Sir Andrew Waugh to try to obtain a survey re-posting. He then sailed to Karachi to join the First Battalion at Mooltan (Multan), now the capital city of the Punjab province of Pakistan. This Sufi city lies in a bend created by five rivers on the east bank of the Chenab, more or less in the geographic centre of the country. (The source for most of this journey is, apart from the military service *Abstract*, the ever-inquisitive Phillimore.) Here, Haversham received his answer from Sir Andrew and found that 'Gen Waugh, the Surveyor-General of India, had applied for his services as topographical assistant in the Trigonometrical Survey.'

Waugh provided highly complimentary reasons for re-appointing Haversham:

> He ... has exhibited a high order of talent as a field draftsman in delineating difficult ground. The mountainous ... character of the tract in which he has been employed is such as to tax severely the physical strength of all engaged in the survey.

Re-appointed assistant in the Kashmir area, Haversham then exchanged to the First Battalion, and by the middle of June, (that personal information glutton of a military historian, Phillimore, says 1 June 1860) he had again reached Srinagar. More advantageously, in the short term, Haversham's sick leave of one year and 14 days was condoned, i.e. recognised as genuine for pay and pension purposes.

At this point, his career in the Survey, which lasted another 18 years under the toughest physical and mental conditions imaginable, became fixed. When he joined the Kashmir Survey, Haversham began to leave his youth and youthful indiscretions behind. He was ordered to head to the land of its highest mountains and its greatest glaciers, to commence the Survey of Baltistan, an undefined, though culturally fairly distinct area, lying between Ladakh and Gilgit, to the north of the Vale of Kashmir. Its highest point is K2 (Mt Godwin-Austen), the world's second-highest peak. This new task, blazing the trail to the world's hardest mountain, required him to do what he is most famously known for: to compute K2's height with unarguable accuracy and survey its position on what was thought to be the watershed of the Great Snowy Range, on the outermost frontier of the Empire.

CHAPTER TEN

A PLACE IN HISTORY

Explorer

The sub-assistants of the Great Trigonometrical Survey would ... make every personal sacrifice for the honour and success of the Department ... No class of men in India go through more hardships and privations, and none are more cheerful, well-disposed, obedient and loyal.

Sir Andrew Waugh

BETWEEN 1860 AND 1865, HAVERSHAM undertook five expeditions; they rank among the very highest achievements by any British mountaineer-explorer. Only Eric Shipton[1] and Bill Tilman,[2] Sven Hedin,[3] Sir Martin Conway,[4] Sir Chris Bonington[5] and perhaps Tom Longstaff[6] stand any sort of comparison. These expeditions simply represented his carrying out his 'orders', as he modestly says on several occasions, although his adventurous streak ignores, over and again, the order not to go above 16,000ft/c.5000m. These expeditions naturally developed from Haversham's well-received work, outlined in Chapter 8, on the Kashmir Survey. From the lower hills of Kashmir, he had reached out and travelled towards the high crags and Buddhist cliff-face *gompas* (lamaseries) of Zanskar; he had approached two 7000m/c.23,000ft peaks, Nun and

---

[1] Eric Earle Shipton CBE (1907-77). As an account of Shipton's legendary exploits in many of the world's major mountain ranges, Peter Steele's biography has some merits, but has now been superseded by Jim Perrin's *Shipton and Tilman: The Great Decade of Himalayan Exploration*, Hutchinson, London, 2013.

[2] Major Harold William 'Bill' Tilman, CBE, DSO, MC and Bar, 1898-1977 (disappeared at sea). His prodigious achievements are well described in his own works or in Jim Perrin's work cited in note 1, above.

[3] Sven Anders Hedin KNO1kl, RVO (1865-1952) was a Swedish geographer, topographer, explorer, photographer and travel writer, as well as an illustrator of his own works. During four expeditions to Central Asia he discovered the Transhimalaya (once named the Hedin Range in his honour) and the sources of the Brahmaputra, Indus and Sutlej Rivers, Lop Nor, and the remains of cities, gravesites and the Great Wall of China in the deserts of the Tarim Basin. The posthumous publication of his *Central Asia Atlas* marked the conclusion of his life's work. His leanings towards Fascism meant that he was eventually stripped of his British honours. Although he conducted only four Central Asian expeditions, they were of a length and duration to excel all others. And, like Haversham, he worked in two distinct climatic and topographical circumstances, in his case, mountain and desert.

[4] William Martin Conway, 1st (and last) Baron Conway of Allington (1856-1937), English art critic, politician and mountaineer. His 1892 Karakoram expedition superseded Haversham's second Karakoram expedition.

[5] Sir Christian Bonington, CVO, CBE, DL, b.1934 in Hampstead. Our greatest contemporary mountaineer-explorer, in particular for his leadership of the Annapurna S face, the Everest SW face and the Kungur (Qungur) expeditions.

[6] Tom George Longstaff (1875-1964) was the first person unarguably to climb a summit of over 7,000m, Trisul, in the Indian Himalaya, in 1907. He also made important explorations and climbs in Tibet, Nepal, the Karakoram, Spitsbergen, Greenland and Baffin Island. He was president of the Alpine Club from 1947 to 1949 and a founding member of the Alpine Ski Club in 1908.

I have excluded Wilfred Thesiger from this list, because he was more of a traveller than a bona fide explorer.

Kun, and he had penetrated some very remote high country, much of it for the first time by any Westerner.

Kenneth Mason, in that definitive early Himalayan history *Abode of Snow*,[7] correctly called Haversham 'a great explorer and probably the greatest mountaineer of his day'. And not only, as Sir Roderick Murchison, the President of the Royal Geographical Society, declared, had Haversham scaled any number of peaks and crossed any number of passes above 5000m/c.16,000ft,[8] but he had, amazingly, achieved these feats before the first ascent of the Matterhorn, (4478m/14,692ft) by Edward Whymper on 14 July 1865. He succeeded, moreover, on all of these icy and rocky climbs – at least four of which reached above 20,000ft/c.6000m – without crampons, ice-axes (he appears to have used garden hatchets for step-cutting) or rope suitable either in length or design, or indeed any kind of formal rock- or ice-climbing training; his men used 'long staffs, with a short cross-piece of ibex-horn at the end, which the *shikaris*[9] carry to aid their steps along the steep mountain-sides'.[10]

Haversham had been elected as a member of that original band of mountaineers, the Alpine Club, according to its *Register*, in 1860. His 'qualifications' for membership are listed rather puzzlingly as the 'Mountains of Tartary and Thibet' (sic), none of which Haversham had visited by 1860. (His experience at that date was confined to Kashmir.) Even by 1860, however, his qualifications for membership stood out. (The qualifications for joining the AC now are given on its website as, 'we ask that you have climbed a minimum of twenty respectable alpine routes or peaks, or the equivalent of this in other ranges and wilderness areas. Hard alpine routes and serious climbs in the greater ranges will carry more weight than easy *voies normales*.') He was obliged to resign his membership in 1892 either because of financial pressures or because he had done no mountaineering since the 1870s and could see little prospect of it happening in the future, given the volume of scientific work he was pursuing. The compiler of the *Register*, however, notes that he attended the Alpine Club Jubilee, i.e. 50th anniversary dinner, in 1907 and that he was made an Honorary Member in 1912. The AC President had complained at a meeting in 1905 that Haversham was, 'not so often now to be seen in town'; presumably he enjoyed some sort of informal or associate membership of the club at that time.

More important than these historical bare bones are the human qualities which made Haversham deserve Mason's description. It is a testimony to his exceptional mental toughness – just as much as to his exceptional physical hardiness – that in the four Karakoram-Ladakh-Zanskar-Tibet expeditions, he was the sole Westerner in the party. We do not know the total numbers for each expedition, but he does mention that the 1861 expedition comprised 66 men apart from himself. On that expedition he had, '*Three servants, two Government natives of India, 2 Dogra Sepoys of the Maharajah of Kashmir's Army; the rest, carrying loads, were Baltis of Shiga and Askoley.*'[11]

Haversham understandably asserts in the same letter the '*great deprivation*' of not being able to '*discuss the nature of the country and admire its magnificent scenery*' with a fellow European on the spot.

His strength and stamina were amazing: Godwin-Austen held the world summit altitude record from 1862 until William Johnson's possible ascent in 1865 of a Tibetan peak which was downwardly assessed to 6710m/22,014ft from Johnson's optimistic original estimate. (The mountaineering record in 1862 was probably held by Robert and Adolf Schlagintweit, who estimated that in 1855 they had reached about 6700m/22,000ft on Abi (Ibi) Gamin, next to

---

[7] *Abode of Snow*, Diadem, 1987, p78. Generally recognised as the most important history of the pre-1953 phase of Himalayan climbing and exploration. Now superseded by Isserman and Weaver: *Fallen Giants* (2008); Yale University Press.

[8] Haversham's ascents are summarised in Appendix B.

[9] *Shikari*, a big game hunter, or (here) a native guide.

[10] *On the Glaciers of the Mustagh Range*, pp37-8.

[11] Letter to the *Surrey Advertiser*, 7 September 1909. Dogras are an ethnic group from the Jammu region of Kashmir. Sepoys were native soldiers within the British Army or armies of territories under British rule or protection. The modern placename spellings are 'Shigar' and 'Askole'.

Kamet.) Godwin-Austen's undoubted ascent of the Rupshu mountain, Mata Peak, 6283m/20,613ft, (now more usually called Mentok Kangri) in 1862, marks his record. If Johnson's claim is incorrect, and the improbable claims of William Graham on Kabru, 7412 m/24,318ft in 1883 are discounted, then Haversham's record must stand for 29 years until Sir Martin Conway and Matthias Zurbriggen's 1892 ascent of Pioneer Peak, 6501m/21,329ft (although Conway's original overstated measurement of this peak had to be revised downwards, also). In sum, this mountaineer-by-accident appears to have made, during his five years in Kashmir and the Karakoram, at least four ascents above 6000m, the highest being 6283m; at least 21 ascents above 5000m (at least ten of which were above 5500m), as well as at least eight ascents of lower hills of high alpine height (3500-4500m); he crossed at least 14 passes above 5000m, at least seven passes above 4000m and at least 26 passes above 3000m (these figures being conservative estimates).

In his letter to the *Surrey Advertiser* of 7 September 1909, however, Godwin-Austen claims that his highest survey point was 21,600ft (6584m). This height lies about 305m/1,000ft higher than any position he records in any of his scholarly articles. He goes on to claim, '*I may have reached 22,000 feet once or twice.*' There is no direct evidence for this. If one reduces these three height references by 1,000ft/305m, however, then this statement from memory made, after all, when Haversham was 75, becomes accurate.

EARLY EXPLORATION OF THE KARAKORAM

It is not my intention to offer even a short history of the exploration of this area; that has already been accomplished by John Keay in his wonderfully readable (if, in Haversham's case, not entirely accurate) work, *Where Men and Mountains Meet*; instead, I should like to note how the tracks of earlier explorers connect to Haversham's own journeys, to show just how much more thorough, determined and systematic his expeditions were, compared to any of his predecessors.

Haversham should be remembered as the first Westerner to explore, with any kind of depth and thoroughness to his plan, the Karakoram, the greatest mountain and glacier system in the world and, despite a lack of technical equipment, reach its very heart high on a spur of Masherbrum. Until he worked his tireless way through it, this remote region had received very few Western visitors (and the Baltoro Glacier none at all). Those who did penetrate this far comprised an interesting mixture of merchant-adventurers, religious obsessives, dilettantes and scientists. The great surveyors such as Montgomerie, Godwin-Austen and Johnson represented the second wave of Western Himalayan exploration, before their place was taken by dedicated semi-professional mountaineers.

The earliest European wanderer through this area was probably François Bernier (1625-88), who accompanied the sixth Mughal emperor of India, Aurungzeb, to reach Kashmir in 1663. In 1786, George Forster travelled across northern India to Kashmir disguised as a Turkish merchant. A much more important traveller, William Moorcroft, in 1812 reached the sources of tributaries of some of the main rivers in this area, for example the Ravi, the Chenab and the Beas. In 1819 he penetrated Kulu, from where he crossed the Rohtang La, a relatively low (13,051ft/3978m) but demanding pass. To reach it, when on the way to Zanskar in 1992, our bus had to ascend an apparently vertical cliff by a series of exciting zigzags, the thrill of which was as nothing compared to the descent on the other side. Near the top, the luridly painted bus had to pass through an ice tunnel as the mist mercifully hid the vast abyss to the side of the vehicle. This pass provides access to Ladakh, the capital of which, Leh, Moorcroft reached via the Baralacha La, the route of the present main 'road' between the Rohthang and Leh, with his fellow traveller George Trebeck (1800-25) in September 1820. Moorcroft remained there for three years, during the second of which he met the Hungarian scholar Csoma de Körös en route for Zanskar. (See Chapter 12). These travellers ventured only a little way north of Leh, to the Shyok and Nubra rivers.

A more enigmatic early Western explorer to arrive at Gilgit, probably in the late 1820s, was Alexander Gardiner. It is thought that he crossed the Pamirs and reached Yarkand – something of a theme of 19th-century exploration in this area – then made his untraceable way down to Leh and from there to Srinagar (and beyond).

Other early visitors included Victor Jacquemont (1801-32), a French botanist who left Simla in 1830 and reached Ladakh. He may then have travelled as far as Tibet, although Tibet's boundaries were hardly clearly fixed at that time. In 1831, he crossed the Sutlej, a common access point for the Karakoram and Ladakh, a river basin which played a very significant part in Haversham's story, then crossed the Pir Panjal mountains bordering the Vale of Kashmir where Haversham began his surveying career, to reach Srinagar. In the four months he spent there he possibly unintentionally opened up the way to the High Karakoram, having explored as far as the Gurais Pass, a journey extended by Vigne when making the first-ever visit to Skardu.

At almost the same time, Dr Joseph Wolff, a man whose religious passions consistently irritated those who met him, whatever their faith, came out of Afghanistan and crossed the Sutlej at Bilaspur, to reach Srinagar on October 11 1832.

One of the oddest of the early Karakoram visitors was Old Harrovian, Geoffrey Thomas Vigne (pronounced 'vine'). This botanist-geologist first crossed northern India (where he met Sir George Everest in 1834) as far as Simla. From there, he traversed the Pir Panjal no fewer than five times, then made his way via Kishtwar up the Muro-Wurdwan Valley crossing, it is said, over 20 passes in the Chamba and Kulu areas. He plotted the courses of the Ravi, the Chenab and the Jhelum rivers. It is the word 'plotted' that is important: Vigne's significance lies in his carrying out and recording his explorations with some sense of system. From the Vale of Kashmir in 1835, he established the 'yak route' over the Gurais and Burzil passes to reach Skardu, the first Westerner in a town which annually now accommodates thousands of them. On the way, he viewed Nanga Parbat, although he grossly underestimated its height. From Skardu, Vigne undertook the first Western expeditions to penetrate the Karakoram in any depth. He followed one direction to reach Astor, then daringly attempted to make for the Hispar Glacier, the first sighting of which was Haversham's privilege in 1861. Vigne walked as far as the Basho Valley, about 40km/25 miles north of Skardu and saw the snout of the dog-legged Chogo Lungma Glacier, although once again it was Haversham who later made the first survey of this enormous glacier. Vigne finally planned to discover the source of the Shyok. He reached the lower end of the Nubra Valley before porter problems and bad weather forced a withdrawal; his re-attempt the following year, 1838, up the Saltoro Valley, again because of bad weather, met with no greater success.

Exploration was temporarily halted between 1841 and 1842, as the Dogra Wars between northern India and China-Tibet ran their course. It seems that no Westerners saw much of the region during this time of military and social turbulence.

The early scientist-explorers are well represented by the Scots botanist Thomas Thomson, who arrived in Simla, the other main access point after Srinagar, in 1847. He was supposed to survey the Empire's northern border with a Lieutenant Cunningham, who had crossed the Sutlej in 1846 to meet him, and with Colonel (eventually Lieutenant-General) Richard Strachey (1817-1908), a relation of the great biographer Lytton Strachey – and another man with Guildford connections. They proceeded from Simla to Lahaul, Spiti and Ladakh. At Leh, the party divided. Cunningham and Strachey followed the Ladakh-Tibet frontier before eventually returning to Simla. Thomson headed for the Nubra and Shyok areas and claimed the distinction of being the second European to reach Skardu, the main settlement of Baltistan. Thanks to bad weather, Thomson was then obliged to winter there, the first European so to do, despite an attempt to escape via the freezing uplands at Dras and the Zoji La. In 1847, however, he succeeded in leaving and crossed the Zoji La to reach safety in Jammu.

Thomson's next journey, more interestingly, took him to the Pir Panjal range then to the Chenab, one of the Indus's mightiest tributaries. By following the Chenab he reached Gulabgarh, from where he made the first European crossing of the crest of the Himalaya over the helpful cleft of the Umasi La (5342m/17,526ft) and so into Zanskar. From there, he reached Leh by what appears to have been an extremely arduous route, then crossed the Ladakh Range via the Khardung La and reached the Nubra and Shyok rivers, from where he probably penetrated as far as the not very difficult but remote and high Karakoram Pass. He therefore covered much ground which Haversham explored to a much larger extent and with a much finer degree of scientific accuracy about 15 years later.

The remaining serious explorers were the elder Schlagintweit brothers, Hermann and Robert. They were protégés of the great German naturalist Alexander von Humboldt and specialised in glaciology and the study of geo-magnetism. They crossed the Karakoram Pass in 1856, and eventually reached Khotan, on the southern Silk Road fringes of the Taklamakan Desert in what is now the westernmost Chinese province, Xinjiang. From there, they traversed the Kun Lun, still probably the least explored of the Earth's larger mountain ranges. Adolf (1829-57), their younger brother, repeated the route and reached Yarkand, then Kashgar, where he was beheaded without trial as a suspected spy by the amir, Wali Khan. In 1856, he had visited Askole, via the Shigar Valley, and was the first European to be seen there before Haversham appeared in 1861.

HAVERSHAM'S EXPEDITIONS: THE OVERVIEW

The information about these predecessors helps us to contextualise Haversham's expedition plans. (Several letters between Haversham and his father mention GT Vigne.) But it is Haversham's blow-by-blow articles in the Royal Geographical Society's *Journal*, his lively private memoirs and various letters and maps from his archive and his family papers lodged at the Surrey History Centre, as well as snippets of information derived from his quite extensive obituaries, which make it possible for the first time to reconstruct his routes and define the extraordinary dimensions of his exploration. Most of the placenames along his routes, with a little imagination in finding their modern equivalents, can be linked. For three of his five expeditions he not only produced his own maps, sometimes objects not just of topographical exactness but of considerable artistic value, but he also marked precisely his campsites (he mapped hardly any other way). Only the Rupshu and Zanskar expedition of 1862 leaves some puzzling gaps.

These expeditions had two purposes: the scientific one attempted to define the watersheds of the Indus and the 'Great Snowy Range', i.e. the Karakoram, and survey this geologically complex area. The second was political: to establish the northern limits of British India and to look at military and trade access routes to Chinese Central Asia.

THE FIRST KARAKORAM EXPEDITION

5 August to early September, 1860.[12]

Apart from a rather limited descriptive vocabulary which gives you little clear idea of what Godwin-Austen could actually see - views are almost invariably 'magnificent' or 'splendid' - Godwin-Austen also keeps you guessing in several of his expedition reports about how he arrived at his starting and finishing points. The first Survey expedition into Baltistan is no exception:

> I shall not here detail the march of 20 days from Srinagar in Kashmir to Skardoo in Thibet. It is a well-known road, long since described by Mr Vigne and Dr Thompson,[13] and is now very frequently travelled over.[14]

---

[12] Much of the public information in reconstructing the two Baltistan or Karakoram expeditions comes from Haversham's paper *On the Glaciers of the Mustagh Range,* read in absentia to the Royal Geographical Society on 11 January 1864.

[13] Dr Thomas Thomson (the preferred spelling), b.c1817, botanist, eventually (1855) director of the Botanical Gardens, Calcutta.

[14] *On the Glaciers of the Mustagh Range*, p19.

The most probable route took in the Deosai Plateau, often referred to as 'plains', but described by Haversham as, '*A curious undulating region of round-topped granite hills, encircled by lofty snowy peaks.*'[15]

From Haramukh, one of the plateau's high points, Montgomerie had obtained the first sighting of K2 and Masherbrum in 1856. From the plateau, the most likely route to the high peaks led from Astore over the Banak La (4964m/16,286ft). We know that Haversham did not travel via the Burji La because in 1861 he mentions wanting to visit it, which he then did, to ascertain its height. He was accompanied by Kudikji for the whole of this expedition, although beyond that we know nothing about her presence on it.

Although Haversham had received his orders from Montgomerie '*in the spring of 1860*',[16] the expedition proper did not begin until 5 August thanks to the long walk-in, which actually began in Sialkot, quite late in the expedition season which normally lasts from about May to September, during which time the passes could be expected to be free of snow. When he '*effected an exchange back into the 1st Battalion, 24th Foot*', on his return to India following his recovery year back in England after the '*enraged Sepoy*' attack in November 1858, he found that Sir Andrew Waugh, the Surveyor-General of India, had applied for his services. Haversham was re-appointed to the Kashmir Survey with a view to surveying Baltistan and the Indus watershed; he also gained his promotion to captain. He was deputed to go to Srinagar to receive his orders, which he had done by mid-June. The walk-in to Skardu began immediately after that.

The beginning of this expedition (see upper map, p237) began dramatically by crossing the fast, wide and deep Indus and Nubra (called the Shayok on some maps, but not to be confused with the main Shyok river further east) rivers in a *zuks*, a native raft made of inflated goatskins to which poplar stakes are lashed and which is propelled by the current and the power of *insh'allah*.[17] Having survived this hazard, by legend said to be connected to the time of Alexander the Great, and with no serious thoughts about acclimatisation, a sub-science of mountaineering which appears to have meant nothing to this super-fit adventurer, Haversham set out next day, 6 August, before sunrise to ascend the crumbling rock and snow tower of Bianchu (4875m/15,994ft), a little higher than Mont Blanc and lying just above the Indus-Nubra confluence. The day's ascent from the valley floor required '*most laborious work*' of a hefty 9000ft (c.2750m).

From there, the next day, he headed north, first over a peak he calls '*Kon-Kon*'; he then crossed a high pass, '*over 16000ft*', the Tuggo La, into the Thalle Valley which he ascended as far as its head, where it split into three valleys or glaciers leading to the Thalle La, the Tlathlung Lungma Pass, which he calls the Tusserploa La (16,679ft/5084m), and the Thalle Glacier, all lying at around 5000m. On his expeditions, Godwin-Austen worked relentlessly: if he could help it, not a day was to be wasted. Halts occurred only in weather so severe that surveying (which required reasonable visibility and a steady plane-table) became impossible.

Having then descended the Thalle to its confluence with the Nubra, three demanding ascents are completed, the first, Monlong, recorded as 14,905ft/4543m, the second, Kauchinokla, about 17,000ft/5182m, then Chungoksi Go, rising to nearly 5750m/18,865ft. It is not just the heights which are extraordinary, but the nonchalance and matter-of-factness of this mountaineer's attitude to his climbs. Haversham provides a fairly detailed description of the ascent of Chungoksi Go, not untypical of many others he made to that height:

> Our way was at first over a broad snow-bed, and then across a slope of loose debris of slate-rock which lay at an angle of 45°. The ground was continually sliding

---

[15] *Ibid*, p22.

[16] Letter to the *Surrey Advertiser*, on the subject of The Duke of the Abruzzi in India, written from Nore, 7 September 1909.

[17] This term, meaning, 'If Allah wills it', I first learnt in 1989, after an odyssey of terror being driven in the dark down the Karakoram Highway by two Afghans in a rickety lorry, thousands of feet directly above the murky Indus.

> away; the whole mass seemed full of snow or ice in a melting state. However, we managed to reach the summit about 1 o'clock and crossing over I set up the plane-table, overlooking a ravine on the other side, down which went a small glacier. There was a sharp point of rock some few hundred feet higher than the ground we were on, and as I had sketched in all that I could see below me, we determined to go to it. Our way was over snow, and somewhat difficult. After a while it became so steep that we had to cut steps for the remainder of the way. Finally we reached the point; it raised us altogether clear of the ridge, and I was amply repaid for the labour by the magnificence of the view. Below was a glacier with all its crevasses and varied features laid out – as on a map: to the left was Zoah, 21000 ft,[18] right ahead up the Hushe valley was Masherbrum, rising to 25000ft, with its pink-tinted rocks shining out in warm contrast with the masses of snow which rested on its sides. Our descent was at first much more expeditious than our ascent had been; lying on our backs, we shot rapidly down the snow-slopes, but the subsequent journey over the loose debris was far worse than the labouring-up had been.[19]

The Hushe Valley, while perhaps more scenically attractive, turned out to resemble the Thalle in having a three-part division at its head, below the colossal south versant of Masherbrum. Haversham began the exploration of the head glaciers with a six-mile- (9.6km-) clamber up the trackless, bouldery moraine of the Atoser Glacier until broken ice made progress impossible. The next day, he followed the Hushe River to a further division, one branch of which he named the Masherbrum Glacier, and proceeded up it until he met a '*cliff of vertical ice, a great mass of which broke away … and went down with a bang and a splash into the smooth lake beneath.*'[20] He completed his survey of this section by attempting to ascend a third, western glacier but was beaten back by the steepness of its sides.

The first phase of his expedition having been completed, Haversham then returned to Skardu. On the second phase, he started out from Gol, a few kilometres to the south-east, and from there headed south. He climbed several peaks, the first of which was called Dindasgo (15,882ft/4841m), scarcely mentioned despite being higher than Mont Blanc (4810m/15,781ft). The second peak in this group appears to have been Gonmathanmigo, probably 5335m/17,503ft; he praises it for its view of the high Karakoram, and contrasts this with the sight of Skardu lying deep in the valley below. The ascent of Thurigo, height unknown, then followed. Having obtained his plane-table perspectives, Haversham promptly returned to Skardu.

The third and final phase of this expedition began with Haversham's first failed attempt on a mountain. Once he had re-crossed the Indus, he tackled a peak ahead of him, almost certainly Marshakala (5153m/16,906ft), but was beaten back at an unspecified height, '*owing to a recent five days' fall of snow*', a remark which in its understatement must hide terrible physical and mental trials. Nothing daunted, Haversham then resolutely headed up the Shigar valley, a vast basin almost directly north of Skardu; it was, however, by this time, September 18, and approaching the end of the climbing 'season'. From the village of Shigar he ascended a peak two days' march up an eastern tributary of the Shigar, this almost certainly being Marpo Chungi, (4695m/15,403ft), a mountain significant for provoking Haversham's first comments on its fossil character. The technical terms used in his paper tell us that even by this early stage in his career (he has just turned 26), his geological knowledge is considerable and his interest has been deeply engaged in the subject for some time.

Another terrifying crossing – though not to Haversham, who describes it with characteristically tight-lipped understatement – on goat-skin rafts at Chutrun, higher up the Shigar, led to two ascents at the furthest limits of this phase of the expedition: Busper (4663m/15,298ft) came first,

[18] Probably Honboro Peak, 6459m/21,191ft.
[19] *On the Glaciers of the Mustagh Range*, p20.
[20] *Ibid*, p21.

then, on his final excursion to the other (west) side of its valley, he reached the Gunto La (4606m/15,112ft) apparently without difficulty (but see the paragraph in Chapter 11 on the inconsistencies in Haversham's recollections about the traverse of the Gunto La in 1860 and 1861). From here, the explorer in Haversham cannot resist a mention of the view north-west towards Nagar (just east of Gilgit), which is known 'only by native report', this view having come about through the ascent of Munbluk (c.5600m/c.18,370ft) from the Gunto La. He finds, the following year, one of the main access routes to Nagar via the colossal Hispar Glacier and might already have been forming the idea of that journey in mind; there is no doubt that Haversham had thought out a clear strategic overview of the aims of his four Karakoram expeditions.

What he would not have known that following year as he gazed down the Hispar Glacier in the direction of Nagar was that another Guildfordian, a fellow-pupil from the Royal Grammar School, Colonel Algernon Durand, educated there about 20 years after Haversham, was in 1891 to direct the British capture of Hunza, which lies across the Hunza River from Nagar, in one of the most spectacular and heroic military campaigns in the whole of British imperial history, vividly described in *Where Three Empires Meet*, by EF Knight, *The Times* correspondent attached to the military force. Other minor campaigns in this area are also described in this very readable and well-illustrated work.

From the Gunto La, Haversham descended its far (south-western) side to the stony hamlet of Rondu in the Indus valley. At that point, we are bluntly told that 'the season for surveying is now over', and we are left to find our own way back to Srinagar. This abrupt ending concludes this extensive exploratory and scientific adventure, but its remarkable achievements provided merely a taster for the next expedition, the one that would leave Haversham's mark on mountaineering history.

CHAPTER ELEVEN

THE SECOND KARAKORAM EXPEDITION

K2 (Mount Godwin-Austen)

TEN MONTHS, THE LOSS OF his first wife, and a hasty wedding later, Haversham returned to Skardu to begin one of the most remarkable expeditions in British mountaineering history. Not only did he open the way to the world's second-highest mountain but, beyond Askole, he became the first Westerner to explore the awesome glacier system in its neighbourhood. (Haversham's original route via the Braldo River then along the Baltoro Glacier was extended up the Godwin-Austen Glacier almost to the summit of Windy Gap by Jonathan and Stephen Godwin-Austen, the sons of Haversham's two eldest living descendants, in a magnificent trek completed in July and August 1988.)

Haversham's reconnaissance of the world's second-highest peak was achieved with the utmost determination, flair and precision (digital measurements of K2's height have not altered the accuracy of Haversham's original brass-and-mahogany theodolite measurement of 28,265ft/8615m). He completed this feat exactly 60 years before the equivalent expedition opened the way on the Tibetan side to the arguably more accessible Mount Everest. (Access to Everest on the Nepal side came only 30 years after that.)

The original sketch of K2 had been drawn by Captain TG Montgomerie of the Royal Engineers on 10 September 1856, while he was working for the Survey of India in Kashmir. He was taking measurements of surrounding mountains from the high altitude Survey station at Haramukh, on the Deosai Plateau above Srinagar, when a break in the clouds enabled him to catch sight of the Karakoram mountain range, an amazing 220km/137 miles away to the north-east. The Karakoram had not yet been properly surveyed, so Montgomerie took the opportunity to draw a quick sketch of the distant mountains. The two prominent peaks in the sketch he labelled 'K1' and 'K2'.[1] Giving labels like this to unsurveyed mountains was common practice at the time although the labels would normally be replaced with appropriate local names once the peaks had been properly surveyed and maps drawn of the area. Montgomerie, when in charge of the Kashmir Survey, instructed his surveyors to continue the 'K' labelling for all of the snow-covered peaks in the range. Thirty-two peaks were identified in this way and as the survey progressed local names were found for almost all of them.

During the ten months between his first and second Karakoram expeditions, much had

---

[1] 'K' for 'Karakoram'.

changed in Haversham's emotional life. Kudikji, his devoted companion (in his account) on the 1860 expedition had probably died (see Chapter 9) at some point in the months after September 1860. On 5 April 1861, he had married Pauline Georgiana Chichele Plowden; only two months after their marriage he set out on this arduous and dangerous enterprise, by far the most demanding he had yet undertaken. Although he was assisted by 66 coolies, *shikaris* and a *bhut moonshi* (interpreter/secretary), Haversham was the sole Westerner in the party. While frustrated at not being able to discuss exploratory matters with a fellow English speaker, Haversham characteristically never complained of loneliness.

The expedition comprised four aims and four phases, each roughly overlapping. First, he intended to cross the Mustagh Pass (5370m/17,618ft), a well-established trade route to Yarkand in Chinese Central Asia, a wild and politically unstable country which nonetheless promised considerable trade advantages for the British. This was political mountaineering; a throwaway remark shows that in approaching the pass, Haversham was thinking about its practicability as a military access route to Central Asia.[2]

Second, Haversham aimed to discover where K2 lay in relation to the watershed of the 'Great Snowy Range', which marked the northern limits of British Indian territory – defining it and what lay beyond, especially when coupled with a feasibility assessment of the Mustagh Pass, was an important basis for British political and economic thinking in the Great Game.

Haversham pursued his third aim by heading a long way north-west, to reach the Nushik La, the access point for the immense Hispar Glacier, which leads to Nagar, then Hunza, at that time both unknown territories.[3] Knowledge of these lands was vital to British political strategy: their mountain gorges were wrongly suspected of providing a possible invasion route for armies, most probably Russian, from Central Asia, and their inhabitants were also sufficiently warlike to prevent intelligence getting back to British India after being gathered in Kashgar by Indian traders and *pundits*.

Haversham's final aim seemed more modest by comparison: he was to try to find a new and quicker route back from Skardu to the big British army base at Udampur, outside Srinagar. While interesting in exploratory terms as country unsurveyed by Westerners, there is a hint that this new route might also help speedier army mobility.

TO THE MUSTAGH PASS

Haversham first speaks of '*preliminary work*'. This entailed visiting '*several peaks*' to the south of Skardu, hardly any details of which are forthcoming. The only peak mentioned by name, Thyarlung, (almost certainly Thalanka, 5130m, which Haversham records as 16,844ft/5134m) was described as visible from Skardu but requiring '*a long round*' to climb it as Haversham had travelled via the Burji La in order to assess its height (15,879ft/4840m). He left Skardu on 7 July, the day after his 27th birthday, though we are not told when he left Srinagar or by which route Skardu had been reached. On the way up to the Burji La, he records that his pulse beat was 138 a minute, while for one of his Balti porters it was 104. From the summit of Thyarlung, he could see K2 '*of an airy-*

[2] He failed to cross it; it was only first traversed 20 years later in an attempt of hair-raising danger by Lieutenant-Colonel Sir Francis Younghusband, KCSI, KCIE (1863-1942), an army officer, explorer, and spiritual writer. Younghusband held positions which included British Commissioner to Tibet and President of the Royal Geographical Society. His life is fascinatingly described by Patrick French.

[3] Hunza is now an adventure tourist destination, lying on the Karakoram Highway, a great road-engineering project carried out jointly between the Pakistanis and the Chinese and completed in 1988. Because of the great geological instability of the region, round-the-clock improvements to the dirt surface and landslide clearance continue. The road links Rawalpindi-Islamabad and Gilgit then traverses the Khunjerab Pass (c.4730m/c.15,518ft), one of the highest surfaced roads in the world, before descending through the Chinese Pamirs to Kashgar. For my own experience of this terrifying road, and in making first ascents in China to the east of the Karakol Lakes, see *Alpine Journal* 96.

blue tint' in contrast to the yellower colours of Masherbrum (K1). A rough descent led him back to the plains, west-north-west of Skardu; he was now ready to begin the expedition proper.

Having once more crossed the Indus just short of its confluence with the Shigar in the exciting little goatskin ferry, Haversham declared that Marshakala (5153m/16,906ft), the mountain which had beaten him in 1860, '*had to be ascended*', an interesting comment on his compulsive behaviour and sense of professional duty. While waiting for the weather to clear, he had time to describe the river-level areas in close detail. His description is quite Romantic:

> ... the village of Kuardo is beautifully wooded and green; fields rise in terraces for a distance of two miles along the valley; the houses are scattered about them, surrounded by fruit-trees - apricot, mulberry, walnut, apple; willows are common. The vines ... hang in festoons from the branches ... The field cultivation consists largely of wheat, coxcomb, tromber[4] and barley. Excellent gourds, melons, cucumbers, turnips etc are grown ... The houses are built ... in two stories *(sic)*: the ground-floor walls are of the rounded stones from the ravines, with mud, or of sunburnt bricks of large size, cut out of the hard lacustrine clay. This lower story *(sic)* ... in the winter ... serves ... as a residence for the cattle, sheep and goats. The walls of the upper story are made of strong wicker-work ... and well plastered with mud ... A portion is left with a flat roof ... where they clean their grain. The ascent to the houses is by a ladder from the outside, so that the inmates in a measure are secure ... The women clean the grain, and, with the exception of ploughing, do as much of the out-door labour as the men, even carrying heavier loads. When cattle are scarce, it is not unusual to see a couple of men harnessed to a plough ... During the apricot season, the large rocks and the roofs of the houses are covered with the fruit and in two or three clear days become sufficiently dry to be packed in skins; they are then called 'khobanies' and form a very considerable article of consumption in the country; the better sort are exported to Ladak, Astore and Kashmir. The process of making khobanies is by half splitting the apricots and taking out the stones: several are then inserted, one inside the other, with a sweet kernel in the centre, and they are thus dried in the form of small balls. When required for use they are put to soak in water overnight, and eaten with chupatties next morning.

It took two days '*in intense cold*' to reach the summit of Marshakala. A narrow ravine, then steep pillars of alluvial clay were followed by '*extremely steep*' rock. On the summit, Haversham stoically recorded typical conditions which a surveyor had to put up with:

> We made ourselves as comfortable as it was possible to do; the plane-table was set up, and as much as could be seen cut in at once. I got into a snug corner out of the wind, and then the day was spent in sundry rushes to the plane-table, whenever, on looking round, some peak would show out from amongst the heavy clouds which hung about. Many such days as these - days of long, dreary hours - have to be passed among the Himalaya. Our altitude was 16919' (5157m).[5]
>
> The frost in the course of the night was severe, and at sunrise next morning the thermometer showed $32^{\circ}5'$.

He was rewarded with an unforgettable view of K2 and Masherbrum in one direction, and Nanga

[4] *Tromber*: a Kashmiri word for cultivated beehives and the honey thereby supplied. This word has not made it to the *OED*. I am consequently grateful to Ms Pamela MacGregor for her explanation.
[5] The modern height is 5153m/16,906ft.

Parbat (26,630ft/8117m; the accepted height is now 8126m/26,660ft), the world's ninth-highest peak, in the other.

The achievements of the 1861 expedition are all the more amazing because after this ascent, Haversham was held up for 12 days by bad weather, a significant proportion of the seasonal 'window' open to him. When the weather improved, he began the ascent of the Skoro La. At his first campsite on the way over the pass, at 12,552ft (3826m), he came in for a shock:

> Whilst lying in my tent, after finishing up my work, I heard an unusual rumbling sound, and on going out I found all the men were wondering what it could be. After a few more seconds of suspense, some Balti coolies, who were cutting brushwood higher up the ravine, shouted out that the stream was coming down, and in a few seconds more we saw a black mass coming out of a lateral ravine from the right, and moving rapidly over the broad slope of boulders which formed the bed of the valley. Before the black stream reached us, it divided into two, and we saw that it consisted of a mass of stones and thick mud, about 30 yards in breadth, and about 15 feet deep. The servants by the side of the little rill near the tents had just time to escape before it came down upon their fires. It was a most wonderful sight: a great moving mass of stones and rock, some of great size, measuring 10 feet by 6, all travelling along together like peas shot out of a bag, rumbling and tumbling, one over the other, and causing the ground to shake ... I was almost bewildered by the spectacle.

As usual, there is no excuse for dawdling, and after an ascent of 4,091ft (1247m) up steep grass the next day, Haversham reached the Skoro La, 16,644ft/5073m. Having been enjoyably amazed by his first view of some of the great glaciers he was soon to traverse, Haversham completed a quick descent to Askole, the last village of any importance before the wild country really starts. To reach Askole, he had to cross the Braldo River by a rope-bridge 270ft (82m) long. It was constructed of nine ropes to make a footway, with nine lateral ropes to hold on to: the ropes are made of intertwined birch twigs. I learnt, for example, in Zanskar in 1992, that these bridges are remarkably strong, because the intertwining of damp birch-twig and yak-hair rope contracts considerably when drying out and so produces a long-lasting piece of cable. Given that a collapse of the bridge in most circumstances would cause certain death, their traverse is still something of a trial of faith. As Haversham correctly but understatedly records, '*the passage across was by no means pleasant.*'

Perhaps to commemorate his survival, Haversham carved 'HHG-A 1861' on a nearby rock, not just a piece of vandalism, however, but of use in the following years, when he added '1862' and '1863', to measure variations in the river height, an important piece of information for a hydrographer. These initials were spotted on their walk-in by the ill-fated American K2 expedition of 1953.

Less easy to justify was apparently carving his initials on a rock between Dras and the Zoji La, although it led to the following curiosity. A newspaper clipping in the archives, probably from the *Daily Telegraph* judging from its font – and although undated almost certainly from the 1930s – tells us that Annesley Godwin-Austen, Haversham's nephew, received a letter from his cousin, Charles Chenevix-Trench, an officer in Hodson's Horse.[6] Chenevix-Trench was returning from a shooting trip in Baltistan when, on sitting down to eat a sandwich between Dras and Matayan, he noticed a rock with initials carved on it, probably with a hammer and tent-peg. Next to it was carved 'HGA 1882', proving that Harold Godwin-Austen had passed that way. (The Zoji La is one

---

[6] Chenevix-Trench (1914-2003), became the author of a wide variety of popular historical works after serving as an Indian army officer in the 1930s, winning an MC during the Second World War and then becoming a District Commissioner in Kenya. His cavalry regiment, Hodson's Horse, had originated as part of the Indian army. It was raised by Brevet Major William Stephen Raikes Hodson during the Indian Mutiny of 1857, and exists today as the 4th Horse Regiment in the Indian army.

of only two generally-used routes to Leh, and I believe the other, the Baralacha La, was not in regular use at this time, certainly not by Westerners.) Harold would have left his mark when returning from his expedition to Yarkand with Ney Elias (see Chapter 3). We know it was the return journey because Elias had already reached Kashgar by 1880.

While Haversham modestly tells us that his arrival was *'an event'* in the Askole area, this understatement draws attention to his being only the second European to reach there, the first having been Adolf Schlagentweit, (see Chapter 10) in 1856.

With the help of two local guides, the party of 67 quickly left Askole and soon reached the snout of the Biafo Glacier. By 2.30 the same afternoon, Haversham, evidently suffering in no way from the effects of altitude, had climbed to a point 2,500ft (c.800m) above it, in order to set up a measuring station. From there, the party pressed on to the junction of the Biaho and Braldo Rivers, then ascended the vast Punmah Glacier to a height of 10,318ft (3145m). Haversham made sketches of the bleak ice- and boulder-strewn wilderness to produce a fantastic panorama watercolour (see plate 26).

He was now heading for the Mustagh Pass. While moving up one of the largest glaciers in the world, Haversham still found time for some natural history observations: the character of the local birch and juniper (both very small shrubs in the locale) is described as are the differences between the local yaks and *dzo*, a half-yak, half-cow (Haversham spells it 'Bzo'), as well as the local cows, lower down. The party was moving up the lateral moraine of the glacier and living off its nerves – everyone was constantly on the lookout for blocks of stone falling without warning off the top of the moraine; sometimes the party had to run past particularly dangerous sections, at least an indication of how well they had acclimatised. After a night squashed together in a place where little room could be found for the tents, Haversham was moved to describe the prospect from his tent:

> ... as extraordinary a scene as the eye could picture; it was the desolation of desolation ... There was not a speck of green to relieve the great precipitous crags of grey and ochre ... Small pools of emerald-green water, with cliffs of ice, filled many of the hollows, while in some parts flowed streams, which generally ended abruptly ... On every side were heard the noises of falling stones as they rolled down the ice-slopes or dashed into the pools.

After another three miles, they were obliged to halt, because of bad weather. Then an unexpected diversion occurred:

> About 6 in the evening we were surprised by the sudden appearance of four men from Yarkund, who turned out to be Baltis of Shigar and the Braldoh, who had emigrated to Yarkund some years back, and had now come over to see their friends on this side. I soon got into conversation, and learned from them a good deal about the country they had come from. The poor fellows had suffered a good deal while traversing the mountain portion of their route, having to travel by night and hide away during the day, on account of the robber tribes. These men wore the sheep-skin cap and long-skirted coat of Yarkund, with voluminous sleeves padded with cotton; thick leggings, and stout leather boots or pubboos, completed their somewhat stout appearance. The goodness of their clothing contrasted strongly with that of their brethren of Baltistan, and showed that emigration had been advantageous.

While such a friendly meeting in such a wild and remote spot was welcome, and gave Haversham the opportunity to practise his Balti (having reached army certificate level in Hindi not long after first arriving in the Punjab), this conversation really concerned intelligence-gathering: Haversham was assessing the accessibility, the economic and the implied military strengths of the land beyond the Mustagh Pass.

The demands made by the terrain continued the next day (8 August): after a long time on the march, the party could see their likely campsite only a few miles away across the glacier, but it was getting dark before they reached it. Miniature icebergs, some measuring 15x18x10ft (approx. 5x6x3m) were observed on a dried lake bed. Despite these inhospitable surroundings, the party also came across a few *chikor*, a grouse-like bird, a favourite trophy of game-hunters in north-west India, three of which Haversham shot as a welcome variant to their diet. (This moment reminds me of perhaps the most enjoyable meal I ever had in the mountains. It was eaten at an Advance Base at about 4200m in the Chinese Pamirs, in 1989, by way of celebrating the Glorious Twelfth. We tucked into Baxter's Grouse in Port Wine Jelly. The memory of its deliciousness is still with me; and the bones made an excellent soup the next day.) The details of the daily diet are otherwise not recorded. Despite the desolation, however, Haversham's artistic responses are still very much alive:

> The sun setting behind the line of snowy peaks ... lit ... up with a beautiful pink tint; whilst the rocks of the moraines, red, yellow and green, heightened in colour by the wet, sparkled in the sunlight.

On a surveying excursion next day, Haversham and his assistants found it impossible to get back to their camp until a 'coolie' discovered a way through an ice-cliff via a kind of crevasse filled with rocks. They got off the ice and into camp with the last glimmer of twilight. Haversham remarks dryly, with characteristic understatement, '*A little longer, and a night on the glacier, without covering, would have been our portion; with the cold wind we should have been frostbitten to a certainty, and perhaps laid up for some time.*' At that point, little did he know how a real version of that imagined concern would soon be dealt out to him and his men on the Baltoro.

Having reached a camp on the Chiring Glacier by 11 August, the party was in a position to attempt the Mustagh Pass. Haversham took eight men with him, with ropes (they had, altogether, the miserable amount of 162ft/49m, for the purposes of river-crossings!) and 'other appliances', presumably sticks, and hatchets for step-cutting. They soon encountered crevasses, the like of which they had not previously seen:

> For the first three miles the crevasses were broad and deep in places only, and we could avoid them by making detours. They soon became more numerous, and were ugly things to look into, much more so to cross – going down into darkness between walls garnished with magnificent green icicles from 6 to 20 feet long, and of proportionate thickness, looking like rows of great teeth ready to devour one.

Although, in Haversham's expression, they looked like a chain of criminals, they soldiered on through a crevassed snowfield, with high crags sometimes visible through the swirling mist, until they were about a mile short of the pass. The weather turned: the clouds rolled in. They lit a fire and found that their boiling-point thermometer indicated an altitude of 17,300ft/5273m. The pass was no more than 500ft/c.150m above them. This was not the occasion to dither, though; everyone made their way as quickly as they could back down the noisy glacier, which was making terrible crunching and splitting sounds as its two-mile breadth was forced through a channel only a quarter-of-a-mile wide. Back in camp, Haversham expressed his disappointment at having failed to reach the pass, just as Adolf Schlagentweit had done in similar circumstances a decade before. But he recognised that he had done his surveyor's duty by 'determining the ridges' on the northern side of the Chiring Glacier. He then gave out his real purpose:

> This *(pass)* is quite feasible, and with a small guard the survey might be carried into the Yarkund country for a considerable distance, as, from all the accounts I have heard of the tribes, their numbers cannot be great, nor their matchlocks much to be dreaded.

It was pretty much this sort of thinking which gave Sir Francis Younghusband his reasons for the murderous invasion of Tibet in 1904. In two days, Haversham *et al* had returned by their outward route to the junction of the Braldo and Biaho rivers at Tsok, and were immediately ready for phase two, the uncovering of the arduous way to K2.

K2 AND THE BALTORO

To gain access to the Biaho Glacier proper, another fragile-looking rope-bridge had to be crossed and gullies and slopes of stony clay clambered up: some were so steep that steps had to be cut. Thorny scrub, inhabited by many hares, followed. No-one, not even the Baltis, had followed this route to Yarkand – in Haversham's rather Biblical-sounding phrase – 'since the days about the time Ahmed Shah became Rajah of Skardo', presumably in the early 1830s. Soon, after climbing more shingle, they reached the surface of the Baltoro Glacier, which according to Haversham extended itself at an angle of no more than 4° in a straight line of 36 miles (58km). While sensitive to colour, whenever Haversham comes across a new topographical feature, he describes it first scientifically, then second, if at all, in terms of its beauty. The occasion must have been not dissimilar to the first close sight of the Western Cwm opening out above the chaos of the Khumbu Icefall (although the proportions and technical difficulties are dissimilar) during the 1951 Everest reconnaissance expedition. This glacier's flatness was deceptive, however: the terminal part was full of pressure ridges and hollows which were a great labour to clamber over. There were other hazards: constant rockfall from the lateral cliffs, and apparently bottomless pools of water into which large blocks of ice would tumble with a loud bang. In one day, the party's laborious daylong scramble covered only four-and-a-half miles.

The compensation for this miserable progress, as ever, was the scenery. The glacier's sides were split by tributary glaciers, each divided from the other by towering walls of spiky rock rising to about 22,998ft/6705m. In front of them, the spire of Gasherbrum IV (8080 m/26,509ft) drew the party on. After a push ahead of his men, Haversham was rewarded with a view of Masherbrum (7821m/25,659ft):

> ... one vast mass of ice on all sides falling into the glacier. The pale ochre-coloured rocks showed here and there only, and in beautiful contrast to the pink-tinted snow at sunset, giving an appearance of much greater distance than the reality, for its summit was only 6 miles from me in a direct line.

This time, of course, he was looking at the opposite face from the side he had clambered up the previous year. The daytime, therefore, comprised a mixture of frustration, danger and reward. Night-time on the glacier, on the other hand:

> ... was dreaded by us all; for in the evening the wind from the east sprung up again, and the cold became intense. The ice was much exposed ... At sunset, every puddle and lake in the glacier began to film over with ice, which in the morning was more than an inch in thickness. The wind blew all night, getting stronger towards morning; and I pitied the poor coolies with only the rough, sharp stones to lie on, and separate them from the ice. The Balti, who carry only one rug, huddled together two or three under the same covering. Mahomed told me next morning that he heard them saying, 'Would that the Sahib felt it as cold as this! He would soon go back!' Few, if any, got any sleep; and all were right glad when the sun rose over the peaks ahead.

The following day, 19 August 1861, is one of the great landmarks in British mountaineering history. Haversham was anxious to see K2, to discover if it lay, as he suspected, on the main Great

Snowy Range watershed. He realised that to resolve the matter he would have to ascend a top on an outlying spur of Masherbrum above the area now known as Urdukas. He made for this minor peak across the moraines and ice ridges of the Baltoro:

> The foot of the ascent reached, we found good footing for another 1000 feet, and then came some steep rock; after this another slope was gained. When we reached the ridge of the spur, Masherbrum came into view. Eagerly I had looked, whenever we stopped to take breath, along the line of snowy mountains to the north in search of the Great Peak; and now, fixing the position on the plane-table, I showed those with me the mountain behind which lay the peak I had toiled up so far to see. Following up the ridge another 1000 feet of elevation was gained, when a distant bit of rock and snow could be seen just peering above the nearer snow-line. After another sharp push up to a point where it was impossible to mount further, there no longer remained a doubt about it. There, with not a particle of cloud to hide it, stood the great Peak K2, on the watershed of Asia!

Despite his excitement, Haversham found time to make a sketch, subsequently converted into a magnificent charcoal and watercolour painting (see plate 29), a valuable record of the first close sighting of K2, while the 'coolies' toiled at setting up the plane-table and built a cairn in the foreground. The painting is a real piece of history: the figures, despite Haversham's relative weakness in drawing people, animate this scene in a way which reminds us that there is no equivalent for Everest. While he looked down on the Baltoro, he would, because of the intervening ridges of the outlying heights of Biange (clearly visible in the painting), have seen nothing of the wide and winding K2 access glacier named after him. He did, however, view the length of the Baltoro, with '*fifteen distinct lines of colours*'. The guides completed their cairn, then they descended.[7]

On the return down the Baltoro, Haversham reflected on his experience and suggested that further visits to the region could only be accomplished '*by some man of influence*', a thought which must have struck Sir Martin Conway and the Dukes of the Abruzzi and Spoleto. By this, he meant someone rich enough to keep the porters motivated, as he suggests that were it not for his *bhut moonshi*, Mohamed, the porters would have bolted some time before. By the next day, they had reached the Biaho, where the thickness of the ice was estimated at 400ft/122m. From there, Haversham resolutely ascended the Biafo Glacier for about five miles to a '*low knob*'. The glacier's length was estimated at '*upwards of 40*' miles (64 kilometres). Haversham digresses briefly into the history of the area and describes a raid on the Baltis, one of many by the Nagar men, the last raid having been carried out by 700 or 800 Nagaris descending the Biafo and carrying off about 100 men and women with as many cows, goats and sheep as they could lay their hands on, in 1837.

The weather again turned bad. Haversham was obliged to keep to the banks of the Braldo until regaining the upper Shigar valley, despite needing to use two further rope-bridges and having to traverse a very dangerous broken track above the torrent. He took in a quick ascent, up to about 5300m/c.17,400ft, above Chokpojong, before moving on to Dassu for the river crossing by *zuks*.[8] Haversham dwelt on a description of his crossing in some detail:

---

[7] I am very much indebted to John Cleare for working out with quite extraordinary precision exactly where Haversham ascended to get his K2 sighting. See plates 27 & 28.

[8] This river crossing sounds like a very nerve-racking experience. Having almost lost my life in a glacier river in the first European expedition to the Dzjungarskiy Alatau, north-east Kazakhstan, in 1998, I find it very difficult to distance myself from the idea that water represents the biggest single objective hazard on any exploratory expedition at these sorts of altitudes. Of course, common sense would help. I wrongly reasoned that the crossing-point for our ponies would do for me, having forgotten the old principle of 'four legs good, two legs bad' and was doubly foolish in crossing on my own. I was in a state of shock by the time I reached the far bank, having thought, when I went under, that my last moment had come. The effect of the shock was to cloud my judgement completely, to the extent of attempting a return crossing a few hundred metres upstream from where the same near-approach to Nirvana occurred.

> The men expected that we should reach the opposite bank near a large rock; but though they exerted themselves to the utmost, we soon saw that this was quite impossible, and we were swept past it some 50 yards distant, and went down the river at a mad pace, causing that curious feeling of excitement which a sense of some coming danger always produces; nor were we long kept in suspense. The ... raft-men behaved with great coolness, and steered safely close to the edge of a bank of boulders ... we passed also some ugly waves beneath the cliff on the other side ... Nothing could be seen ahead but curling waves and foam, with great black rocks here and there: into the midst of this our raft glided. Mahomed and the manjis[9] repeated the Kulmah or Mahomedan Creed in a rapid whisper, as we were carried over the crests and down again into the troughs between the curling waves. At one time I did not expect that we should come out of them; for whilst in the trough a great wave broke right over our frail craft, and completely buried us ... For a moment we came to the surface, the manji shouted, 'Ya, Ali, hold tight!' ... When at last we came out upon the smooth water below, and looked back up the rapid we had just come down, every one breathed freely again, and the 'Thank God'of each was never more sincerely uttered.

As this second phase of his expedition was now ending, Haversham took time to describe the countryside and economy of the Askole area, and its relationship with the 'robber tribe' areas of Hunza and Nagar. These are linking thoughts: his next expeditionary aim was to head north-west, to reach the Nushik La, in order to assess its accessibility as a route to these unvisited tribal areas.

THE THIRD PHASE: TO THE NUSHIK LA

On 29 August, as he headed up the Shigar River, Haversham came across great changes from the previous year. A glacier lake had burst in July; the ensuing flood of debris had covered a whole village and part of one other. The party had to pass over difficult steep ground, by using ladders and planks, because some of the riverside area had been washed away. At the river crossing place – Chutrun – Haversham noticed a bungalow, for the use of visitors; his remarks here suggest his awareness of the possibilities of tourism (though not, of course, mass tourism) in the future. This is not such a strange idea since the prospect of tourism in Kashmir and the Himalayan foothills had been considered much earlier in that century. Above Doko, he briskly ascended a peak to get a view back to the Shigar area then a forward look over the Chogo Lungma Glacier towards the Nushik area (which was hidden by intervening ranges). This peak was probably Berginsho, (5820m/17,323ft), situated about 1,500ft/c.450m above its access glacier. This latter height he achieved in an hour, an excellent rate, and the same as that which Frank Smythe set himself on the north face of Everest in 1933. It is a demonstration, if any by this time were needed, of Haversham's superb physical fitness and perfect acclimatisation.

From the herding hamlet of Arandu, Haversham began his ascent of the unexplored Kero Lungma Glacier, which leads to the Nushik La. He passed an area of dead birch wood, killed through the formation of a now (naturally) drained lake formed by a glacial dam – it was a haunting sight. Further on, he passed some dwarf willows, and shepherds' huts, both in their turn dwarfed by the desolate vastness of the main and subsidiary glaciers. On 4 September, quite a risky late point in the season to be so far from safety, Haversham and his party cut some steps up a steepish snow slope then crossed an almost crevasse-free ice-bed with steep cliffs on either side to reach the Nushik La, the height of which he must have measured but for once forgot to record.[10]

---

[9] The oarsmen, from the Manji tribe of Sindh, now in Pakistan.
[10] It is 4990m/16,371ft.

As they approached the top, further evidence of Haversham's outstanding physical fitness became apparent:

> Several of the men felt the height, and had to remain behind from sickness and headache. I scarcely felt it; and I think that the state of the stomach has a great deal to do with these sensations.

He thus became the first European to see the enormous Hispar Glacier, 12 miles (19km) of which was visible, that being only about a third of its actual length. He remarked that 'no glacier in the whole of the Himalayas can exceed this in magnitude of all its features', and he was (with the possible exception of the Siachen Glacier) probably right. Having satisfied himself about the difficult 1000m/3,300ft descent to the Hispar (*'it is necessary to let a man down with a rope round his waist'*) – and therefore its military unsuitability – he retraced his steps and with astonishing speed had not only arrived back in Arandu next day but by 6 September had made his way about five miles (8km) up the Chogo Lungma Glacier.

On this extremely long double dogleg of ice he spent five days working his way along its true left bank and making minor surveying ascents before reaching Arandu once more. On 12 September he left for Skardu, having triumphantly completed his aim of surveying the whole Shigar basin.

There is a little inconsistency in Godwin-Austen's account of the next section. In his 1864 paper, likely to be the more accurate as being closer to the event, he says he struck west to the Tormik River and the Stok La, the western extremity of his Karakoram travels; he does not say precisely by which route. In an 1894 letter to the RGS on the subject of the wrong titling of a recent woodcut, he claims it was by the Gunto La. According to the 1864 paper, however, he had first traversed this pass, then descended to Rondu in the Indus valley, having taken in a peak, Munbluk (c.5550m/c.18,209ft), next to its summit, in 1860. The 1864 paper asserts that his 1861 starting-point for the Tormik valley is the hamlet of Matuntoro Klas (*'klas'*, he explains, is a summer hut), in which case it is much more likely that the 1861 route lay up the side valley from Matuntoro Klas and across a pass to the north of the Gunto La.

The trail became well-defined only in the Tormik area, however, where the villagers were cheerfully cutting their crops and tending their bullocks, a pastoral scene which Haversham revelled in after the bleak and near-lifeless worlds he had been living in for the previous two months. The weather, unsurprisingly by this late stage in the year, turned nasty and prevented him from reaching the Stok La, although he could only have been a few kilometres, perhaps even only a few hundred metres, short of it.

Recognising that he now had to move swiftly to reach Kashmir before the serious snows set in, Haversham (no doubt unwillingly as it was almost all familiar ground) led his party back down the Tormik to below its confluence with the Indus at Strongling, some way downstream from Skardu, in order to obtain a safe crossing on the rope-bridge.

From there to Skardu, the road is not long, but it is tricky:

> We reached another bad part of the road ... consisting of a series of ladders placed against the nearly perpendicular face of the cliffs; but the ladders were more dangerously placed, and some of them very rickety. This route should not be attempted by any one liable to get dizzy on looking down from great heights. The ladders often rested only against pieces of wood driven into cracks in the rocks, and on looking through the rungs as you go up, the view presented is that of a great river rushing along like a foaming torrent, at the base of vertical cliffs, which descend 300 feet sheer beneath one's feet.

THE FINAL PHASE: BACK TO SRINAGAR

Having crossed various low passes, such as the Katch Bore La (3000m/c.10,000ft), Haversham and his party arrived at Kachura, a few miles west of Skardu, their turning off point for the expedition's final phase, the crossing of the Alumpi La in order to reach Srinagar. They ground their way up to the bleak plateau of Shigarthang, not far from the Burji La, at 10,200ft/3109m. At least, however, the way up to the Alumpi La had remained open.

Although Haversham measured its highest point at the not-too-dangerous height of 15,200ft/4633m, the party came across the bleached skeletons of several men, a not uncommon sight in this part of the Karakoram. On top of the pass, by way of reinforcing the potential danger of their situation, they discovered a further 50 bleached skeletons. This sight reminded the party that they had to move on quickly as it was 29 September and they were cutting their time very finely indeed – large clouds were rolling in from Nanga Parbat and over the Deosai Plains. ('Plains' is a misnomer: the Deosai is a high upland plateau.)

Once the pass had been crossed, Haversham wound up his expedition in characteristically brisk but uncharacteristically imprecise fashion. He mentions the Dore La as the best (i.e. most straightforward) pass in the area, and also the Kashmiri character of the Astore valleys: this statement implies that he descended from the Alumpi La to Gudai, south-east of Astore, before heading along forested hillsides over the Dore La to join the track descending from the Burzil La, just south of it, at Kamri. From there, a straightforward walk over the Raj Junghan Pass takes one readily in to Bandipur and the large British base there, just outside Srinagar. The whole expedition was, however, merely one part of the Kashmir Survey; it was –

> ... immense work – that of weeks, months, and years ... the altitudes climbed, the long, wearing descents ... the days and nights spent taking observations from the many hundreds of trigonometrical stations ... and the large number of men who share in the work, from the officers in charge to the native signalers.

This expedition was probably the greatest single mountaineering expedition undertaken by any human being up to that point. It had fixed the height of K2 and it had defined the structure of a large part of one of the world's greatest mountain ranges. Haversham's place in mountaineering history is assured by these achievements; but much more, in even remoter parts, remained.

CHAPTER TWELVE

BEYOND THE KARAKORAM

Summiteer

IN 1862 HAVERSHAM EXTENDED HIS work beyond the Karakoram - each expedition ventured into successively more remote country - by visiting Rupshu, Ladakh, and Zanskar. This surveying was carried even further afield the following year, to Lake Pangong, in the forbidden land of western Tibet, then to the Changchenmo river basin in the wild Aksai Chin, east of Leh. Information about this 1862 expedition is very uneven. There appears to be no reliable scholarly papers to describe it. In his unpublished, typewritten abstract of his army service we are given only a sketchy outline of his route and achievements, apart from one short detailed account copied from his Journal (now lost) of a section of the expedition and the '*hard walking and climbing it involved.*' This is a regrettable loss to mountaineering history, and to me personally: of all the areas explored by Haversham, this ranks as my favourite, not least because a good part of it has still not been much visited by Westerners. The information we do possess is not exactly comprehensive: undoubtedly, quite a number of unrecorded ascents of peaks and passes, most of them above 5000m/c.16,500ft,[1] have gone unrecorded. Haversham also keeps up an old habit, at least in part: we know how he reached Ladakh, but we do not know by what route he returned. The dates and placenames supplied by Haversham are complemented by dated watercolours at Papplewick Hall, but these are not enough to reconstruct the whole journey in satisfactory detail.

His Rupshu, Ladakh, and Zanskar expedition is also notable for Haversham breaking the world mountaineering summit altitude record[2] on his ascent of Mata Peak (Mentok Kangri, 6283m/20,614ft), west of Tso Moriri, now a fairly well-established trekkers' peak, but with difficult technical sections, depending on the route.

RUPSHU, LADAKH AND ZANSKAR

Early July - early September, 1862 (start and finish dates unspecified).

Haversham started his expedition proper from that beautiful and isolated lake, the bird sanctuary of Tso Moriri, in the high country of Rupshu. We know that he reached it from Leh; of the two probable routes, that over the Namshang La seems the more likely since it is the more direct. We

[1] In this chapter, heights given first in feet then in metres are from Haversham's own measurements in feet, followed by the metric equivalent at 3.28 metres/foot. Heights given first in metres then feet are from the Karakoram and Indian Himalaya sheets published by the Swiss Foundation for Alpine Research.

[2] There are of course records of higher altitudes gained, probably for religious and/or burial purposes, in the Andes, but I doubt that these should be designated 'mountaineering'.

do know that he reached Leh via the Zoji La (he spells it Zogi La, 3528m/11,575ft). This high pass is now one of the main military and tourist highways into Ladakh and Zanskar. Its *z*-bends, however, are so narrow that traffic is regulated to allow either ascending or descending columns one at a time only.[3] This was not Haversham's first crossing of the pass, as a watercolour at Papplewick of Mulbekh Gompa, showing that he visited it in August 1858, proves. (A *gompa* is the Tibetan word for a lamasery. In this area, *gompas* are usually built on hilltops or on inaccessible ledges of high, steep rock faces. Mulbekh lies near Kargil, the administrative capital of Zanskar and well down the road to Leh.) Haversham's 1858 painting of the nearby 7000m/c.22,300ft twin peaks of Nun and Kun from their northern versant confirms that this wild upland was Haversham's furthest north-westerly point on the survey of Kashmir proper.

Haversham climbed Mata Peak (Mentok Kangri) on 23 July. From there, next day, he wove his way along the west bank of the lake (Tibetan: *tso*) to a '*Red Knob*', 17,000ft/c.5200m somewhere near the lake's south end. He then swung west, to head for Zanskar, 'the land of white copper'. He did climb to '*near*' an 18,000ft/c.5500m peak (unnamed) on his way to the Pangpo La (17,000ft/c.5200m, although the Swiss 'Indian Himalaya' map gives 5305m/17,405ft). The Pangpo La lies usefully close to the watershed between Rupshu, Lahaul and Zanskar-Ladakh. Haversham stopped there for several days. Recognising its watershed importance, he re-crossed the pass twice. On 6 August, he ascended one of the highest peaks in the area, Kato (now usually written Khating), 5810m/19,062ft, before making his way almost certainly along the Tsarap Chu, but including one side-valley diversion as far as the Lachalung La, c.5100m/16,732ft, to the Marang La, 5250m/17,224ft, a pass familiar to all those who have driven along the magnificent, if hair-raising, road from Manali to Leh.

From here, Haversham's trail becomes harder to follow. He states that he climbed a 19,500ft/5944m peak but does not name it. It may be an unnamed top, marked at c.5900m, near the Baralacha La-Zingzingbar district. To have reached it from the Marang La in the time available, by heading south down the Leh-Manali road past Whisky Bridge, would fit. He further tells us that he reaches the Surichan La on 18 August, i.e. a week after leaving the Marang La. His arrival on 22 August at one of the wonders of the Asiatic world, the cliff-face Phuktal Gompa, consequently makes sense. We know from the date of a Papplewick watercolour when he reached Phuktal; the drawing has been made from the Purne (west) side, its most usual line of approach. Haversham does, however, refer to reaching a summit called Lankar, precisely recorded at 19,273ft/5874m, on the day before reaching Phuktal; I have not been able to trace this top – it is most likely one of those high peaks to the west of the Tsarap Lingti Chu on the section between Kargiakh (to where one descends after the Surichan La) and Purne, the turning-off point for Phuktal.

Haversham's route to Phuktal followed the line usually taken by today's trekkers going over either the Shingo or Phirtse La from Darcha to Padum. They make the brief diversion from the main trekking route at Purne to see this *gompa*'s fantastic buildings, seemingly hacked out of an all-but-vertical cliff above the Tsarap Niri Chu and topped by its golden temple roof, above which lurks the cave where the water level never changes. I visited Phuktal in 1992, from the Purne side, having staggered somewhat breathlessly over the Shingo La en route for Zanskar. The abbot told me that he loses two lamas a year, on average, from falling out of their cell windows down the cliff. It is a fabulous sight and roughly comparable to, but to my mind surpassing, the sensational Tiger's Nest (Taktshang) in Bhutan.

Haversham was only the second Westerner to reach this remote spot. In 1826, on his second trip into this region, the Hungarian scholar Csoma de Körös studied there – his carved name can still be seen.[4] The only remaining hard information we are given about this expedition is that

---

[3] The road also boasts the infamous 'Captain's Corner' where a military officer drove off a hairpin, followed lemming-like by 33 other vehicles behind him. This story's status lies somewhere between legend and truth, although I am sure a great many variations exist of the numbers involved.

[4] Sándor Körösi Csoma (1784-1842) also known as Alexander Csoma de Körös, was born Csoma Sándor in Körös, Transylvania. He was a Hungarian philologist and orientologist, author of the first Tibetan-English

Haversham climbed Shingrugo, 17,208ft/5245m, just east of Padum, the historic capital of Zanskar, on 29 August, i.e. a week after leaving Phuktal. Whether he reached Padum via Purne and the trekking route down the Tsarap Lingti Chu, or via the Stonde La, we do not know. The former is much more likely as it offers greater cultural and geological interest, a better margin of safety and better possibilities for a livelier diet. The Stonde La route would also probably be slightly too short for the dates given, assuming the weather was reasonable.

How Haversham returned to Srinagar we also do not know. The track from Padum out to Kargil, which would complete a route he had almost certainly covered in part, from just below Ringdom gompa onwards, in 1858, seems the most likely possibility. The alternative routes via Zangla back to Leh seem unnecessarily long diversions, given the lateness of the season.

For this powerful piece of surveying, therefore, it is most regrettable that all we know about it are the bare bones of most of the route. No records of incidents along the way, the nature of the people or the countryside, or the character and size of his party are forthcoming. The very remote country east of the Marang La remains not much visited even today, and some description of it would therefore have been very welcome. Fortunately, such deficiencies were more than made up for in Haversham's final expedition in this part of the Himalaya, an achievement which provides the most emphatic evidence yet for the singularity of this extraordinary Western explorer.

LAKE PANGONG AND THE CHANGCHENMO BASIN

14 July - September, 1863.[5]

This is a well-documented expedition, both in the scholarly papers in which Haversham describes his progress and observations and in the degree of detail provided by the private sources in his archive. It should also be considered Haversham's highest expedition, if we look at the mean height of his ascents, three of which exceeded 6100m/20,013ft. The vast area surveyed was the logical eastern extension of the Karakoram and Ladakh surveys he had undertaken in 1860 and 1861. His political purpose was not made explicit, but it was quite clear: to examine where Tibetan resistance to foreign invasion might start. As he remarked, it '*taxed my powers of endurance to the utmost*.'[6] Despite these demands, very shortly after returning to Sialkot (near Jammu) in early October, he received orders to proceed on his next expedition, this time at the other end of the Himalayas in Bhutan.

Godwin-Austen began his expedition at Leh, having arrived there by the usual route over the Zoji La: he mentions bypassing Spituk Gompa, which lies close to Leh, just off the main road between it and the Zoji La. Moving south-east up the Indus, Haversham chose the Chang La, 17,470ft/5325m, as his route to Lake Pangong. He preferred it to the slightly higher Kay La, just to the south of the Chang La, by which he returned. He concedes that when they crossed the Chang La on 15 July 1863, almost everyone suffered from altitude-related illness, not helped by having to wade through deep, late snow. Haversham notes that on the return over the Kay La (now more usually written 'Che'), only one man suffered from altitude sickness, despite this latter pass being 780ft/238m higher.

From this pass, he made his way relatively easily down to the hamlet of Durge and then along '*almost level ground*' to Tankse. He was much taken by the valley in which he now found himself appearing at one time to have been a lake. Soon, however, he had reached the '*low pass*' called the

---

dictionary and grammar book. He studied for 16 months in Zangla, the second village of Zanskar, and on his second Zanskar expedition lived and studied in Phuktal, although his teacher-lama appears to have disillusioned him by his unhelpfulness.

5 Most of the detail for this expedition comes from Godwin-Austen's personal abstract of his services, then from *Notes on the District of Lake Pangong in Tibet, JRGS*, Vol 37 (1867), pp343-63, and from *Explorations in the Glacier Tributaries of the Shayok* (sic) *River, Kashmir Territory*, in *Nature*, 17 March 1910.

6 *Abstract* of Army service, 1863, unpublished.

Surtokh La, from where he obtained his first view of one of the highest and longest lakes in this part of Central Asia, its elevation having been '*determined trigonometrically*' as 13,931ft/4246m. As with almost every traveller after him, Haversham was struck by the range and intensity of colour in his view:

> ... *(the water's)* colour is of an intense blue, as clear as crystal ... knobs and peaks were seen for many a mile along the spurs ... These all appeared comparatively low, only on the highest lay a few small patches of snow, thence to their bases was one succession of shades of yellows, purples, reds, and browns, the invariable colouring of Tibet, not a scrap of green.[7]

Having made some local enquiries, Haversham discovered that, after a few days' walk, the lake narrowed to a neck where it could be crossed. On 22 July, therefore, he set out along the southern shore. His task was made easier by finding plentiful supplies of wood and fresh water, the lake itself, like most of those on the Tibetan plateau, being too salty to drink. He observed a ruined fort, evidently an ancient border post, at Tuggu Nuggu, although his men could tell him nothing about it. Next day, he crossed the boundary between the '*Kashmir Rajah's territory*' and the '*Chinese district of Rudokh*' (although the present, disputed, boundary lies a little further east). At this point we come across Haversham's first reference to upland shells: he identifies sub-species of the *Lymnaea* and *Planorbis* genera, as well as one small bivalve of the *Pisidium* family. Apart from being near-conclusive evidence that the lake was once much deeper, he mentions only his regret about finding no living species there. The geological peculiarities of the area struck him so forcibly that he devoted a further three sides of his *Notes* to it. At the same time, he also recorded details of the local bird life: Brahmini geese, teal, a red-headed diver and a '*black-plumaged duck*' among the water birds. He recorded only sandpipers and wagtails among the land birds, and a sea eagle which was attracted by the plentiful fish in the less brackish water of the second lake (i.e. the section beyond the narrows). Like a true Victorian sportsman, Haversham had brought his rod with him and caught 25 fish (unidentified, but resembling tench) in an afternoon, the largest weighing about 4lbs (1.8kg). He gave some lines to his men and, not being Buddhist, they fished enthusiastically. One afternoon, a squall blew up; the sight of the waves on the lake reminded Haversham of, '*the beaches in old England*', an uncharacteristically sentimental note hardly to be found elsewhere in the surviving journals.

Haversham was able to make good progress along the north bank of the second section of Lake Pangong because the place was effectively uninhabited; Khampa herdsmen could be seen and, while they are normally associated with eastern Tibet, here on these uplands these nomads were spread out in small numbers. The Khampas grazed their animals in the higher valleys, partly because they produced a better freshwater supply for the grass, and partly because they could stay out of the way more effectively in disputed territory. He perhaps unsurprisingly opined that, because of the fort near Tuggu Nuggu, and despite it being in ruins, the territory belonged to the Rajah of Kashmir. Permanent inhabitants might also be daunted by bad weather – even in summer, apart from a few fine days, it was '*miserably cold*'. I can go along with this assertion: on my last visit to the Tibetan plateau, north of Saga, we established a base camp at 5000m/c.16,500ft for three weeks, in August 2003. While some days were comfortably mild, it snowed at least three times. Chilly winds in the afternoons and the occasional frost did not help. Fortunately, we had trained in Scotland and nothing in my Central Asian travels so far has equalled the scream of a proper Cairngorm blizzard.

By 1 August, however, Haversham's party had been spotted. They pitched camp at Paljung, which was not visited again by Europeans until Sven Hedin in 1901 and Captain Rawling in 1903. Next day, they headed for Pal, well down the northern shore, en route to which they met their first

---

[7] *Notes on the Pangong Lake District of Ladakh*, *JRGS*, Vol 37(1867), p346. See also plate 32.

nomads, whose giant, black, guardian mastiffs made a fearsome impression. They were told that word had been sent ahead to the Tibetan authorities to alert them to the presence of *feringhi* (foreigners) – Tibet was very much the 'forbidden land' during this period. Diplomatically, Haversham got his *bhut moonshi* to write to the local governor to request permission to continue.

After two days of enforced delay in bad weather, Haversham sat his first real diplomatic test when he was met by the *Zimskang* (under-governor) of Rudok and 20 followers who, clearly intending to bar further progress, pitched their tents on the stream bank opposite Haversham's camp. The Zimskang appears to have come across as an amiable if nervous fellow; he brought presents of the traditional white scarf, two bricks of tea and some sheep and goats. He did, however, also bring a letter which said it was beyond the *Zimskang*'s power to allow Haversham to proceed to Rudok (the provincial capital). The *Zimskang*, a native of Lhasa, understood the politics of the occasion very well. Subtly, he added that he knew that Lhasa did not permit journeys across the frontier and that, since he could not resist Haversham's party by force, he would rely on Haversham's sense of decency in not making him (the *Zimskang*) lose his job (and probably his head) by proceeding further. In his turn, Haversham stated that he had received orders not to confront 'Chinese' officials, and knew that he therefore could not go further. He did, however, one suspects as more of a gesture than anything else, '*show anger at their absurd wishes*.'[8] It got him his way: he was allowed one more march eastwards. In the evening, the *Zimskang* returned to repeat his request not to proceed further, but Haversham held him to his original agreement: Haversham was about to arrive at the most easterly point of his Karakoram and Ladakh survey.

As the weather had cleared on 5 August, Haversham set off in the direction of Noh, on the way to the far end of the lake. The *Zimskang* stuck to him '*like a leech*'. He describes the Zimskang's men as, '*A curiously-dressed rabble they were, with their enormous flat mushroom-shaped hats, and all mounted on little shaggy, but sturdy ponies*.'[9]

The atmosphere felt much less tense than on the previous day: Haversham stopped at the hill where he promised he would, much to the *Zimskang*'s relief. On the journey from Pal, Haversham showed the Tibetans his maps and drawings, which they '*thoroughly understood*'. He was thanked profusely for not causing any further trouble. Before leaving on the following day (6 August) Haversham was invited to the *Zimskang*'s tent for some traditional butter tea to help him on his way. After distributing some presents, Haversham's party departed northwards in the direction of the Dingo La and the wild high country of the Aksai Chin. Again, after this encounter, the myth of Haversham's '*tetchiness*' is exposed. His brief anger directed at the *Zimskang* sounds no more than a negotiating ploy; this is the way I interpret the event, from my own experience calculated crossness with those in supposed authority, such as liaison officers, has helped my party out of fixes in Kashmir, Pakistan, Xinjiang and Tibet. By the standards of British imperial politics of the time, he acted with propriety.

Along the first part of their route, now north-westerly, they were accompanied by a group of Khampas. The Godwin-Austen eye spots:

> One of the Champa girls *(who)* was very nice-looking ... I had the young lady brought over to my tent, where she sat for her portrait, and was delighted at the drawing made of her.[10]

On the following day, they passed over the watershed between Pangong and Changchenmo at the Dingo La, 16,270ft/4959m. Haversham '*determines*' to ascend a hill about 1000ft/c.300m higher. On the way up, he passed a form of rhubarb: this is not as odd as it sounds; I have come across giant rhubarb in Kazakhstan at 3000m/c.10,000ft. He also noted the limestone nature of the rocks. On the summit he was disappointed that owing to haze he could fix the heights of only one or two peaks.

---

[8] *Ibid*, p357.
[9&10] *Ibid*, p358.

Haversham's group then parted company from the Khampas, who were to stop for a few days' grazing, also no doubt because they had seen their fellow-travellers safely away from Pangong and Rudok territory. As they advanced, the countryside increasingly resembled an Alpine desert: water became harder to find although fuel could be obtained by burning the local furze-like scrub; bird, animal and plant life become distinctly less evident. On 8 August they reached the Deenjor (Domjor) La, which the boiling-point thermometer indicated lay at 17,464ft/5323m above sea level.

At the summit, two Argali sheep appeared and Haversham took a pot shot but missed both; one wonders why as Haversham was a good shot and Argali sheep are not small – perhaps his eyesight was affected by the altitude and the cold. Another hill, however, rose above the pass; as is often the case in very clear atmospheres, the distance to the summit was further and the gradient steeper than it looked. Haversham was very fit, however, and soon reached the summit, at 20,240ft/6169m, his highest ascent on this expedition. Rather revealingly, when contemplating the view, he says he, '*longed to go in the direction*' of the great peaks above the Shyok; this is the mountaineer, not the surveyor, speaking.

At this spot, Haversham caught sight of a number of *kyang*, the Tibetan wild ass; several Tibetan antelope were also spotted; these sightings helped to distract the party from the bitter cold. He was now aiming for another high point, Kepsang: he took a roundabout route which bypassed his caravan as they crossed the gravelly plateau. He made the summit attempt the following day. A freezing early start: the yaks' backs were covered in frost. He came across a hare, benumbed by the cold. It could not move and was killed by one of the 'coolies' with a stick. After traversing much irritating debris, he reached the summit, the view being '*a recompense to the labour of legs and lungs*'. It was about 3,199ft/975m from the campsite to the summit; Haversham described the climb with schoolboy enthusiasm as, '*a delightful little pull*'.

The summit lay, once more, above 6000m: he had reached 20,033ft (6106m). His men told him that he was looking over country occupied by the Kyrghiz, a Turkic ethnic group found mainly in Kyrgyzstan and Xinjiang, China. If they were Kyrghiz, they had wandered rather a long way from home which traditionally is a good deal further west, in the Kashgar area and westwards. This was evidently one of the coldest summits Haversham had ever reached: he complained that he could hardly hold his pencil. Nonetheless, with his usual considerateness and practicality, he got a fire going, no mean achievement in itself, and brewed a pot of tea for his men. By the end of this long day they had reached the Kiung Gang (Chungan) La, a cold place for a camp, at 17,259ft/5261m. (On my 2003 expedition to the Tibetan Plateau, in the Longpo Kangri range north-west of Saga, we made three first ascents of peaks all around 6200m and a number of lesser hills above 5000m. Various advance bases and assault camps were placed at about 5400m and 5800m. During that period, it snowed on three days, once, heavily, and was stormy on several others. We did not, however, experience particularly low temperatures or severe windchill.)

Because it marked the political border between Kashmir and Tibet (Rudok), a small guard was stationed on this pass. The guards were not expecting Haversham's party and were caught unawares, becoming the victims of some chafing by Haversham's porters. The only other life on the pass consisted of hares, a few of which Haversham managed to shoot for the pot, and seven wild yaks, massive but nervous animals, rather more intimidating than their domestic counterpart. One other human appeared, however: a Mr Turner who was trying to retrace Adolf Schlagentweit's routes. He admitted to being in the wrong place because his porters had led him in the wrong direction, thanks to his inability to speak Hindi. Mr Turner accompanied the party for two days, then, mysteriously, pottered off again on his own.

Haversham meanwhile declared, '*I longed to explore but could not do so*', as time and the weather were beginning to work against him, a sign of his interests developing from the purely objective work of surveying towards a more subjective interest in investigating the unknown, where his instincts, as opposed to his orders, led him. On leaving Mr Turner, Haversham made a '*short ascent*' above the valley of the Changchenmo, but did not specify the height; this could, however, mark the third spot where he claimed he went above 6000m/c.20,000ft. This altitude would only lie 3,000ft/914m above his campsite, a perfectly reasonable '*short ascent*' in his state of fitness.

At this campsite, Haversham also touched on one of the curious sub-plots of his travels. He mentioned that in this area many fine agates and cornelians could be unearthed. He collected about 250 semi-precious stones, the bulk of which were cornelians. Various carved medallions or stamp-seals, Buddhist and Moslem, were added to the collection. Sadly, the bulk of them had to be sold as his bankruptcy approached: the sale catalogue's date is 13 May 1897 and they appear to have been bought by the British Museum. Some of the remainder were donated to the British Museum in 1921; the remainder were left to the museum by Haversham's niece Cicely in 1943. An enquiry by Dr Richard Godwin-Austen in 1968 elicited a grudging offer to view the collection. He was told that they had only been taken out two or three times, and had never been put on public display.[11] In 2010, however, a more enlightened attitude obtained: a request to the British Museum to view the 111 surviving stamp-seals was met with every helpfulness.

Helen Parkin undertook to bring the collection, now spread over the Roman and Greek, Middle-Eastern and Eastern departments, together, for the benefit of a viewing by Robin and Richard Godwin-Austen (Haversham's grand-nephews) and myself. Having also been joined by Richard's son Jonathan, we enjoyed a fascinating couple of hours in a research area of the museum where we could examine the collection at leisure and were expertly guided through it by Dr St John Simpson, the whole event being an example of constructive public service which was deeply appreciated.

From this point onwards, Haversham followed a downward course off the gravelly and snowy high ground, past the hot springs at Kyam, to Boomzi (probably Pamzal). One last rise to almost 18,000ft/5486m followed by another long descent took the party back to familiar ground at Mugli, then Tangse. The surveying work was finished, but Haversham's appetite remained unfulfilled. He ascended to 19,672ft/5996m, to a top on a spur of Shyok 2, 21059ft/6419m, from where even he was awestruck by the view of the soaring Shyok and Nubra ranges, rising to at least 22,000ft/c.6700m before him, while thousands of feet below he could pick out the great bend in the Shyok. From there, the party made its way back to Leh over the granite country of the Key La, 18,256ft/5564m. As is his habit, his account of the expedition then breaks off early, on this occasion at the top of the Key La.

He could not have spent long in Leh, or indeed in returning to Srinagar over the Zoji La, and could hardly have timed his return to Pauline better. On 1 September 1863, Haversham received 'the thanks of the Viceroy and the Govt. of India' for the 1857-63 service, and on 14 September 1863, his new son and heir, Arthur, was born. Arthur's birth would have consoled the parents as another son, Alfred, who had been born in Jullundur (now Jalandhar, also the birthplace of Lawrence Durrell and the late President of Pakistan, Mohammed Zia ul-Haq) in the Punjab, in April of the previous year, had died in Lahore only two months later.

At almost the same moment, army orders rather callously required him to go immediately to the other end of the Himalaya, to join Sir Ashley Eden's ill-fated 'diplomatic' mission to Bhutan. Haversham would never see the Karakoram again. And he could scarcely have imagined what utterly different territory now awaited his attention.

[11] Dr Richard Godwin-Austen: private letter, 5 March 2010.

CHAPTER THIRTEEN

POLITICAL SERVICE

Bhutan 1864-5

GODWIN-AUSTEN HAD SPENT BARELY two months back at base in Sialkot after his exhausting expedition through Tibet and the upper Changchenmo when an order arrived on 2 November 1863 from the Governor-General to report to General Walker at Dehra Dun: Godwin-Austen was to prepare himself to join the political mission of Sir Ashley Eden to the Durm and Deb Rajas, the spiritual and temporal rulers of Bhutan since the 16th century. (The Rajas were replaced by a British-supported hereditary monarchy in 1906. This dynasty is still in place and enjoys extraordinary popularity.) Haversham was to act as surveyor and Assistant Political Agent. Having made himself ready at Sialkot, on 18 November he proceeded to Darjeeling, which he reached on 30 November. During December, special preparations were made, as Western Bhutan was largely unknown country: a British trade mission had reached Bhutan in 1774; Turner in 1783 and Pemberton in 1837 had made the only other significant visits. Haversham's route followed Pemberton's; this entailed entering Bhutan by the south-west, a particularly mountainous area of high passes, dense forests and deep gorges, and without an adequate map.

It may symbolise the general misguidedness of the Eden Mission, which consisted of Sir Ashley, Cheeboo Lama the Prime Minister of Sikkim, Dr B Simpson, Haversham, Captain WH Lance and an escort of 15 Sikhs and 15 Gurkhas commanded by the appropriately named Captain Lance, that it left Dalingkot, near Darjeeling, on the not very sensible date of 1 January 1864. (Dalingkot(e) was known by this name while under Sikkimese rule; it is now more usually called Kalimpong. It is still a large Indian army base.) It is difficult to understand why the most mountainous country in the world should have been tackled in the dead of winter, although it might have been an attempt to allow several months to complete the mission before the monsoon set in. (It is said that the longest stretch of straight road in Bhutan, which is about the size of Switzerland, lasts for about nine seconds' worth of driving. Having traversed the country from west to east, in 2008, I am sure this is true: Bhutan, an otherwise wonderful country, should be avoided by the potentially car-sick.) The start had been delayed because of 'political differences' – Eden, who had already successfully led a similar mission to Sikkim, thought that the Bhutan expedition would largely mirror it but, despite being neighbours, Bhutan's climate and terrain are quite different from that of Sikkim. This first of several misreadings by Eden got his mission off to a very bad start.

It aimed to reach the low-lying Punakha, the winter seat of the Bhutanese government at that time, '*about eighteen marches or about two hundred miles away*'. Disaster soon resulted: while the rest of the British party were enjoying horse races at Paro, Eden was being humiliated by the infuriated Bhutanese (an emotion one does not normally associate with them). One report describes Eden's incompetence being rewarded by the local insult of being slapped in the face with a piece of wet barley dough, 'to the accompaniment of much laughter'.[1] Eden's clumsy attempts to bring the Bhutanese under British 'protection' (Bhutan's northern border marches with China) failed because they were not backed by military force and because the Bhutanese possessed their own defence capability. The Bhutanese quickly repudiated the agreement signed with the British and this repudiation, along with Eden's vengeful response to his loss of face, as well as a history of aggressive incursions by Duar Bhutanese into British Assam (which had been formally annexed in 1838) combined to produce the *casus belli* for the following year's Anglo-Bhutanese spat, the effects of which are still seen today in the human rights problems which Bhutan keeps hidden in the Duars. ('Duar' means 'door' or 'gateway', in other words a low-altitude pass. Somewhere between 18 and 24 duars can be found along the southern frontier of Bhutan.)

Remarkably enough, a map of the first section of the mission's route, on something not much thicker than tracing paper, has survived among the Godwin-Austen papers at the Surrey History Centre. Helpful as ever, Godwin-Austen has even marked each campsite on it. On 16 January, following orders and taking a sensible escort of 15 Sikhs, he headed for Julpygori, now Jalpaiguri, a city of North Bengal which still retains many colonial characteristics, to buy pre-arranged provisions, a sign of the trust placed in his self-reliance and sense of diplomacy. Julpygori was developed by the British in the 19th century as a centre for the tea industry. Haversham reached it in two days, but the supplies were refused by the exotically titled Subah of Dalingkot. (A *subah* is a province and provincial governor in Sikkim.)

In that cold season of 1864, the first bar to the mission's progress arose with a storm on the Tegong La (12,200ft/3719m). To reach it, several other passes at an only slightly lower altitude such as the Pango La and the Tule La, needed to be crossed. From Sipchu, at the border, the party had passed over the Tsangbe La then carried the survey up to the Tegong La; Godwin-Austen and 'some coolies' remained behind the main mission party to carry out survey work. On the Tegong La summit, the party were forced to make an emergency bivouac under a rock for three days and nights in a blizzard. On 15 February they then decided to make a break for it, in what must have been a terrifying ordeal, approaching exhaustion for the survivors. Two 'coolies' did die and nearly all the survey party's baggage was destroyed. The survivors persevered, however, and crossed the Chi La, 12,490ft/3807m, now known as the Cheli La: the present metalled road rises to 3780m/12,402ft and is a contender with the Thrumsing La in Central Bhutan for the highest road pass in the country.

The whole mission then reached Paro, the neat, fertile valley containing one of the few flat pieces of land in Bhutan (and now the site of this hilly country's only major airport). On the way, Haversham remarked, as modern travellers do, on the magnificence and variety of the rhododendrons and the unusual height of the tree-line, which rises in places to well above 4000m/c.13,000ft. On this section of his journey Haversham almost lost his life: a fellow-officer accidentally knocked over a barrel of gunpowder, which blew up. The officer was killed, but Haversham, who was sitting only a few feet away, escaped because he was protected through having seated himself behind a grassy bank below the level of the detonation. Once again, he recounts this story of a near miss with death nonchalantly and with extraordinary understatement.

The mission rested at Paro (see map on page 239) for 17 days, presumably while negotiations continued with the Bhutanese. (The party had, however, now lost its taxidermist, who had been sent back early on, at Sipchu, in order to make supplies spread further.) After moving on, the party crossed the Bie La, 11,166ft/3403m, although they were not quite able to see Tashichodzong, at that time an impressive but isolated fortress, despite its position having been fixed by some houses which could be made out nearby. Now, it has become Thimphu, (2800m/c.9,000ft), the agreeable modern capital of Bhutan. One final ascent was needed to traverse the Dokim La, now the Dochu

---

[1] See Michael Ward, *Geographical Journal*, Vol 132, No 4, pp491-506 and his unpublished manuscript: *Bhutan: A Short History of Exploration*, lodged in the Royal Geographical Society archives. The 'dough' would have been *tsampa*, barley flour, mixed with water or butter tea, the staple fare in Bhutan and Tibet (at the time).

La, 10,000ft/c.3000m. It lies a one-hour, characteristically winding drive away from Thimphu. It is covered in 108 *chörtens* (burial monuments); acclimatisation walks much favoured by trekking groups start here. The descent from the Dochu La follows a never-ending series of hairpins, to bring one down over the course of about two hours' drive from a sub-Alpine environment to a tropical one (a malarial zone with pineapple and banana palms and something approaching jungle), to the winter capital of Bhutan, Punakha (4,500ft/c.1400m) which the mission reached on 15 March 1863.

The Punakha *dzong* is, to my mind, one of the most attractive buildings in this part of Asia. A *dzong* is a Bhutanese castle which contains a lamasery, royal courts of justice and various town and county hall functionaries. The capital of each province contains its own *dzong*, often the finest building in the province. The ruler, who lives in or near the *dzong*, is called the Penlop. The pro-British Penlop of Trongsa, Sir Ugyen Wangchuk, became the first of the modern hereditary kings of Bhutan in 1906. Several *dzongs* date from the 17th century. The Punakha Dzong sits at the confluence, an auspicious place, of the Pa-chu and Ma-chu rivers (representing the masculine and feminine forces in nature) like a vast elegant houseboat. Haversham was evidently very struck by it: he painted one of his finest watercolours of it, and also produced a very precise military surveyor's plan of this strategic fortress (see plate 34). His military notebooks contain very finely drawn sketches of the *dzong* and its surroundings, from the perspective both of its native architecture and as a military site; the Bhutanese tell various colourful stories of the way they defeated invading Tibetan forces who had penetrated Bhutan from the Punakha Valley to the north.

After negotiations broke down, the mission returned by its outward route and arrived back in Darjeeling by April. Godwin-Austen's follow-up suggestion was that success might have been more readily obtained if at least one of the party had been able to speak Tibetan (of which Dzongkha, the Bhutanese language, is a variant). I agree with him: the value of learning your neighbours' languages is enormous. I had a letter published in *The Times* on August 31, 2010, to that effect, and it brought out a strongly positive response.

An interesting consequence of Haversham's Bhutan fieldwork was to draw attention to the accessibility to an army of the Teesta and Chumbi Valleys, which lead to Tibet. This information probably defined the access route for the Younghusband 1904 Tibet invasion (which proceeded via the Chumbi Valley); in its turn, this route was followed by the first Mount Everest reconnaissance and mountaineering expeditions in 1921, 1922 and 1924, a matter on which Haversham wrote at some length in his later years (see Chapter 24).

SECOND BHUTAN EXPEDITION
THE DUARS

1 October 1864 - April 1865

It is a curious piece of exploration which begins in warfare. Haversham had recessed at Mussoorie over the summer of 1864 in order to complete his map of Western Bhutan (see plate 35). Mussoorie is situated about 34 kilometres from Dehra Dun in the state of Uttarakhand. This hill station contends with Shimla for the title of *Queen of the Hills*. Various versions of this map exist, but the original remains unpublished and uncatalogued by any public institution, and constitutes as beautiful an example of the mapmaker's art as one could hope to see.

On 1 October, Haversham was detailed to leave Dehra Dun to join the Right Column of the British Army stationed at Guwahati (Haversham spells it Gowhatty), which lies on the Brahmaputra with a population of just under a million today. It remains in the memory of my own Bhutan Expedition in 2008 as the first sizeable place we reached in India following our departure via Samdrup Jongkhar in south-east Bhutan. Within hours of arrival, most of our party were affected by sore throats from the pollution. The city has a grim fascination: the Brahmaputra here flows widely, but filthily. Perhaps the most memorable public building, the Temple of the Nine Planets, has a floor which appears to be carpeted in a mixture of congealed soot and animal blood.

To visit this dark place, you need to walk around in bare feet: when you realise what the floor is covered in, you quickly acknowledge that this is not a place for the fainthearted.

From Guwahati, Haversham had travelled to Gwalpara, in modern Bihar state, to complete some observations, when he was suddenly ordered to join Brigadier-General Dunsford's Left Column, a fine piece of Army 'left hand-right hand' muddle. By '*riding day and night*' through Koch Bihar, Haversham arrived in time to assist at the siege of Dalingkot (Kalimpong). He acted as ADC to the general, just as he had done for his grandfather and then General Reed 20 years before, and also found time to map the fortress and surrounding area for General Dunsford's report. He then assisted during the operations which led to the capture of Chamurchi, a tea-planting centre and important junction, a campaign necessitated by a recent successful surprise Bhutanese attack.

From there, he single-handedly took up the survey of the Duars and '*the line of the outer hills from Dalingkote to the Longitude of Buxa*'.[2] He surveyed '*by chain*', an extremely laborious and finicky process, up to the Gorun and Paro[3] rivers '*as far east as Alipur near Chichacotta*'. In time, although it is not clear exactly when, he was joined by a small party of fellow surveyors and they surveyed until April 1865, when work was abruptly halted by the '*sudden and complete prostration of the whole party and Sepoy guard by jungle-fever*'.[4] This illness, possibly malaria, was Haversham's first experience of a disease which laid him seriously low on several future occasions and which he said '*broke his health*' (although he recovered quite quickly after his return to England in 1877). The effect on the party was severe: the majority were carted as far as Caraghola Ghat, from where they could be more comfortably transported to Calcutta – several died. Others were transported to Allahabad, to the large British garrison from where they were invalided out to their homes. It is noticeable, however, that from the time of his first appointment to Bhutan, Haversham had begun again to work with groups of other Englishmen, the first time since his early days in the Kashmir Survey up to about 1858.

TRANSITIONS

On his return from Bhutan, Haversham was awarded the Indian Frontier Medal, with clasp, and was Mentioned in Despatches. The despatch citations[5] have survived:

> Extract from the despatches of Brig Genl. HF Dunsford[6] CB*(?)* Co Left Column Bhutan Field Force[7]
>
> No 88 At Camp Dalinkote 7th Decr 1864
>
> 'Captain Godwin-Austen who arrived in camp only the evening before, at once volunteered to accompany me, and his previous knowledge of the place proved invaluable.'
>
> No 141 At Camp Chamurchi[8] 12th *(?)* Jany 1865
>
> 'I would beg to be allowed to bring forward to the notice of the HE the very gt. assistance I have at all times received from Captain Godwin-Austen who accompanied the force for the purpose of making a survey of the country, his long experience of hill territory makes him an unerring judge of the line of road and drainage.'

---

[2] *Abstract of service*: unpublished. Buxa is near Jalpaiguri.
[3] Not to be confused with Paro in Bhutan proper.
[4] *Ibid*, p7.
[5] Unpublished document.
[6] Not known, other than as commander of the Left Column in this campaign; he was replaced after some successful Bhutanese counter-attacks.
[7] The British invasion force in the Doar War.
[8] Dalingkot and Chamurchi: captured by the British in October 1864.

Extract from Letter Book[9]

Brigade Office, Shillong[10]

These copies of the original citations are preserved in Godwin-Austen's private papers. To them, he has added in his own hand:

True copy Sd HHGA

PS General Dunsford subsequently recommended me for Brevet Majority for my service under him but I am not in possession of his letter and the reply of the C-in-C.

GA

(In the United Kingdom the brevet commission was awarded only by courtesy. The term stood for a promotion given on such occasions as a coronation or the end of a great war, and originated during the reign of James II (1685-88), but it was abused so frequently that in 1854, during and after the Crimean War, it was strictly limited to cases of very distinguished service in the field and on the principle of seniority. In the British Army, brevet commissions were confined to grades from captain to lieutenant-colonel. The Brevet conferred rank in the army, but not in the regiment.)

Haversham remained ill, however, for six months, which he spent on sick leave during the winter of 1865-6. Malaria, assuming that is what is meant by 'jungle-fever', was much harder to shake off than the smallpox he had caught as a young man. Part of this leave was profitably occupied with geological examinations and fossil-collecting in the Siwalik Hills, near Dehra Dun.[11] An intense work-rate, maintained until a few months before he died, is one of Haversham's most noticeable characteristics.

He also occupied himself at Dehra during his recovery in producing the first map of Central Asia, compiled by his old friend Colonel Walker, Superintendent since March 1861 of the Great Trigonometrical Survey in succession to Sir Andrew Waugh. He also organised the mapmakers' photozincographic office at Dehra. (Photozincography, known commercially as 'zinco', is the photographic process developed by Sir Henry James FRS (1803-77) in the mid-19th century. This method enabled the accurate reproduction of images, manuscript text and outline engravings, which proved invaluable when originally used to create maps for the Ordnance Survey of Great Britain during the 1850s.) The geological and fossil specimens which Haversham gathered were presented to the Geological Survey Museum in Calcutta. These activities, and the new kind of terrain covered in the second part of his Bhutan expedition, all indicate that a transition was occurring in Haversham's life: the mountain-explorer was being left behind, the jungle-explorer and Natural History scholar (in many different forms) was taking his place.

As ever, the Army provoked his next move. Haversham had been promoted to major,[12] to take charge of Topographical Party No 6, which was to survey the unexplored Khasia (Khasi), Garo, Cachar and Jaintia Hills on the borders of northern Burma and Assam, at the same time as he had been appointed to the Bhutan Mission. The subsequent war and his illness prevented that appointment being taken up until November 1866. Off and on, because of illness, Haversham was

---

[9] Record of official correspondence.

[10] Shillong is the capital of Meghalaya, one of the smallest states in India. It is also the district headquarters of the East Khasi Hills District and is situated at an average altitude of 4,908ft/1496m above sea level.

[11] The Sivalik or Siwalik Hills or Outer Himalaya are the southernmost and geologically youngest east-west mountain chain of the Himalayan System. They crest at 600m/1,968ft to 1200m/3,937ft and have many sub-ranges. They extend 1600 kilometres (994 miles) from the Teesta River in Sikkim, westward through Nepal and Uttarakhand, into Kashmir and Northern Pakistan.

[12] Gazetted in the *Illustrated London News*, 18 November 1868.

about to embark, as the area's first serious Western explorer, on six seasons in the humid, malarial jungles of northern Burma and Assam, populated by hostile, resistant tribes who attacked him and, on another occasion, killed one of his best assistants. When combined with what we now know of his work in Kashmir, his four Karakoram, Ladakh and Tibet expeditions and his two expeditions into Bhutan, we can readily ask ourselves if there was ever a more resolute and versatile British explorer. Kenneth Mason can be given the last word:

> The ideas of Godwin-Austen dominated the Survey of India for the next forty years.[13]

[13] *Geographical Journal,* 74, p276, RGS publications, London, 1929.

CHAPTER FOURTEEN

PAULINE

Family Background

WHILE THE EXPEDITIONS TO K2 and thereafter were taking place, and during the first part of his service in Burma and Assam (see the following two chapters), Haversham was going through his second marriage, this time to Pauline Georgiana Chichele Plowden.[1] It was not a happy marriage, perhaps not helped by Haversham living apart from his young wife in extremely remote country for months at a time as well as suffering from bouts of severe illness. Haversham married Pauline, seemingly rather quickly, in April 1861, only seven months after returning from his first expedition into the Karakoram, on which he had been accompanied by Kudikji.

Our knowledge of Pauline is uneven and, in some aspects, very uncertain. Her plaque in Shalford church gives us bare but rather inaccurate details of her life:

PAULINE GEORGINA CHICHELE PLOWDEN 1845-1869 First Wife of HH

In memory of PAULINE GEORGINA daughter of LT COL ARTHUR WELLESLEY
CHICELE PLOWDEN 1st Light Cavelry 1st wife of LT COL HENRY
HAVERSHAM GODWIN-AUSTEN born 1845 married 1861 died in Calcutta
1869 aged 24 years and 6 months her son MAJOR ROBERT ARTHUR
GODWIN-AUSTEN places this brass

We know Pauline was in fact born at Firozepur on 9 October 1844 to a military family with Hampshire roots. (Firozepur was founded by Sultan Firuz Shah Tughluq (1351-88); it lies on the banks of the Sutlej and holds an historic place in Indian history as it was here that the Raj established control over much of North-West India, as well as what is now Pakistan, through the Anglo-Sikh wars. Its population in 2001 was about 95,000.)

THE CHICHELE PLOWDENS

Pauline's father's tombstone, and the family history (see below) provide rather more reliable information about Pauline's background and ancestry. Her father, Lieutenant-Colonel Arthur

[1] There are a number of variant spellings, e.g. Chichele-Plowden, Chicheley-Plowden etc. The mis-spellings and wrong dates in the Shalford Church plaque are transcribed as they occur there.

Wellington Chichele Plowden, of the First European Light Cavalry, was born in Calcutta on 1 November 1814 and died on 3 January 1861 at Dehra Dun only four months before Pauline married Haversham. Was Pauline's acceptance of Haversham a kind of expression of freedom from her father's influence? (If we go by the Chichele Plowden family history, the Shalford church plaque (as it is in a number of details) appears to be wrong in calling him Wellesley, no doubt through muddling the Iron Duke's surname and title.)

In the family history, Arthur was the son of Richard Chicheley *(sic)* Plowden (1782-1825), a salt agent at Hijili, written Hidgellee, Bengal, where he is described as a 'Collector', and Sophia Fleming, from Calcutta. He began his career as a cadet, in 1832, became a cornet in the Third Bengal Light Cavalry on 19 June 1833, was promoted to lieutenant in December 1841, then captain in October 1852, brevet major in October 1852, and brevet lieutenant-colonel in October 1860. We are told that he 'arrived in India in December 1833', so presumably he had been educated in England. (There is a family tradition of the Plowden boys being educated at Westminster or Cheltenham, although which was Arthur's school is not known.) On 7 May 1840, he married Caroline Charlotte, the fifth daughter of Charles Mackenzie, BCS (Bengal Civil Service); she died on 19 November 1848, when Pauline was only four, a significant event for his daughter's development in freeing her – as her father was absent on duty much of the time – from learning the high Victorian arts of self-discipline, self-restraint, and even sound housekeeping. Arthur fought at the Battle of Ghazni (medal): during the First Anglo-Afghan War, the city had been stormed and taken over by the British on July 23, 1839, and then at the Sutlej in 1845 (medal and two bars). The Sutlej Medal was approved in 1846, for issue to 'officers and men of the British Army and Honourable East India Company who served in the Sutlej campaign of 1845-46' (also known as the First Anglo-Sikh War).

Arthur was also a grandchild of Richard Chichele Plowden, who was a Director of the East India Company. According to the same family history,[2] Arthur's branch came from the Ewhurst and Liphook areas of Hampshire. At one time, they owned Ewhurst Manor, although this was later (but not by them) sold to the Duke of Wellington. Arthur's branch was descended from the 'great Elizabethan lawyer', Sir Edmund Plowden (1517-84), whose grandson, Sir Edmund of Wanstead (1590-1659), not only married a Chichele and thereby founded the main line of the family, but was famous for his unusual 'New Albion Charter', given to him by Charles I. New Albion is situated in modern-day New Jersey, Pennsylvania and Maryland. Colonisation was unsuccessfully attempted by Sir Edmund Plowden under the authority of a charter granted by Charles I in 1634. The first successful settlement was made by Plowden in 1642; it ended, however, in an attempted mutiny. After that, Plowden managed New Albion from the Colony and Dominion of Virginia, selling rights to adventurers and speculators, until he returned to England in 1649. Both Sir Edmunds were descended from a Roger the Crusader, who was known to have been at the siege of Acre in 1191.

Some members of this family had a difficult time in the Mutiny. Pauline's father, for example, was present at the moment the Mutiny began:

> His regiment was the one which prematurely opened the ball in Meerut, in 1857, on that memorable Sunday, the 10th of May. It is probable that, had the Mutiny been delayed till the day fixed for a general rising (the 1st of June), matters would have gone very hardly with the few British troops then maintained in India. As is well known, the ostensible reason given for the Mutiny was the greased cartridges served out to the native troops, and as this grease was said (though falsely) to be composed of pigs' and cows' fat, which would have defiled Hindoos and Mussulmans alike, it was naturally strongly objected to. Only some eighty-four

---

[2] *Records of the Chichele Plowdens, AD 1590-1913: with four alphabetical indices, four pedigree sheets, and a portrait of Edmund the great Elizabethan lawyer.* Walter FC Chicheley Plowden (late Indian Army); printed for private circulation, Heath, Cranton & Ouseley, Ltd, London, 1914.

troopers of a cavalry regiment were then armed with carbines, the rest having swords and pistols. Of these eighty-four as many as were present in Meerut of the 3rd Light Cavalry, about sixty or seventy, refused to use the cartridges and were tried by court martial, and were all condemned to some years' imprisonment. They were first confined in the old jail, near the present Sadr (Principal) Bazar, and on the following day, Saturday, the Bazar people refused to hold any communication with the troopers of the regiment, who, they said, virtually consented to their comrades being imprisoned for conscience sake. This was the immediate cause of the rising on the following morning, when the British troops were at church, unarmed. Want of combination among the native regiments, cavalry and infantry, as also the absence of any definite and projected plan, was, under Providence, the salvation of the garrison of British troops and their families. The writer heard many anecdotes from eye-witnesses, when he was in Meerut forty years after, as cantonment magistrate. His principal informant, Khan Bahadur Karim Baksh, CIE,[3] was positive that the greased cartridges caused the outbreak, though he was quite confounded, and unable to supply any explanation, when asked: 'Why, then, did the mutineers sack the magazines and take away with them for use at Delhi these very same cartridges?'

Anyway Major Arthur escaped slaughter, and survived till 1861, leaving at his death a son, Cornwallis Alfred, and a daughter. Cornwallis joined the Punjab Police, in 1867, and married, in 1872, his cousin, Mary Plowden. He died at Ferozepur, in 1894, leaving two children ...

Arthur's cousin, George, became the Commissioner and Agent (political representative) of the Governor-General in Nagpur, in the state of Maharashtra, the largest in central India and also the third largest city by population in that state. He had an even luckier escape.

During the outbreak the residency was one evening burnt to the ground by his own guard of native troops, and he had considerable difficulty in escaping with his wife and four younger children to Kamptee, thirteen miles off, where there were British troops. The writer was one of them, and among his earlier recollections, this is probably the earliest, the long drive in the dusk, on the coach-box of a four-in-hand carriage, with one or two horsemen as escort, and the tall building burning behind.

George played an important part in putting down the insurrection in the Central Provinces, and in assisting Sir Hugh Rose (afterwards Lord Strathnairn and Commander-in-Chief in India) in his Central Indian Campaign.[4]

(A more distant cousin, Walter, was murdered trying to assist King Theodore of Abyssinia, (c.1818-68), Emperor of Ethiopia. Walter's adventure has some similarities to that of General Reade Godwin-Austen in the 20th century: see Chapter 2.) The only other person of note was her uncle Augustus, (1805-52), whose obituary memorably declares:

We regret to notice in The Mofussilite an account of the sudden death from apoplexy of Mr. A.U.C. Plowden, Magistrate and Collector of Bolundshahr. The deceased gentleman was in perfect health on the morning of the 30th April, and had been walking in his compound, when he called for a cup of tea, immediately

---

[3] Commander of the Indian Empire
[4] *Ibid*, p180.

> after drinking it he felt giddy, and was placed in a bed from which he never rose. He died at 8 o'clock in the evening of the same day, and was it is said attended to the grave by hundreds of the inhabitants. Aged 47. Augustus was six feet seven inches in height, and weighed twenty-seven stone, which, if correct, probably accounts for his apoplectic stroke. He was a very kind-hearted man.

'Mofussilite', this arch term for the writer's source, is an Anglo-Indian expression introduced about 1845 for a country person, presumably here used disingenuously.[5] The writer also added that Augustus's nickname was 'the Babe', because he had once gone to a fancy dress ball dressed in swaddling clothes, seated in a child's perambulator which required several 'coolies' to push it.

MARRIAGE TO PAULINE

We do not know how Pauline and Haversham met, though it is tempting to think that the Colonel MacKenzie buried next to General Godwin was somehow related to Pauline's mother Caroline MacKenzie, especially as Pauline's father and Colonel MacKenzie were in the same regiment. General Godwin's avuncular interest in Haversham's wellbeing, which has been noted since his Sandhurst days, suggests that he might have had some hand in this matter.

Whatever the connection, the marriage at Nowsherah, in the Punjab, came as a surprise to the Godwin-Austens, who seem not to have known the girl or her family. The following rather defensive letter confirms that Haversham told his family while he was either stationed or on leave in Rawalpindi; we do not know exactly when he stayed there, but it is most likely to have been in autumn of 1860, as the Regiment is likely to have moved down to the lower ground (where Rawalpindi is situated) once the heat of summer was over, and because we know Haversham was in Dehra Dun over the winter. If the time of the end of his relationship with Kudikji is correct, (see Chapter 9), Haversham must therefore have proposed, been accepted and informed his family of his engagement within a very short period indeed. He wrote to them from Nowsherah, now usually Nowshera, in the North-West Frontier Province of what is now Pakistan.[6] The letter was written not only a week before his marriage, but only a few weeks before leaving on his 'K2' expedition:

> Naosheirah near Peshawar
> 'Good Friday'. March 29th 1861
>
> I am now able to answer yours of the 10th my dearest mother. And I was very glad to get it. You will have received mine from Pindee[7] when you get this, with the news of my engagement, and I shall be married on the 5th. I *(illegible)* and then start for Kashmir, how the family will like the news I don't know. But I see very plainly that I have far better prospects out here in the Staff Corps and shall be far more independent than coming home with the 24th. I could not live in England without an allowance from my father and I see very plainly that that is what he could ill afford without taking from my other Brothers and Sisters on whom my allowance will be much better spent. I shall if I can try and get enough together to get one of the sisters out to me who would like it, and be also a companion to my wife, during my work in the field. You ask in your letter who Miss Plowden is – I

[5] Source: *Routledge Dictionary of Historical Slang*.

[6] The town lies on the Grand Trunk Road, 27 miles/43km due east of Peshawar. The population in 1901 was about 9,000.

[7] Rawalpindi (formerly Rawal Pindi, formerly Rawal Pindee), one of the most important cities in Pakistan, now almost joined to the capital, Islamabad.

> told you in my last and Princess[8] will know more about them than I do – all I know is that they are all descended from a Richd Chilchley[9] Plowden who was an EIC[10] Director. Geo. Plowden with whom I am now to stay is a son of Geo. Plowden who was resident[11] at Nagpur.
>
> The marriage will be rather hurried as we have just come off the march not far from here and not certain whether the Rg't will go on to Peshawar or not. I have to go back sharp to my work with *(?)* the new arrangements about the Su: Lt C*(?)*.[12] It is very uncertain where Plowden will go to, I could not wait till the end of this season and perhaps better when it is all settled and over. I don't know what you good people will think of it and would I dare say have liked to have seen her at the beautiful day. All I know is the little girl is very fond of me and that I saw more of her than most men do of their wives before marriage in this country.
>
> I am glad that Jip[13] is going on better and only know that he will get his commission all in due time.
>
> I am so very, very sorry that Sophy[14] could not be sent out to India. It would have been such a capital arrangement but I know Sis was looking forward to it so much and asked me to write home to press it – I am determined I will get out one of them, and will do all I can for it. When married I shall have less than before, but again she would be no extra expense to me. I will come down £100 – but this will better be talked over hereafter in a mail or two hence.

Unsurprisingly, Robert Alfred Cloyne commented on the unseemly speed with which the marriage had been arranged. The rather stiff but finally forgiving letter, dated Chilworth, 14 May 1861, began rather sternly:

> Perhaps had you waited for a few months, you might have had the benefit of my notions respecting marriage as applicable to your case in particular, but you have conducted matters with such speed that there has not been room for anything of this sort. You may rely upon it that whatever I might have thought on the subject proceeded from the conscious interest I feel in you, and in the part you will take in life. And knowing well how much that depends on the Wife a man takes, it is only natural that I should wish to have had some acquaintance with the lady of your choice. There is another matter too, and to speak of which I would wish there had been some little delay. You stand the eldest of a generation and the heir to a property which if not large has been kept together for the last 860 years[15] solely by the prudent arrangements of those who have been interested in it. And I could have wished that there could have been made those settlements which are invariably made when the continuance of a family ... and its Property are at stake. Though I myself have never and may never be in possession of the family property,[16] you will be able some day...to see the sacrifices I have made ... so that it may go to you unchanged and augmented. I would never have interfered in your choice for ... life, and I would not *(missing word, e.g. 'wish'?)* that a single mail should go out without

---

[8] Presumably one of his sisters.
[9] Curious that he misspells 'Chichele'.
[10] East India Company.
[11] The 'Resident' was the British imperial administrative officer for a particular area.
[12] Abbreviated: Sub-Lieutenant C – (but not a surname). Possibly 'Cartography'?
[13] Evidently a nickname for Haversham's brother Alfred, who would have been 17 and at Sandhurst at the time of writing the letter.
[14] Haversham's sister Sophia.
[15] An indication of Robert Alfred Cloyne's interest in his family's genealogy.
[16] In fact, he inherited the estate in 1871 and died in 1884.

taking to you my most beautiful hopes and wishes that the important step you have taken may conduce to your happiness.

I am very sensible of the considerate thoughts contained in your letter to your mother. It has been mainly owing to this that I have been able to keep your brother at Sandhurst. At the present this battle of life is somewhat hard to fight with the wants of so many to attend to ...

... I shall send up your last sketches to Longman to exhibit at the next monthly meeting of the Alpine Club ... [17]

... You will convey to your wife every kind and earnest feeling for her future happiness.

Your affectionate father,

We know, from Phillimore, and an unpublished set of Godwin-Austen family history notes, that Haversham and Pauline's first son, Alfred, was born on 20 March or, according to Phillimore, in April 1862, most likely in Mussoorie,[18] where they may well have set up home. We know that Pauline witnessed her sister's wedding to one Melville on 27 March 1865 at nearby Dehra Dun (home of the Survey of India and therefore central to Haversham's work) but according to an unspecified but clearly scrupulous Austen family historian (probably Robert Alfred Cloyne himself) the birth was at Jullundur – some 186 incompatible miles (300 km) away from Mussoorie. This son lived only for fewer than three months; he died at Lahore (according to the family historian), on 16 June 1862, or in Mussoorie, again according to Phillimore. The son and heir, Robert Arthur, was born (all sources are happily agreed) in Srinagar on 14 September 1863. (Arthur's adult story is dealt with in Chapter 19.) A third son was stillborn.

Other than that, we know, factually, very little. The marriage was not a very happy one; money seems to have lain at the root of the problems. Several sources of the couple's financial difficulties quickly arose. Pauline was extravagant – this trait is referred to on several occasions; Haversham, moreover, would have been funding Eddy's upbringing (see Chapter 9) and it appears that the Survey expeditions had to be funded out of army pay (though it is not clear to what degree these expenses would have been reimbursed). These difficulties, combined with Haversham's frequent absences which made it trickier to resolve such problems, produced strains on the marriage which became near-intolerable.

As early as during his engagement, Haversham asks for advances on his military pay:

> ... With the heavy travelling expenses in the field ... I am now very much pressed ... by different tradesmen with whom debts have ... been contracted since the sudden reduction in my allowances has ... prevented my own servants receiving their pay ... I do not wish to involve myself ... by borrowing money at ruinous interest.[19]

By late 1863, only two years into the marriage, Phillimore quotes Haversham's superior, Montgomerie, who declares that Haversham was, 'the only officer in India who has actually lost pay by joining the Staff Corps.' Haversham at this time was, however, promoted to a higher scale at Rs450 per month. (His next promotion in rank, to major, was gazetted on 18 November 1868.)

From the surviving letters, the first hint of real trouble appears four years into the marriage, in 1865. Robert Alfred Cloyne, in his characteristically interventionist way, discusses at some length

[17] The world's oldest mountaineering club, founded in 1857, now at Charlotte Road, London EC2. It should never be referred to (for which I was once chastised by one of its stuffier members) as the *British* Alpine Club.

[18] Mussoorie is a town and a municipal board, about 34km (21 miles) from Dehra Dun in Uttarakhand. This hill station, situated at 2000m/6,562ft in the foothills of the Himalaya, is also known as the *Queen of the Hills*. The population in 1901 was about 6000.

[19] Quoted by Phillimore.

the interest which Haversham is paying on a loan which has been funded by his mother, it seems for the very substantial sum of £1200. He also refers to bills being run up in Guildford and London: 'The things sent to Pauline about £30 I think will clear it.' The letter, however, is still signed off 'with love to Pauline'. Matters have, however, taken a serious turn by only a little later that year. (Internal references to Haversham's 'absence in Bhutan' and the 'Mustagh Paper', published in 1864, make 1865 the most likely year.) Writing in July, Robert Alfred Cloyne (who rarely puts the year on his letters) refers to Haversham's 'wretched letter' of May, where it becomes obvious that Haversham has been complaining bitterly about his wife's extravagance and its consequences (having been invalided out of Burma in April, after his whole party was struck down by 'jungle fever'). Haversham's father, as ever, takes quite an intrusive role in the marriage, as he had done in his son's career, although it is hard not to accept that, in doing this, he is acting for the best:

> I do feel for you in all your disappointment in discovering the extravagance which had been going on in your absence. The most unsatisfactory part of the matter seems to me is that ... P will not take any part in household affairs. I am sorry to say this, but it looks like temper, whereas what we would have liked to see would be an acknowledgement of error and a disposition to put things right. I had my suspicions as to extravagance from a letter received during your absence in Bhootan containing orders which if executed would have amounted to near £90. I said that it was not to be attended to, as many items were unnecessary ... such as two riding habits. I know that the wife you had married was very young and inexperienced and with disadvantages of education and bringing up which would throw *(?)* a good deal of her training ... I had heard too that some of her belongings were worthy people, there were others the less she saw the better. I am sorry you put the sisters in this list. But marriage, my dear Haversham, is a contract 'for better and worse' and if things are not quite as you wish, do not make them worse by any errors in judgement and treatment, as so many do. You are quite right in this, 'that P must be brought to see the effects of squandering money in the way she has' and if you find that she does do and feels this, then let bygones be bygones – looking happily to the future. I and all here have tried to form a favourable opinion of P as your wife. I hope that it never will be otherwise and that your next *(illegible: 'mail'?)* tells us that matters are more satisfactory.

The crisis appears to have arrived in 1866. (This idea does not quite square with Pauline's letter to Haversham, on pages 127-8, however, unless one believes, from the tone and content of Robert Alfred Cloyne's letter – this time, unusually, the year is given – that Haversham, unknown to his wife, has been discussing the possibility of separation.) The letter is full of pain. The father's sometimes tetchy tone is, moreover, not helped by him having been unwell:

> I have been ill for the last 8 weeks, with that I had at times but little hope – I am now getting better, but slowly and am so weak that a few lines is all I can manage ...[20]

There is a near-collaborative tone to the first part of the letter:

> I need not tell you how much you occupied my thoughts during this last month. The thought of it all makes made me miserable. I was not angry with you – I never meant to be – I pitied you ...

[20] RAC Godwin-Austen to Haversham, Chilworth, 1 November 1866. In fact, RAC G-A's letter stretches to four quite closely written sides.

He goes on to talk about 'the disappointment ... of all the hopes I had formed', as though he were the husband, and refers somewhat cryptically to, 'that unhappy fancy which has afflicted your position.' He accuses Haversham of 'being an enemy to your own position' – it is not clear what is meant, although it may allude to a wish to give up the Survey or simply to bring about the disgrace for someone of Haversham's social standing which would have been attached to a separation from Pauline; either that, or, Haversham has been passed over for promotion because of tensions within his marriage having become public. More worryingly, Robert Alfred Cloyne then goes on to talk about, 'The whole of that wretched affair has been imported here – and such things are viewed more seriously here than out in India.' What Robert Alfred Cloyne means by 'that wretched affair' is also unclear. There are several possibilities: either Eddy's existence has become known (which, given the secrecy surrounding that part of Haversham's life, seems unlikely), or that Pauline's extravagance has somehow become a kind of public scandal, or, that she herself has had an affair (which has become publicly known).

We do know that Haversham was allowed official sick-leave 'during the winter of 1865-66', (see Chapter 15) after his work in Bhutan was brought to an end by the, '*sudden and complete prostration of the whole (Survey) party and Sepoy guard by jungle-fever.*' Between completing his work in the Duars and setting out on the Topographical Survey of Burma, Haversham divided his leave between the Survey Headquarters at Dehra Dun and collecting fossils while making a geological examination of the Siwalik Hills, the main range of Himalayan outliers. Given that this period also coincides with the most difficult period in his marriage, it is not unreasonable to suppose that this 'sick-leave' might have been a form of compassionate leave also.

Haversham's father then continues:

> Your letter shows much want of harmony between your wife and yourself. And when you closed it *(Haversham's previous letter)* it was not clear whether you were to go down *(?)* alone to your new district[21] or she with you. I know of all your grievances then but what can you promise to yourself? She may well say – it has been your own conduct which has lost you a good position and I of all others am the last who should be a sufferer. I trust, however, that the next letter we get will say that she has joined you. Such as your married life is – you will do well to make the best of it – it was your own doing

With this rather cheerless thought, Robert Alfred Cloyne turns to the future:

> I think that just now you have had a sufficient spell of Eastern habits and ideas. England will do you good after which you might return to India, but if you come all that I can offer is a home here and with that you and yours must prepare to be contented. It will not do to see your wife over by herself. Many reasons forbid this. But you may send your boy over whenever you like. We will take every care of him, and the sooner he is removed from India the better.

Only one letter from Pauline survives. It is to Haversham and is dated Wednesday 11 April 1862. This is a puzzle, however. The year looks clearly enough to be written '1862'. But the only Wednesday on which 11 April fell between 1861 and 1869 was in 1866. 1862 is also suggested by the reference to the 'poor little baby', presumably their sickly first child, Alfred, who only lived for less than three months and died in June 1862. With such an uncertainty, therefore, this letter is best viewed as a comment on Pauline's character rather than as a useful document in describing the history of the marriage. Unfortunately, although it stretches to seven A5 sides of rather scrawly

---

[21] This would have been the Topographical Survey of the Khasi and Garo Hills, on the Burma-Assam border (see Chapter 15).

writing in two different pens, the second part of which begins with Pauline writing that she has just had her breakfast, much of the letter is illegible. Haversham is chastised because he 'forgot to kiss the poor little baby before you left'. He is, however, addressed as 'dearest Haversham' on many occasions throughout the letter, and the signing off is to 'my very dearest Haversham from your very affectionate wife, Pauline Godwin Austen'.

In the other intelligible fragments of this letter, Pauline does mention catching 'fever' – 'I know I would be ill'. The frequency of 'and' throughout suggests a chatty, rather unreflective and perhaps rather self-absorbed – or maybe just simply bored – person. The only other phrase which is readable, because it too is frequently repeated, is, 'I hope I shall get a letter from you tomorrow.' Pauline's anxieties at this point may well have stemmed from the loss of her mother when she was only four, her father's absence on military duties, then her father's death in 1861 just before she married Haversham. Indeed, her extravagance does probably derive from these causes combining to provide a lack of firm parental guidance though also, as the insistent hope for a letter from Haversham indicates, from boredom or unhappiness in love. And it is probably also relevant that Pauline was only 17 when she and Haversham were married.

The illness theme is developed in a letter from Robert Alfred Cloyne to Haversham, dated 9 January 1869, in which he alludes to Pauline being seriously unwell:

> Your place as to your wife I imagine to be that you will leave her under the charge of her brother, the young man we knew when he was at school in Richmond. I conclude that all of her family know of her infirmity. From what I have heard only very lately, it would be only annoyance and mortification to bring her to this country. The boy[22] of course will come with you. It is high time that he should be at home, and that he should be setting to work at his English education.

The word 'infirmity' may suggest a lingering illness. The records strongly suggest that Haversham was at work in Burma between May and December 1869, so it appears that Robert Alfred Cloyne's wishes were not acceded to for two years, by which time Pauline had evidently died.

Pauline's muddled life extended into a muddle about her death. While her plaque in St Mary's Shalford tells us that she died in 1869, aged 24 years and six months (having got the year of her birth wrong), other sources, and particularly the usually precise and detailed Chichele Plowden family history, tell us that she died in 1871, at the age of 27, this figure in turn being complicated by a handwritten family history in Haversham's papers which states '1870'. Unfortunately, this history in turn compromises itself by adding 's.p.' (*sine prole*: without offspring), which is self-evidently wrong. All sources are, however, agreed that she died in Calcutta, having spent her whole life in India. The further fact that Haversham went on furlough from March 1870 to October 1872, however, leads us to believe that he might have used this period not just to recover his health (which had been '*broken*' in Burma) but to see his wife through her last illness, then, in England (where he spent most of the furlough), recover from his wife's death and make arrangements for Arthur's education. 1871 seems therefore the more probable date for Pauline's sad and premature death.

The return to England discussed between Haversham and his father in 1869, described by Robert Alfred Cloyne as a 'mortification' – a rather extreme term, used to that effect, for example, about Lydia eloping to live in sin with the bounder Wickham in *Pride and Prejudice* – implies that the marriage remained in a rather unhappy state until Pauline's final illness; such a return was perhaps being contemplated because Haversham, Pauline and Arthur could no longer afford to live in India – Pauline's extravagances could be better curtailed in Surrey (although that would have been a rather suffocating existence for her). At any rate, Pauline does not come across either as Haversham's intellectual equal or domestic and emotional partner in the way that Jessie Robinson, the real love of his life, so convincingly and consistently did two decades later.

[22] Arthur, his grandson, who would have been just under six at the time of the letter.

While this painful period in Haversham's life was evolving, he had left the mountains to start new survey work, in the face of opposition from the 'impudent' Raja of Munipur, among the headhunters in the Assamese and northern Burmese rainforests, easily equalling the Karakoram as a land of the severest physical and mental tests. Haversham's personal history pivoted on this change: he was about to be freed from the constraints of a difficult marriage and was to undertake exploratory work which could hardly be more different from what he had already experienced. Haversham the mountaineer-explorer was metamorphosing into Haversham the scientist.

CHAPTER FIFTEEN

NORTHERN BURMA, MANIPUR AND ASSAM (I)

Among the Headhunters

WHILE THIS MAY SOUND LIKE a rather sensational chapter heading, it is nonetheless accurate. The tribes of Manipur, and those of northern Burma and Assam, particularly the Nagas, are described in too many unimpeachable works of that time for it not to be entirely believable that they would decapitate, and in some cases, scalp their enemies, then display their skulls on door lintels in their villages. Haversham himself experienced an unpleasant encounter with some Naga villagers (see below), in which shots were exchanged and from which political repercussions ensued. Many such difficulties were compounded by the Raja of Manipur, a minor potentate of Mafioso character who was foolish enough to display 'impudence' towards British imperial ambitions and power. Haversham lived through a dangerous and difficult time in other ways – he was caught in a violent earthquake in Cachar in 1869, although, like the good scientist he was, he had the presence of mind to make notes as it happened. He also suffered the loss of his wife, Pauline, (see previous chapter) and underwent several destructive attacks on his health from 'jungle fever'.

A paper[1] published only a year after one of Haversham's main 'seasons' (the non-rainy period of the year, from about December to April or May) in Nagaland, 1872-3, confirms the character of the Nagas, and other similar tribes, such as the unhappily-named Kukis:

> A very common and conspicuous feature of the Nagas, Garrows, Kukis, Lushais, and other hill races of this part of SE Bengal is their custom of taking human heads, either by regular warfare, raids, or casual surprises. Not only does the custom seem almost universal among them but it has obviously existed for some ages in its present form ...
>
> ... It is often supposed that (even here) these raids and murders are due to revenge, disputes, or blood feuds, and that there is always a more or less personal element in the affair, than which there could hardly be a greater mistake. Occasionally there are causes, but by no means always; and among the tribes near here, probably not more than half, if as many, have any grudge personally against those they kill.

---

[1] SE Peale: *The Nagas and Neighbouring Tribes*, *Journal of the Anthropological Institute of Great Britain and Ireland*, Vol 3 (1874), pp476-81.

During the discussion that followed this paper, Haversham spoke up with first-hand authority:

> ... With reference to head-taking, I can testify that it is not always caused by the existence of blood-feuds; they will take human life whenever a safe opportunity offers, and one of our dak-men *(letter-carriers)* was thus killed, perhaps more for the sake of his musket than aught else. A party lay in wait for him and speared him as he passed ... The wearing of rows of cowries on the kilt is permitted to those who have killed an enemy ... In one village of Gaziphimi, we counted no less than fifty-three skulls, stuck up on poles.

On another occasion, Haversham reconfirms this headhunting practice:

> Skulls here to be seen in most of the houses, generally 2 or 3 over the front of the entrance just under the eave

... and in his *Journal*, he gives his reasons for the headhunting continuing, despite British disapproval:

> The fact is that the Government through its district officer is trying to stamp out this abominable custom in our territory, *(the Raja of)* Munipur directly fosters the natural keen desire of the Kookis to accumulate them (the heads), by employing them in his raiding expeditions against the hill people.[2]

The Nagas were, in every way, warlike people, perhaps unsurprising where clean water was quite hard to find and limited agricultural land was fought over because of the ecologically-destructive practice of 'jhum' (spelt 'joom' by Haversham):

> Here, forest is felled and the site used for two years only, when, in consequence of the growth of rank weeds, fresh forest is again joomed, and a system of permanent culture of one spot is impossible.[3]

Haversham observes this widespread custom at one village where they camped:

> Butler had been exerting himself for good in his usual way & got camp pitched in the village in an open position that had been recently burnt.[4]

Indeed, the anthropologist cited above (Peale) goes so far as to assert (and none of the other eminent anthropologists at the meeting disagreed with him) that:

> The inhabitants of this tract *(Manipur and Nagaland)* are divided and sub-divided to such an extent that few parts of the world, I should suppose, can present such a minute tribal segregation.

ETHNIC CHARACTER

What, then, were these 'Asiatic friends' like? Haversham is evidently fascinated by their

---

[2] *JIB*, entry for 16 March 1873.
[3] Peale, *ibid*, p477.
[4] *JIB*, entry for 27 March 1873.

appearance, manners and culture, to a degree which bore no comparison with the thinner ethnological pickings on the sparsely populated Aksai Chin or Karakoram. He begins by telling us that even by 1870, no European officer had penetrated far into the Northern Burmese interior. (Haversham and his assistants were undoubtedly the first Westerners to explore the local hill ranges in any serious and systematic way. Japvo unquestionably constituted a first ascent, as did Kylas. The main tribes in the area, the Khasis and the Garos, either keep themselves to themselves, or oblige smaller groups called the Migams or Langams to occupy buffer country in between.) The Garo are described in greatest detail. One warrior, for example:

> ... was wearing a kind of crown ornamented with tinsel, the honourable distinction of a man who has taken a life in a fight or otherwise.

Collectively, where a village turns out to observe them, Haversham notes:

> ... some of the men were got up in a very gorgeous way with feathers in their hair and earrings with pendant red dyed goats' hair that is attached to boar's tusks that is passed through the ears. Our guide from here was got up in this way and was a most picturesque savage as he walked ahead, his parti-colored sheets thrown over his shoulders and his spear adorned with red goat's hair – large armlets of wood on his arm above the elbow, his legs bound round below the knee with a long coil of thin cut cane – this I notice leaves this part of the leg most rough and unwholesome looking.[5]

Other tribes wear variants of the above:

> ... the men of the Aphromai *(a subdivision of the Garos)* wear earrings like those of the Burmese: cylindrical pieces of bamboo $^3/_4$" in diameter put into the extended ear lobe. The kilt is not worn but only a short *(illegible: local word?)* which scarcely covers the private parts.[6]

We are also told that the men wear their hair short, giving it the appearance of a cap. As we know from so many observations, going back to his first sight of the local women in Burma in 1853, Haversham much appreciated female attractiveness. His diary entries about the beauty and elegance of the (southern) Burmese women whom he meets as a young man occur fairly frequently and are lingered over. Here in the north, though, he seems less sure of the ladies' attractiveness.

> The women ... are tattoed although it is not a general custom. The tattoing is confined to the upper part of the body in a pattern of treble lines and crosses. The lines commence in the hollow of the chin below the lower lip *(and)* are carried straight down under the chin to the Kiroah.[7]

Rather ungallantly, Haversham goes on to tell us that the women are 'by no means good looking in one part of the Anghami Range (part of the Burrail). Not far from Imphal, they receive an equally bad press from him, but in a different way:

> ... he walked along with a tangkol (headman) and his two wives, they carrying hoes and baskets ... I was rather amused at the abrupt way he stopt on arriving at a

---

[5] *JIB*, entry for 30 December 1872.
[6] *JIB*, entry for 6 February 1873.
[7] *JIB*, entry for 28 March 1873. It is not clear what the 'Kiroah' is, though the word may be connected to the *kira*, the maxi-length Bhutanese pinafore.

> field on the roadside, & motioned to them with a grunt to commence their hoeing which they quickly began without saying a word. Their lord and master apparently not having the slightest intention of doing anything in that line himself ...

Haversham dwells on the women's dress:

> The ... women wear the short skirt extending to just above the knee similar to that worn by all the other Naga tribes – but is generally broadly striped white and black – a blue cloak like a scarf crosses the breast from over the left shoulder ... No strap (for a basket) crossing the shoulders and breast ... transferred to the head when heavy.[8]

Jewellery and clothing aside, the other memorable aspect of these peoples' appearance could be seen in their weaponry (and tools). Virtually all the men carry a spear, the longest one observed being 10ft 9in (c.3.28metres). Their axes, which can fell the greatest trees rapidly, Haversham asserts, are made of chisel-shaped iron ('*the Khasi iron is excellent*') inserted into a hole in the close-grained root end of a piece of bamboo. While very useful as tools for clearing ground, the Garo axes are also used as weapons –

> ... we were shown the grave of one of their number who had been killed. They said his head had been taken off & this had been found but minus the scalp ... [9]

– as well as for hunting and fishing. With their axes (the common hatchet having been Haversham's own substitute for the ice-axe, at least as far as step-cutting was concerned, in the Karakoram), the natives sharpen stakes to make a *kote*, a rough enclosure or stockade of trees near one of the many brine springs where elephants and deer come to drink. Smaller versions of such stakes are also prepared for the *pongis*, the animal traps in the form of a camouflaged pit. And even the children catch fish, very often by trapping them in the brackish lagoons and filling their wicker baskets with them.

One of the most remarkable uses to which the natives' tools (and other forms of ingenuity) were put was house building. Generally speaking, their upland longhouses, which held 50 to 80 people, (see plate 37) were built of bamboo and placed on stilts, with a thatched roof. These features protected against the frequent and intense rain, flooding and earthquakes. In Kohima, Haversham (who has an outstandingly sharp eye for architectural detail) noted a curious decoration on some house fronts resembling rather heavy wooden broadswords, (see plate 38) although he could not explain why some houses had them and others had not. (They perhaps reflected the cowrie-shell necklaces worn by males who had killed someone. Such males, particularly chiefs and headmen among the Khasi tribe, after death were preserved in honey; others were cremated, though male and female ashes were kept separate.) In some of the smaller villages, many of these houses had been burned; Haversham suspected the hand of the Manipuris, directed by the wicked Raja. In swampy ground, the villages were often built on an artificial mound, surrounded by rice terraces and marsh.

Because this terrain was broken up by watercourses, dugout canoes became an essential part of everyday transport. Forest trees of the right size were selected then felled and the canoes hacked out and burnt into shape on the spot. Straight round poles were laid on the ground to lift the canoe and carry it to the nearest water. Haversham estimated their average length to be about 16ft (just under five metres). Much of the British Topographical Party's baggage was brought in by large such dugouts on the 1869 expedition.

[8] *JIB*, entry for 4 March 1873.
[9] *JIB*, entry for 30 January 1873.

Some instances of communal life are also described. The bazaar in Imphal receives quite detailed comment, and Haversham and his party are asked to visit the supposedly exclusive annual *Sarramphang Hath* (market) attended only by Garos. The District Commissioner (presumably with his eye open for future trading possibilities) expresses his keenness that Godwin-Austen's party see it; they were to be the first Westerners so to do. They entered the market area with great care because they were uncertain how the wild tribespeople who attend this market might react to seeing their first, intrusive, Westerners. They need not have worried: despite quite an arduous journey by dugout and on foot, little remained – no market had been held there for some time. Much livelier was the bazaar at Imphal. It assembled daily at about 4pm and lasted until about 8.30pm. Fish, sugar, rice and brass ornaments were all sold from stalls in fixed places. An interesting mix of peoples could be seen: Meiteis (the inhabitants of the valley parts of Manipur), Nagas, Kukis, Sylhets and even some Sikhs all met there.

The remaining large communal event was the Hoolie (now usually Holi) festival which Haversham described in his diary in Burma as a young man in 1853 (see Chapter 6). In India, the festival still enlivens its surroundings with its colour and exuberance, with street bonfires, face-painting and general merriment (as I recall from having negotiated its street-bonfires in Kolkata in 2008). In distantly provincial Manipur in 1873, however, little stalls are set up along the roadsides; they are illuminated by bamboo poles bound with bundles of straw and decorated with white flags. These were set ablaze at night, with people shouting 'Hari bol!' ('Call out Hari!'), followed by dancing and banging of tambourines and gongs.

To investigate this country, Godwin-Austen had actually taken charge of his own mobile international community. He lists his party's nationalities: English, Hindustani, Hazaribagh, (meaning 'a thousand gardens', a city and area to the immediate west of Bengal), Assamese, Bengali, Garo, Cachari, Khasi, Mikir, a small ethnic group from the hills near Nowgong (now Nagaon) in north Cachar, Naga-Angami, Naga-Tangkol, Manipuri, Kooki, Gurkha. At various points, Godwin-Austen quotes words or phrases from most of the languages spoken by these people, further testimony to his unusual linguistic capabilities as well as his sense of cultural diplomacy.

HISTORICAL AND POLITICAL BACKGROUND

Where people gather, however, politics start. When Godwin-Austen was exploring this territory, Manipur made it clear that it was the kind of British protectorate which did not wish to be 'protected'. In 1832, the British (after a previous invasion which failed because of the impenetrability of the country and widespread illness among the invaders) concluded a treaty with the Raja, Gambhir Singh; he was designated the independent ruler of Manipur. A British political agent was appointed in 1835.

The British then began to intrude into the Nagas' territory. The Raja of Manipur gave safe passage to the British to go to Burma (which had been acting aggressively). In return, the British not only tolerated Raja Gambhir Singh but also assisted him with weapons in order to annex the area inhabited by the Nagas. The British imposed a hefty tax on the Raja, who extracted the sums due from his subjects (the extent of whom was rather loosely defined because no one knew where the borders of Manipur actually were) in whatever ways he could. That tax-gathering process appears to have provoked a more systematic subjugation of the Nagas by the Manipuri kingdom. Many oral stories relate how the Meitei (i.e. Manipuri) Rajas invaded the villages of the Nagas these accounts are confirmed by Godwin-Austen's own observations. The last Naga village to be conquered by the Meitei rulers was Liyai, probably in the late 1870s. In spite of fierce resistance the Nagas were defeated by the more advanced and better-armed Meitei troops. In 1890, the Raja of Manipur was, however, deposed by his own general. The British Chief Commissioner of Assam JW Quentin (a civilian), and four others marched to Manipur with a small force but were captured and killed. A strong British retributive force was then sent. The Raja, Kulachandra Singh, wa captured and executed. A boy raja, Churachand, was installed on the throne and, during his minority rule, the British political agent administered the state.

There existed therefore, when Godwin-Austen arrived in 1866, well-established and complex tensions between the Manipuris and the upland tribes (mainly the Nagas, the Kukis and the Cachars), between the Manipuris and the British – Haversham calls Manipur, with true imperial self-confidence, 'this little upstart state' – as well as the squabbles continuing among and within the upland tribes themselves.

THE RAJA OF MANIPUR AND THE NAGAS' RESISTANCE

Godwin-Austen's brief from the Government of India in 1872-73, the best-documented year we have of his time in this part of the world, required him to explore the area from Telizo Peak eastward along the watershed between the tributaries of the Brahmaputra and Irrawaddy which was supposed to mark the northern boundary between Manipur and Assam (and perhaps a small part of Burma). The outcome of this expedition failed to satisfy. The actual demarcation of the border was carried as far as Telizo. But beyond there it remained uncertain in what direction the boundary would continue. For the settlement of this question, a survey party under captain Badgley was intended to follow up the work begun by Godwin-Austen.

The main political problem facing Haversham's little party in carrying out this boundary commission concerned the Raja of Manipur not wanting to have his state's boundaries defined, because indefiniteness gave him greater scope for coercion and robbery. As Haversham states, almost as though quoting from *De Bello Gallico*, the Raja '*was yearly extending his influence, levying tribute from people he could in no way protect or keep in order*'. When in Imphal he first met the Raja (along with Thompson, the British Commissioner) a grand *durbar* (ceremonial gathering), of the kind much loved by the British Raj, was taking place; such a ceremony representing a satisfyingly self-regarding way of theatricalising British superiority over the rulers of her conquered territories. The Sandhurst man casts a critical eye on the proceedings:

> Thompson went in a jampan[10] ... painted with the colors *(sic)* of the Union Jack. Guards were drawn up, very ill-drilled fellows of all sizes in a black uniform, and all was very slovenly done. We were met by the Jubraj (the crown prince) and his two younger brothers, all in green velvet tunics & epaulettes, white stockings and ammunition boots.[11] We then met the Raja at the entrance of his own house ... In the centre (of the audience room) was a round table with a white cloth ... We took our seats in order of rank. The band struck up a march and played a few airs extremely well considering that they are all small Naga boys and have only been under an instructor about 5 months ... Conversation lagged a little at first until the Boundary Question was mooted. Butler explained that he was only carrying out orders laying down the boundary of 1842 & that he had no instructions to go into the rights of this matter. The orders were now clear to make a Survey of the watershed & he (Butler) asked for their *(the Manipuris')* aid. The Raja distinctly gave a refusal saying that to the east of Telizo the country belonged to either

---

[10] A kind of local sedan chair.

[11] Ammunition boots were the standard footwear for the British Army from the Victorian Era until the late 1950s. They were unlined ankle-boots with leather laces, 'pebble-grained' (dimpled) leather uppers and sides, leather toe cap and heel cap, iron heel-plate and toe-plate, and an iron-studded leather sole. They were designed to be hard-wearing and long-lasting rather than comfortable. The term comes from the boots' unusual source: they were procured by the Master Gunner and the Munitions Board at Woolwich (the Regiment of Artillery's headquarters) rather than Horse Guards (the headquarters of the British Imperial Army).

> Munipur or Burmah and that the British Government had no right to go there. Butler said it would be better for him to act with a good grace now ... but that opposition would ruin his case. It was no use. The last thing said by Butler was that he gave his advice as a friend ... and to remember the advice he had that morning given to him. We then shook hands with the Raja, and I walked back beside Thompson ... [12]

A few days later, a frustrated Thompson brusquely draws aside two of the Raja's interpreters because they are not doing their job properly; the cunning Raja seizes his political opportunity and declares in a public letter that the British have acted violently towards these Manipuris. (Godwin-Austen does rather unhelpfully, if interestingly, add that Thompson was known for his violent temper because of a head wound he had received in an unspecified earlier engagement and which had become public knowledge.) Despite Thompson's careful letters to the Raja, a '*decided refusal*' to co-operate followed and a diplomatic stand-off ensued. While the British awaited further instructions from Calcutta, the Raja and Thompson agreed that the low-lying parts of Manipur might be surveyed.

One month later, and despite the government of India's instructions that 'the Survey must go on' (the British Crown having taken over the government after the act of 1858, and the East India Company itself having been dissolved in 1874), Godwin-Austen's party, still frustrated, had returned to Imphal. The Raja called on them. Haversham spots a crafty trick by the Raja, and thwarts him:

> He (the Raja) tried the ruse of drawing me over *(to separate Godwin-Austen from the main negotiations)* by sending twice, once to see my sketches and again the bird skins, but I sent them over by a servant.[13]

Godwin-Austen's final thought comes near the end of the season as this inconclusive expedition is drawing to a close:

> I do not myself consider there is any objection to Munipur pushing on into the hills east of the Shiroifurar range and obtaining a kind of authority, however small that may be, over the Naga tribes on that side, but the Raja should be made clearly to understand that such extension should be made known to the Political Agent who should as soon as possible proceed into hills so subjugated, attended by an officer of the Survey Department to map the country on the $1/4$ inch scale. This Geography, Ecology and Natural History would have a fair chance of being added to...and we would thus by degree gain a fair knowledge of the great drainage area of the Irrawaddy basin.[14]

The Raja resisted the British presence in his state mainly through deceit, subversion and coercion. Godwin-Austen describes how plane-table stations are torn down, or how villagers refuse to give out placenames or help as (paid) porters, or 'mutiny' against Godwin-Austen's orders. Raids, involving murder, theft and burning of property are carried out by the Raja's henchmen to stir up the Nagas' and the Kukis' suspicions of and hostility towards all incomers. While the Survey party is treated in a 'civil' i.e. co-operative way in many villages, the kind of civil disobedience the Raja is encouraging occurs in others.

On one occasion, for example, events take a most serious turn:

---

[12] *JIB*, entry for 17 February 1873.
[13] *JIB*, entry for 14 March 1873.
[14] *JIB*, entry for 30 March 1873.

Next, we heard the 'Assembly' *(a trumpet call)* followed by the 'Double', but we could not get along faster. Then we heard the 'Assembly' again. A short time after we heard shots. Passing on we overtook the coolies. The Naga coolies ... had thrown down their load ... It appears that one of the dobashas[15] ... went on ahead into the village and returned saying they would not let us enter the village. The headman said that the young men would fight and advised him to leave the place or he would be killed. He took the hint, sticking his spear into the ground in front of the door, and said ... he would do his best to save the house ... This spear was still standing ... on our getting into the place ... The inspector with the constables formed the Advance Guard & the Nagas came down in a threatening excited attitude armed with their long spears ...

The gist was ... that we might go to the right or left of their village but not through the middle ... Thompson told them we would go through the place and that we intended no harm & that all we wanted would be paid for. He had sent back for Butler directly when he saw the state of affairs. The Nagas up the hill on & off the path now raised their war cry & came down with their spears at a charge, but at 50 yards received a volley from the constables. We were now all forces together. The halt was sounded as there was some sort of confusion. The police were brought to the front and extended into the jungle on either hand & we advanced up the hill ... I took the right, Butler and Thompson the centre on path, Ogle on the left ... We came out at last in full view of the village where we saw 2 men only near the highest houses. The left hand party cut in at the thorn hedge and cut a way through while the centre and right had mingled. Butler at first thought the 2 men were women and called out ... but their spears soon showed & Butler took a shot at them. They disappeared behind the ridge. We then advanced & got inside the village. 2 or 3 now shewed themselves on the extreme Western end & Butler fired another shot at them & the traces of blood ... told us that his aim had been true. I fired a shot also in that direction at a man I saw behind some planks.

The largest house in the village was selected as our Headquarters, the loads all placed in front of it ... Sundry fowls and pigs were slain and the tents pitched in front of the large shingled house.

Just after dark, fire was seen coming from a house on the further end. The Naga had come in and fired it.[16] We tried ... pulling the thatch of houses in vicinity *(sic)* but only one was pulled down. The winds were then blowing favourably away from our tents and baggage, the houses near us were fired and the place was soon nearly all in flames & continued so all night, for in order to have good light a house near us was occasionally fired. A Naga was seen to *(illegible)* across into a shed near & three shots were fired & it *(sic)* fell. Torches of grass were brought and the hut was searched when to our disgust it turned out to be a poor old woman, the poor creature had received the ball in her thigh which had passed through luckily without smashing the bone. She was brought into the large house. She said she thought her own people had returned to the place & so come out of her hiding place in the jungle. Double sentries were posted & the Kookis looking eagerly out forward from their posts would have made a good picture.

After dinner we settled the matter of watches. Ogle and I agreed to sit out our watches together. The night was mild and rather cloudy with ... rain. Went out

---

[15] From Hindi, 'one who can speak two languages', so, an interpreter.

[16] Godwin-Austen explains in a footnote to this page of the *Journal* that the villagers set fire to the house because they had seen his party acquire all their animals and thought they intended to stay a long time.

> towards houses in village & fired 2 to give a fine light over the jungle and the Nagas who went out and got a supply of their liquor out of one of the houses. Night passed off quietly. Nagas and Kookis firing remaining houses. As soon as they went out pig killing and fowl spearing was going on on all sides. About 10 o'clock 4 Nagas came into the place in a great state of funk and hoped that we were not going to cut up their village. Were immediately quodded[17] but were released in evening.

The night again passed off quietly and the next day the surveying continued. In the evening, a considerable number of Nagas showed up. Their principal man was '*in a great rage*' and declared that, '*having lost all, what was the good of making friends?*' The Nagas were taken to the wounded old woman who, to the Westerners' surprise, was then taunted – when she asked for a drink of zoi (liquor) it was indignantly refused. On the following day, the Nagas returned and made peace, '*in formal way by sacrifice of a fowl which Butler held the head of, the Shipromi men* (the villagers) *the legs and wings & it was then decapitated by Thompson.*' The Nagas then surrendered their spears. Later, the 'Advance' was sounded and the Survey moved on. Godwin-Austen remained, imperturbable, with the rear guard. As the coolies filed out, all the inhabitants returned in a rush and flocked to the largest house. He concludes, in a rather schoolmasterly way, '*It is a serious loss they have sustained but entirely through their own action & it will be a good lesson to them & the neighbouring villages that we stand no nonsense.*'[18]

The Raja, naturally, made much of the incident.

HAVERSHAM'S 'SEASONS' IN BURMA AND ASSAM

Haversham spent six seasons in this area of inter-tribal and international squabbling. Having got over his sick-leave in the winter of 1865-6 as a result of the 'jungle-fever' attack which prostrated his whole party in the Bhutan Doars (see Chapter 13), and probably a crisis in his marriage, he spent the winter of 1866-7 as the newly-appointed leader of No 6 Topographical Party surveying the Khasi and Garo Hills, to the south and south-east of the large and strategically significant, if otherwise undistinguished, city of Guwahati, just above the northern perimeter of Bangladesh. The Survey annals describe the area as, '600 square miles north of Cherrapoonjee.' This was Haversham's first taste of dangerous and hostile jungle country containing inhabitants bent on resisting any outsiders:

> They will have to work over tracts of country into which few Europeans have yet penetrated, covered in dense forests, notoriously unhealthy; and are inhabited by a race of men who are believed to be lawless savages.[19]

It was evidently a successful trip:

> Captain Austen deserves great credit for the very excellent season's work which he has turned out in *(his)* usual characteristic and effective style.[20]

In 1867-8, he returned, and moved slightly east, to the Jaintia and Cachar Hills. In 1868-9, he approached even more dangerous territory when exploring the Burrail, the range of forested mountains which marks the watershed between the Brahmaputra and the Irrawaddy, the very

---

[17] Slang: 'imprisoned', perhaps from quad, a quadrangle. Sandhurst in origin?
[18] *JIB*, entry for 4 February 1873.
[19] *Great Trigonometrical Survey Synoptic Report* for 1865-6.
[20] Major JT Walker, RE.

complicated delineation of which was required to separate and define the then independent (Indian) state of Manipur from Burma and British Assam. His planned work was, however, disrupted, partly by resistance from the Lushai tribe who invaded the South Cachar tea gardens (plantations), thereby requiring Haversham to join General Nuttall (a cavalry veteran of the Second Afghan War (1878-80) where he fought, poignantly enough, in Helmand) to suppress it; regrettably, no descriptions remain of this campaign. Haversham's humid work was also interrupted by the violent earthquake referred to above (and described in detail in Chapter 16), and finally by an attack of the serious illnesses – '*spleen and fever*' – which would lead to his final return to England eight years later.

In December 1869, however, Haversham returned with his usual self-discipline for his second season's work, once again exploring the Garo hills, of which he left a detailed account (see below). That season's work was curtailed in March 1870, when he went on furlough, for an exceptionally long time, until October 1872 (although he would of course have left England some months earlier). He was said to be, 'seriously broken in health and emaciated from the arduous work.' This period of absence, however, is also likely to have covered his wife's final illness, her death and his return to England (Pauline was buried in Calcutta) with his son and heir, Arthur.

The relentless Haversham found himself back in Manipur and Nagaland by October 1872; he had put his bereavement behind him, it seems rather quickly, and was ready to take charge once more of his former party. Thanks to the survival of his only other diary, we can enjoy a detailed account (see below, *passim*) of his activities from November 1872 to April 1873. For unexplained reasons, other than that it was required for '*urgent private affairs*', he was again given leave to return to England from August 1873 to October 1874. The 'affairs' must have been urgent indeed to have spent five months or so at sea on the round-trip to England during that period, with only about nine or ten months there, in between. Simply no evidence at all exists for what brought about this absence. (Robert Alfred Cloyne did not die until 1884, although he had on occasion been seriously ill in the 1860s, and therefore quite probably in the 1870s. Given Haversham's recent health difficulties and his managing only one more season in Burma, he might simply have believed himself to have acquired a potentially fatal, or at least very serious, illness which required treatment in England, although that supposition does not square with his usual candour about his health. Resolving problems connected with his wife's will and seeing to Arthur's education are the other likely possibilities.)

When Haversham returned, yet again, in October, for his final 'season' of 1874-5, he worked for two-and-a-half months in a new but equally exciting area to the north of the Brahmaputra and to the east of Bhutan, alongside a brief and successful British army expedition under Brigadier-General Stafford which was trying to subdue the Dafla hill tribes, a group of clans as hard-headed and aggressive in defence of their homeland as the Nagas. (This assertion of local independence is still there, in a different political guise: Maoist guerrillas killed a soldier of the Central Reserve Police Force in an attack in this area in June 2010. Source: *Times of India*.)

Haversham excitedly claims that the area east of the River Sunkos was unexplored. (This river is not known: if he means the Sun Khosi in Nepal, that seems too far west, particularly as he himself had explored further east of it, in Bhutan, in 1864 and 1865.) From a base at Narainpur (now Narayanpur) on the Brahmaputra, he surveyed with his Topographical Party an area of 1,705 square miles in total, in which a most impressive 50 new peaks were fixed and 37 had their heights determined, although Haversham records that their work was confined to the 4-8000ft/12-2400m asl range. As far as I know, no records remain of which peaks were mapped, although he does mention the as yet unidentified Toroputo Peak, under which he camped, and found some carp in a pool. It appears to lie somewhere up the valley of the River Dikrang, (now more usually Dikrong), which flows from the north-west to just north of present-day Itanagar then into the lower Subansiri shortly before it joins the Brahmaputra. At least nine camps are recorded in this area, although that number does not necessarily imply a wide surveying area as progress was so slight. It was extremely tough going: on occasion, the teak, sal and bamboo forest was so dense that the party was said to have advanced only 100 yards in the course of a day. Haversham leaves a description of this trying country:

> The hill-portion of the Dafla country is covered from base to summit with dense forest, the larger trees being clothed with thick creepers, and the bottoms of the ravines are occupied by a luxuriant growth of bamboo, canes, tree-ferns, screw-pines, plantains, etc. Such luxuriant vegetation renders the scenery on rivers like the Dikrang and the Burroi exceedingly beautiful, but it is monotonous. As one ascends, the underwood becomes thinner, and at 7000 feet the forest is composed almost exclusively of a straight-stemmed bamboo, with oak, rhododendron, and other large trees ... The terai portion between the Dikrang and the Burroi...is covered with a sea of high grass, intersected with sluggish nullas, the banks of which are generally fringed with forest.[21]

The Dikrang appears to be the eastern limit of Haversham's explorations in this area, with the Buroi as the western limit.

During this period, Haversham also carried out substantial work mainly by calculating the course, through the volume of its water, of the Subansiri River, which has its source in Tibet. He was trying to resolve the question of which in its upper reaches was the more likely to be the Tsangpo, by comparing the relative volumes of water produced by the Subansiri and the Dihang (as the Brahmaputra is known in this area): this was one of the Last Great Problems of 19th-century geography. It was also a question of how the rivers might be connected. The discovery of the Great Bend in the Tsangpo-Brahmaputra, which solved the problem, had been given presumptive proof as early as 1765 by Rennell, but was not revealed until the pundit Kinthup's ('KP') heart-rending expedition of 1881-4. For an account of this odyssey, which involved quite astonishing suffering and devotion to duty, and for one of the best-written and most enthralling travel books one is ever likely to come across, I recommend Ian Baker's *To The Heart of the World*, with a foreword by HH the Dalai Lama, Souvenir Press, London, 2006. (Rennell's 'proof' had not prevented a Scots engineer named Gordon from believing that the Tsangpo was actually connected to the Irrawaddy. Other theories even suggested a connection with the Salween.) An earlier attempt, in 1826, by Wilson, to follow the course of the Dihang had been stopped by resistance from the fierce Abors, the Daflas' equally bad-tempered neighbours.

The British expedition, which got itself involved in various armed skirmishes with the Dafla tribes, marked the end of Haversham's military service. After yet another period of sick leave in England, the remainder of his time in India, in a kind of transitional phase, was spent on government work, mostly in Calcutta, where he assisted Dr John Anderson FRS, the superintendent of the Indian Museum, in arranging the collections in the new museum building. Then in 1877, following an invitation from the trustees of the Indian Museum, Haversham re-arranged the Indian Bird Collection and catalogued it. He also mentions spending time arranging and classifying his own unsurpassed collection of bird skins, shells and molluscs, described as '*very extensive*' and comprising '*new species*', an important link to the change of professional identity, from explorer to full-blown scientific authority, which awaited him after his return to England in 1877.

With the same scrupulousness as he applied to his Kashmir and Karakoram Survey work, Haversham records the extent of land he surveyed. In 1866-7, he covered 249 square miles; in 1867-8, 411 square miles; in 1868-9, 98 square miles; in 1869-70, 515 square miles; in 1872-3, 1580 square miles, and in 1874-5, 1,090 square miles – a total of 3,843 square miles. (Like Kashmir but unlike the Karakoram, the area surveyed in total was of course much larger than that because other assistants were working in other parts of the same country.) And it is worth remembering that much of the early work took place at or above a height which made breathing a harder task than usual.

As is often the case in piecing together the different parts of the various phases of Haversham's

---

[21] *JASB*, Vol CLV, part ii, 1876, p65.

life, our knowledge is uneven. For this period, we are fortunate still to possess three detailed accounts of his activities. The first is a long paper entitled *On the Garo Hills*[22] which gives a blow-by-blow account of his work during the 1869-70 season (with some references to previous work in these hills in 1867); the second is a paper, redrafted from letters, describing in detail the Cachar Earthquake of 1869; the third is his personal diary for the greater part of the 1872-3 season. Among his personal papers, we can also see his letters of resignation from the service. As it would be tediously repetitive to recount what Godwin-Austen did during each day of the two seasons of which we have detailed knowledge, I therefore intend to approach these sources thematically, to form a picture of what life was routinely like, and what the exceptional moments were during these times. This picture can then no doubt be seen as an exemplar account of what life would have been like for Godwin-Austen during all the seasons which he spent in this difficult and dangerous area.

MOUNTAIN AND JUNGLE

One of the characteristics of Godwin-Austen as a great explorer is that he could switch without any fuss from hard labour among the highest and least-known peaks on the planet to essentially the same work in country mostly comprising dense forest and jungle. His work ethic had such resoluteness, was so driven by a sense of duty and professional excellence, in the proper sense of rising to a position which towered above others (as opposed to the debased modern sense of 'trying quite hard to be commercially noticeable') that we see a form of greatness in the way that he rose above the physical difficulties of the explorer's world and arrived at an understanding of how universal forms of the knowledge he had acquired might be applied.

The factors common to mountain and jungle exploration were that they involved toiling up '*stiff ascents*' (Haversham rarely uses any other expression to describe ascending steep ground.) At the top, rocks would be cleared in the mountains or trees in the jungle, in order that trigonometrical stations could be set up in a stable way to make use of the clear views. (He describes the Burmese survey stations as '*bamboo platforms built on lofty trees 80 to 90ft high on the hill-tops*'.) Both mountain and jungle expeditions were of a similar '*siege*' structure: long walks and climbs supported by many 'coolies' and usually a translator, with local guides. And both kinds of country offered their own forms of objective danger: crevasses, glacial rivers, rockfall and mudslides, thin air and cold in the mountains; earthquakes, deer-traps with poisoned spikes, cataracts, 'fever' in the jungles.

Then the differences begin. On the Kashmir Survey, Haversham worked most of the time as a sole and sometimes a rather lonely European; in Burma (used here for the whole area in which he worked in Assam, Manipur and northern Burma) he generally worked with fellow officers, and political commissioners (or their deputies), in a way much more likely to sustain good personal morale – many a convivial dinner after a hard day's work is referred to. Haversham's survey officers (as well as several civilians) were usually subordinates. In Burma, Major Godwin-Austen was in command; he was now doing the instructing (he complains of the delays required to instruct young assistants, particularly between 1866 and 1868) whereas in the mountains, he was still a young man, on probation in 1856, who had to answer to Captain Montgomerie and had to show that he was better than his talented and hard-working fellow officers.

Something similar was true in the contrast between the aboriginal peoples of his two exploratory areas. In Baltistan, Ladakh and Zanskar, Haversham moved through countryside of very low population density; it was possible to go for four or five days, on occasion, and not meet any settled locals. Invariably, including his difficult encounter with the Zimskang of Rudok (see Chapter 12), confrontations with the locals were diplomatically resolved; he was treated with courtesy and generosity while he was in the mountains – the word which he comes back to, over

[22] *Journal of the Royal Geographical Society*, Vol 43 (1873), pp1-46.

and over again in his diaries to describe helpful local behaviour is '*civil*'. In Burma, on the other hand, a day rarely goes by without passing through or camping at a village. Here, '*civil*' behaviour – bringing out some rice or fowl for sale, or perhaps providing a guide – occurs only sporadically. Godwin-Austen also meets considerable hostility, in the form of unwillingness to provide supplies, or showing sullenness, occasional armed resistance and various forms of deceit. (The variety of dress and behaviour among the local tribes does revive Godwin-Austen's lifelong interest in ethnology; virtually all of his ethnological studies were published during this period.) And of course, it was not just the people who differed; the climates were also wholly dissimilar – extreme dry cold and heat with powerful winds in the Karakoram or on the Tibetan Plateau; in Burma he was based in one of the wettest and most humid regions on the planet, although, as is seen below, Burma also could be cold, but nothing like to the degree experienced in the Karakoram. Much the same sort of contrast is true of the ecology of the two areas: the high glaciers form part of an alpine wilderness where seeing birds other than raptors, or animals, or anything other than a few stunted or scrubby plants, is unusual; in the Burmese forests, though, he constantly observes new birds, collects shells and molluscs, catches several varieties of fish, discusses types of tree and their uses and mentions a great many animals and plants – the 1872-3 *Diary* concludes with several pages of 'Ecological Notes' on the Naga and Manipur Hills.

TOPOGRAPHY AND ECOLOGY

The climate, of course, is largely responsible for defining the ecology and shape of the country. Godwin-Austen has, unsurprisingly, much to say on the matter. Perhaps the most surprising aspect of his writing on this topic concerns how attractive the scenery is: '*pretty*' is used to describe the countryside time after time. Near the rivers, typically, '*The whole country … perfectly flat is covered with a dense forest as far as the eye can carry one into the misty atmosphere …*'[23]

At an early stage of the 1872-3 season, Godwin-Austen moves further east and finds himself near Kohima, the capital of Nagaland:

> The country east of this assumes quite a different aspect. Forest almost disappears and all the southern slopes of the hills are grass, round and bare with solitary trees here and there.[24]

(This substantial town, in a low-lying gap, was the scene of the great battle from April to June 1944 in which the Japanese in their attempted invasion of India were finally turned back after several days of ferocious hand-to-hand fighting. The battle is as important in modern Asiatic history as Stalingrad or El Alamein was in the western theatre.) When Haversham moves on to higher ground, he discovers that different ecological conditions of course prevail:

> … I spotted the highest point. This part I found covered with a form of small stemmed bamboo from 7 to 8 feet high.[25]

This bamboo is often described: little thorns grow out of the joints of the stem, making it painful to walk or cut one's way through; and the bamboo forest conceals a danger: houses, possibly containing aggressive tribesmen, may be buried within it.

Perhaps more surprisingly in ecological terms, in this area of the Brahmaputra-Irrawaddy watershed, Haversham comes across '*a few solitary crab-apple trees standing here and there*', growing

---

[23] *JIB*, entry for 27 December 1872.
[24] *JIB*, entry for 1 January 1873.
[25] *JIB*, entry for 5 January 1873.

wild, a surprise which could not have failed to remind him of distant, gentler English gardens. As the survey party concludes its work on the watershed and moves on down to Manipur to open negotiations (for further exploration) with the obstreperous Raja, Godwin-Austen's antipathy to the scenery seems to reflect his anxious state of mind:

> The whole way was a dead flat, the valley opening out broadly as one reached ... Munipur. The coloration of this season of the year is particularly ugly - most of the trees bare and the grass of one same tint - of ochre - spread over both plain and hill, shewing black on the hills where burned.[26]

As the season progresses, however, and the weather changes - and with that, the emotional state of the surveyors - a happier appreciation of his surroundings follows:

> This corner is certainly one of the most beautiful we have been in & especially at this season, the varied mixture of coloring from the browns of the old sunburnt grass, the bright greens ... now springing up, and the past leaves of the oak, against which the blackened stems of the trees and a kind of date palm very common on the open slopes tell such a strong contrast.[27]

This clearing of the hills by burning ('jhuming') is another frequently observed feature:

> The good wood ... is used as firewood, to this I attribute the bareness of all the hills within a radius of 12 miles of the Raja's palace, the grass that supplants the underwood & the annual burning at this time of the year that the last undergoes.[28]

Another piece of typical countryside is described during the phase of the expedition which occurs after unsuccessful negotiations with the wicked Raja leave Haversham with no option but to stick to fairly low ground:

> We passed through open woods of oak and well-wooded slopes on either hand extending quite down to the narrow strip of open grass in the bottom of the valley.

Oak and fir, which again sound surprisingly English - although actually widespread in many parts of Asia - are the most frequently mentioned trees, although '*sal*' timber is said to be the '*principal forest tree*'. Deodar (a species of cedar) is also noted; a cutting was sent back to Chilworth Manor, on Robert Alfred Cloyne's request. The India rubber tree is also recorded and, on the low and medium ground, rice terraces with their bright emerald-green shoots are observed pretty well everywhere.

The vibrant ochres and greens (pretty close to his old school's contentious colours) in which this country, outside the forested parts, is rich, are the chromatic emblems of the area. Near the beginning of the 1869 expedition, in the Garo Hills, Godwin-Austen is moved to observe:

> Puna Tith is one of the most beautiful scenes I have ever passed through ... The river here is very deep and perfectly still, and the reflections of the richly coloured dark green rocks capped with trees and foliage, as we saw them about sunset, were most lovely.[29]

---

[26] *JIB*, entry for 15 January 1873.
[27] *JIB*, entry for 2 April 1873.
[28] *JIB*, entry for 4 February 1873.
[29] *On the Garo Hills*, pp3-4.

The ochre, terracotta-like colour is derived from the local sandstone (rather as in Torridon where the pinnacles and outcrops are characteristically bluish- or brownish-red); it occurs in ravines, cliffs and even on summit tops. The hills, indeed, are very often described as plateaux, rather than peaks, with steep, low, narrow ridges leading up to them, quite often terminating in a cliff several hundred feet high. One of Haversham's typical perspectives would be:

> The country to the immediate north of Kylas[30] was covered with dense, dark forest; it presented a flat appearance, the rivers flowing in deeply-cut narrow valleys bounded by steep regular slopes; the somber *(sic)* green of all only here and there broken by a bare bit of light-coloured cliff in the horizontal sandstone.[31]

Of course, its intense annual battering by powerful streams of rain and river water had shaped the soft sandstone and a unique topography for this country. A description of a journey following a river is therefore included:

> On all sides, dense high forest that had never been touched by axe met the view. Through this we marched ... crossing one high flat plateau descending into the level of the Rengchi ... Our path ran with the course, oftener in it than out of it; the fall very gradual, long still shallow reaches, with sandy bottoms, separated by gravelly bouldery falls. Just before reaching the junction with the Sumessary, the roar of whose waters we could hear, we struck up over a ridge, and descended again upon that river ... The scenery was here very grand, and during the height of the rains must be truly so; the whole body of the water is thrown over a steep and sudden ridge of gneiss and sandstone. In this ... gorge, it makes its great fall to the low level at Seju, some 700 or 800 feet, in ten miles.[32]

Such was the land and the people among whom Haversham now found himself. What follows gives us an idea of his daily life among these troublesome but fascinating headhunters.

---

[30] A local hill, not of course to be confused with Mt Kailas in Western Tibet.
[31] *On the Garo Hills*, p17.
[32] *On the Garo Hills*, p39.

1. Haversham's paternal great-great-great grandfather, Robert Austen (d.1718), Lord Commissioner of the Admiralty.
Portrait by Sir Godfrey Kneller

2. Sir Henry Edmund Austen (1785-1871); estate-improver and friend of royalty. Haversham's paternal grandfather.

3. General Sir Henry Thomas Godwin (1784-1853); Haversham's maternal grandfather and formative influence.

4. Haversham's father, Robert Alfred Cloyne Austen, later Godwin-Austen (1808-84).
Courtesy of Dr Richard Godwin-Austen

5. Haversham's mother, Maria Elizabeth, née Godwin (1813-1904).
Courtesy of Papplewick Hall

6. Shalford House and Park, Surrey, in the 18th century.
Courtesy of Papplewick Hall

7. The ancient Dining Room of Shalford House in the late 18th century, showing one of the magnificent early 17th-century fireplaces.

8. Lozenge decoration from one of the 1601 (or 1609) fireplaces in Shalford House.
Courtesy of Papplewick Hall

LOT

232 Mahogany-frame easy chair, upholstered in crimson, and extra cover
233 Ditto ditto
234 A 3 ft. 3 in. mahogany reading occasional table
235 **Set of 8 very elegant mahogany Chippendale chairs with hair seats, and pair of elbow chairs to match**
236 A 4 ft. mahogany dining table with rounded ends, extending ft.
237 Capital 16 ft. bookcase, painted and decorated with gilt edges, with 4 wire doors, 2 drawers and enclosures
238 A 4 ft. 3 in. painted bookcase, decorated with gilt, and 2 drawers
239 Ditto ditto
240 **Very handsome 4 ft. mahogany secretaire, inlaid with tulip and satinwood, with bookcase over, enclosed by elegantly designed glazed doors**
241 Pair of 24 in. well-carved oak figures
242 Pair of bellows with carved face, and barometer in mahogany case, by "G. Adams"
243 A 13 in. brazed inkstand with handles, envelope case, pair of brass candlesticks and ash-tray
244 Three Windsor chairs, seats stuffed in hair, and 3 fringed valances
245 Antique walnut-frame easy chair with ball feet
246 Hassock, paper basket, sundry boxes, and pair of brass and bronze reading lamps
247

Dining Room.

248 **Pair of 20 in. handsome brass fire dogs**
249 **Capital Turkey carpet, 26 ft. by 15 ft., as laid**
250 Persian rug, 10 ft. by 7 ft., and smaller rug
251 Well-made mahogany dining table, in divisions with rounded ends, extending to ft.
252 **A 9 ft. 4-leaf highly decorated, embossed old leather screen**
253 **A 6 ft. by 2 ft. 6 in. Early English handsomely carved oak table with massive carved oak legs**
254 **Antique Sheraton settee with satinwood decorated legs, upholstered in silk damask, with 5 pillows and extra cover**
255 **A similar lot**
256 A 2 ft. 8 in. mahogany side table with 2 drawers, and pair of ladies' high-back chairs, utrecht velvet seats
257 Three antique walnut and mahogany chairs with ball feet
258 A 4 ft. 9 in. mahogany Chippendale frame side table with modern top
259 Set of 12 well-made mahogany chairs with leather seats and extra covers

LOT

260 **A 9 ft. well-made Early English oak sideboard, very handsomely carved, with cupboard enclosures and 4 drawers, the carved figures supporting the two ends and the decorations and embellishments beautifully executed**
261 **An 8 ft. by 4 ft. very handsome carved oak table with 6 massive excellently carved legs**
262 **An Early English beautifully carved dark oak elbow chair**
263 **Ditto ditto, with Cupid**
264 **Smaller ditto**
265 **A 4 ft. 6 in. old Burmese coffer, very highly decorated with gilded figures**
266 Pair of 3 ft. very elegantly designed mahogany dinner wagons
267 Carved oak child's high chair with cane seat
268 Carved oak figure on pedestal, supporting a tray
269 Marble pedestal bust of Washington
270 Ditto ditto of Fox
271 Pair of well-carved high-back Elizabethan chairs with cane seats and backs
272 Ditto
273 Ditto
274 Ditto
275 Finely-carved Elizabethan elbow chair with cane seat and back
276 Ditto ditto
277 Very handsome Elizabethan elbow chair with carvings composed of coronet, thistle and children's heads
278 Carved Elizabethan high-back chair
279 Heathen goddess, highly decorated with gilt and coloured glass
280 Pair of marble heathen goddesses
281 Antique bronze bell, figure and 2 curios
282 Pair of carved oak ornaments, Lions supporting shields
283 Curious carved oak tankard with cover, "Tenasserim Pecu," and antique blue and grey stoneware tankard
284 Ebonized inkstand, 2 hassocks and 4 mahogany stands for china
285 Handsome old brass chandelier with 12 lights, divided into 4 branches
286 Two 21 in. blue and white old china dishes, decorated with Roman chariots and figures
287 A 19½ in. ditto ditto and a 24 in. well dish
288 Indian rug, 9 ft. by 6 ft.
289

Boudoir.

290 A 9 ft. Persian rug
291 Three Indian rugs
292 Hearth rug and 2 skin mats
293 A 6 ft. WELL-MADE MAHOGANY SOFA TABLE, inlaid with brass, and elegant fittings

9. Part of the inventory for the 1899 Shalford House bankruptcy auction.

10. Artistic influences: watercolour by Captain Robert Petley, Haversham's drawing tutor at Sandhurst, titled 'Immigrants at Bedford Basin near Halifax'.
Courtesy of Library and Archives Canada/Robert Petley collection/c115424

11. Artistic influences: Captain Robert Petley, 'Windsor, Nova Scotia, from the Barracks'.
Courtesy of Library and Archives Canada/Peter Winkworth Collection of Canadiana/e002511352

12. Artistic influences: William Carpenter, 'View of houses adjoining a stream, Peshawar, 1855'.
Courtesy of V&A images: Victoria and Albert Museum, London

13. The Shwedagon Pagoda, Rangoon (Yangon), Burma. (1853)
HH Godwin-Austen

14. Burmese landscape at Prome, above the Irrawaddy Delta. (1853)
HH Godwin-Austen

15. General Godwin's Tomb, Simla: 'Fortunate in those he had commanded. Unfortunate in those he had served.' (1854)
Lieutenant-General Fred Maisey (1882)

16. View from the British camp at Nowsherah towards the Khyber Pass. (1855)
HH Godwin-Austen

17. A surveying party at work in Kashmir. The figure in blue on the right is most likely to be Haversham. (1856 or 57)
HH Godwin-Austen

18. The Pir Panjal Hills, Kashmir. (1856 or 57)
HH Godwin-Austen

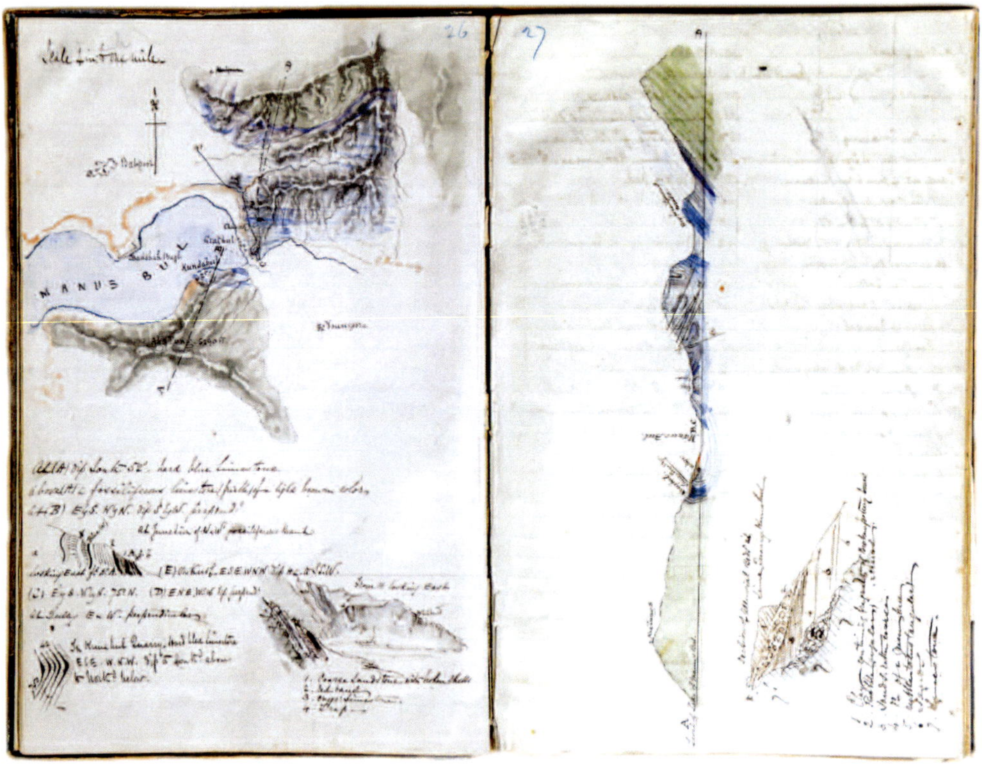

19. Pages from the Kashmir geological survey compiled by Godwin-Austen, 1855-59.
Courtesy of Papplewick Hall

20. The village of Rajauri, Kashmir. Haversham was received by the Maharajah in the tower towards the back on the right. (1854) *HH Godwin-Austen*

21. Zanskar's twin 7000m peaks, Nun and Kun, from a ridge due south-west at about 5300m. *HH Godwin-Austen*

22. Nun and Kun from the north, near the Parkachik La, west of Ringdom Gompa.
HH Godwin-Austen

23. The starting point for the High Karakoram: looking across the Indus Valley at Skardo. *Courtesy of David Hamilton*

24. Looking towards Masherbrum from near Hushe village, upper Hushe Valley.
Courtesy of David Hamilton

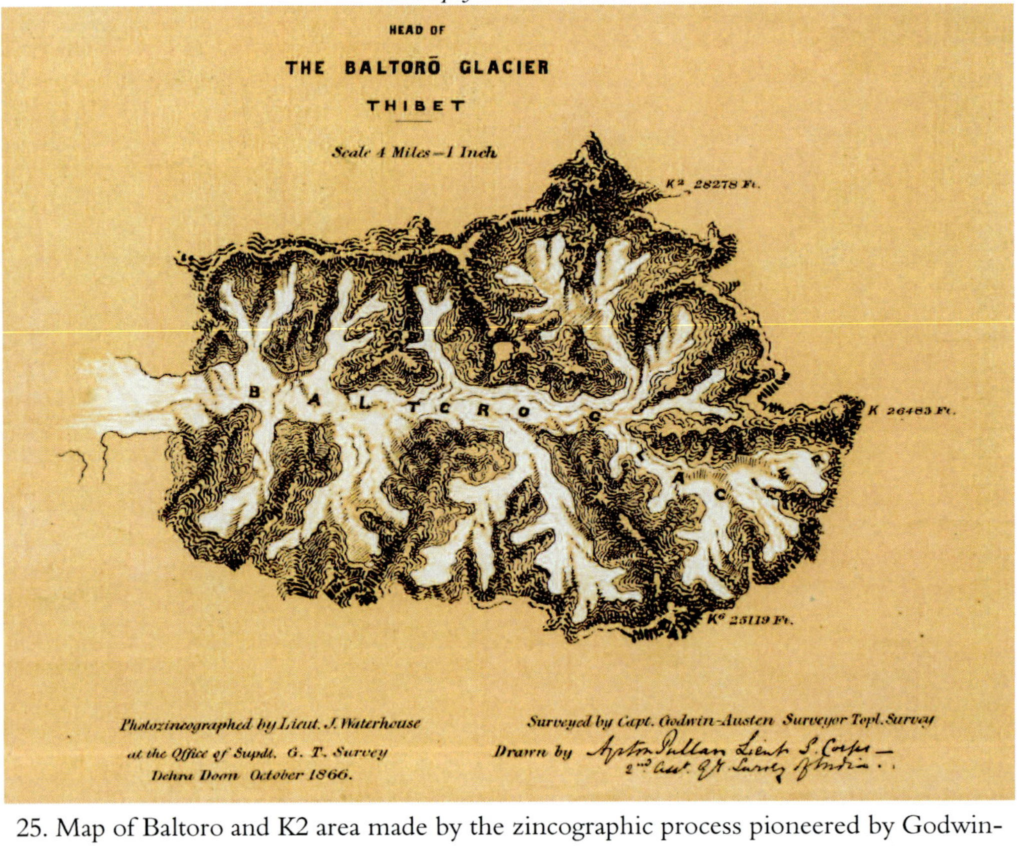

25. Map of Baltoro and K2 area made by the zincographic process pioneered by Godwin-Austen and General Walker, from Godwin-Austen's original survey map.

26. The junction of the Punmah and Choktoi Glaciers, looking north. (1861)
Courtesy of Dr Richard Godwin-Austen

27. Panorama of the Baltoro Glacier from near Concordia, looking west, to show the area of Godwin-Austen's first sighting and measurement of K2.
Courtesy of John Cleare

28. The site of Godwin-Austen's likely ascent above Urdukas, on the south side of the Baltoro Glacier, to reach the point on a spur of Masherbrum from which the first close sighting of K2 was made.
Courtesy of John Cleare

29. K2, to the left at the rear, from a top above Urdukas, about 1000m/3281ft above the Baltoro Glacier, looking over the Biange heights. The first close-up sighting of the world's second-highest mountain. (August, 1861)
HH Godwin-Austen

30. View of the Godwin-Austen Glacier leading to K2, from Concordia.
Courtesy of Jonathan Godwin-Austen

31. View looking south over Tso Moriri, in the remote country of Rupshu. Mentok Kangri (6250m/20,505ft), on which Haversham broke the high-altitude mountaineering record, is the snow-covered peak beyond the lake.
Courtesy of Royal Geographical Society, South Kensington, London.

32. Lake Pangong, Tibet. Part of the area traversed by Haversham during the 1863 expedition.

Courtesy of Royal Geographical Society, South Kensington, London

33. Paro Dzong: administrative, military, religious and legal centre, Bhutan, 1864.

HH Godwin-Austen

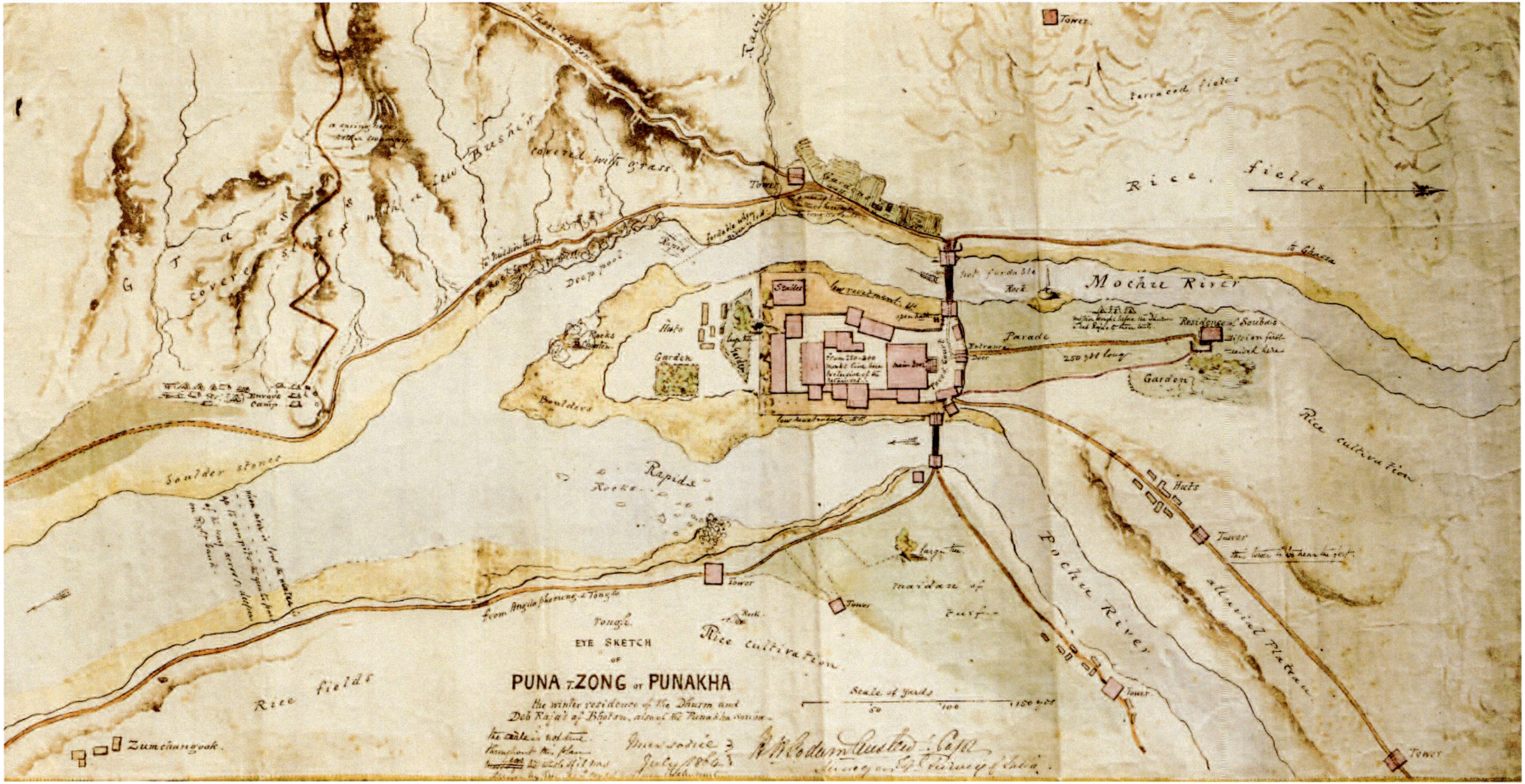

34. A survey map of Punakha Dzong, Bhutan. This kind of illustration shows Haversham's attention to the fort's military significance as well as its architecture. (1864)

HH Godwin-Austen

35. The first map of Sikkim and Western Bhutan, made by Haversham in 1864 when attached to Sir Ashley Eden's failed diplomatic mission.

36. Godwin-Austen outside Colonel McCulloch's bungalow at Cherrapoonjee (Cherrapunji), 1869. Haversham is aged 35 or 36. His sunken cheek may have been caused not only by the physical demands of a soldier-surveyor's life in a trying climate but also indicates that he had already suffered two bouts of malaria.
Courtesy of Natural History Museum, South Kensington, London

37. A murong, or Burmese communal hut, from a diary drawing by Godwin-Austen, 1872. *Courtesy of Dr Richard Godwin-Austen*

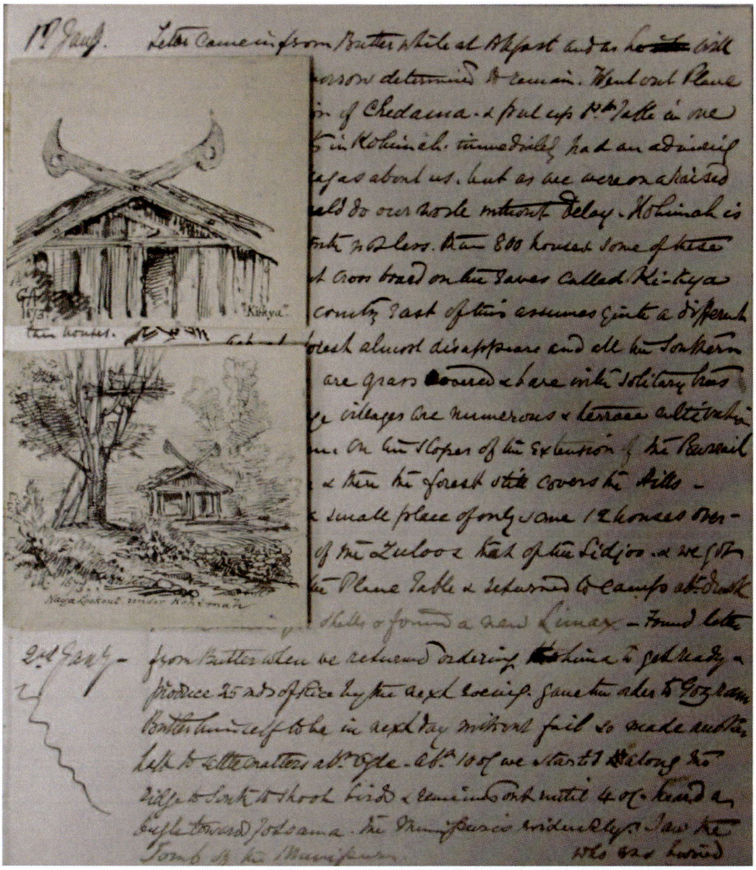

38. Some sketches and a page entry from Haversham's Burma Diary. *Courtesy of Dr Richard Godwin-Austen*

39. The Château de Chillon and Lake Geneva, looking east. (c.1882)
HH Godwin-Austen. Courtesy of Dr Richard Godwin-Austen

40. The Matterhorn from the Zinal Valley. (c.1882)
HH Godwin-Austen
Courtesy of Papplewick Hall

Watford Engraving Co.

41. Haversham's third wife, Jessie Robinson, (c.1910) from the memoir on her life published (1914) by Gertrude Jekyl (sic) and 'MSW' (Margaret Watts)
Courtesy of V&A images: Victoria and Albert Museum, London

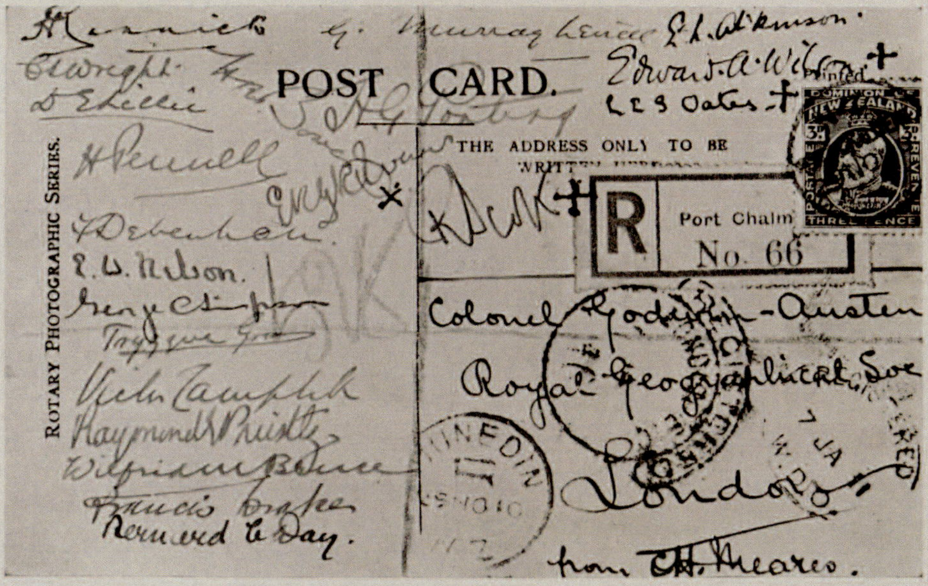

THE POST CARD FROM THE "TERRA NOVA."

Watford Engraving Co.

42. The postcard from the *Terra Nova*, sent to the Godwin-Austens by Captain Scott and his crew from Dunedin on their way to the Antarctic. Jessie has marked crosses against those who died on the expedition, not long before she herself died.
Courtesy of V&A images: Victoria and Albert Museum, London

43. A Buddhist Mandala, or sandpainting, symbolising the order and ephemerality of the universe. This watercolour by Godwin-Austen is one of several pieces of circumstantial evidence for his conversion to Buddhism late in life.
Courtesy of Papplewick Hall

44. The Buddhist temple of Burmese design and probably made from Northern Indian sandstone, discovered at Nore by Dirk Bogarde after he bought the house in 1950.
Courtesy of Mr and Mrs G Kemp, Nore, Hascombe, Surrey

CHAPTER SIXTEEN

NORTHERN BURMA, MANIPUR AND ASSAM (II)

Burmese Daily Life

WHAT KIND OF LIFE DID Haversham lead during his seasons in Burma and Assam? Like any job, despite the unusual working conditions, it had its routines: much of the work was of course based on his previous experience as a surveyor-explorer. It is noticeable, for example, from his *Diary* of the 1872-3 season in Burma that Godwin-Austen's social diary is full. On 3 November 1872, while in Calcutta on his way out to Burma after a furlough in England, he dines with Stolitzka;[1] on 4 November with Blanford;[2] on 11 November, he dines with Stolitzka 'at the club'; on November 17, he spends the evening with Thuillier.[3] These are all important people in their field – Godwin-Austen has become a part of the scientific and military establishment (and, given that he dines out virtually every night for the 19 days he is in Calcutta, he is clearly interested in putting on some weight before the rigours of the Burmese season begin).

MOUNTAINEERING AND SURVEYING

As the previous chapter showed, the nature of the soil and the extremes of the climate combined to form a land which was hilly, even mountainous. But Haversham's ascents have been much harder to work out than his generally well-detailed (see Appendix on page 245) ascents in the Karakoram, Ladakh-Zanskar and Tibet. In many cases, his measurements were taken from the tops or high points of ridges rather than from true summits. Nevertheless, he seems to have ascended many of the major peaks in the area.

Some of the ascents are quite short, the sort of walk which, in Godwin-Austen's state of fitness, could be accomplished in not much more than an hour, for example, '*I ascended the hill above the village for about 1500 feet whence we obtained a magnificent view.*'

---

[1] Ferdinand Stolitzka, (1838-74) was a Czech natural historian who worked in India on palaeontology, geology and zoology, including ornithology and herpetology. He died of high-altitude sickness during an expedition across the Himalayas.

[2] William Thomas Blanford (1832-1905), English geologist and naturalist. He is best known as the editor of a major series on *The Fauna of British India, Including Ceylon and Burma*.

[3] The future Surveyor-General, not the French biologist.

There is a lot of this sort of rather imprecise description. But since we do not usually know the name of the village referred to, or if we do it is not marked on any readily available modern map, or we do not know what the starting height was, the significance of the ascent is lost.

Godwin-Austen's first important climb was Jotsama, '*a stiff ascent*' which '*extends in three separate sections up the slope*'. Jotsomah (modern spelling) appears to rise to just over 7,874ft/ 2400m.[4] A more important ascent, in January 1873, was Japvo (9,888ft/3014m) lying just south-west of Kohima. Its summit lies about 1600m/c.5250ft above Kohima, although the 'valley' Godwin-Austen refers to may be rather lower (since Kohima lies on a ridge):

> Térélu *(the guide)* with some 15 Nagas (he said there were 20) led us up the bottom of the valley over rice cultivation in terraces then crossing the stream that comes down from Japvo. A short but sharp ascent brought us upon a level terrace on the side of the valley and some ____ feet above the river, terminating abruptly in a slope of 45° all composed of sub-angular rock thrown together ... The ridge of Japvo being 9000 feet and the peak just under 10000 feet ...

At that point, in a not untypical arrangement, a camp is set up, with the purpose of finishing the ascent the next day. Godwin-Austen had, with his usual thoughtfulness – there is at times something of the Boy Scout about him – '*brought up a tarpaulin and under this I spread my bed*'. On the following day, once some men had been sent downhill for fresh water and food:

> I took them *(his Naga assistants and 'coolies')* further along the ridge until I spotted the highest point ... The forest consisted of rhododendron, at least three species, two growing to large trees, & yew & box pretty common, another tree now leafless completed the list ... Having got the top of the hill, I laid out plane-table and laid out rays ... [5]

Various ascents were completed along the watershed near Kohima, such as the Napoo Ridge on 14 January. On 17 January, Godwin-Austen determined to go and '*clear*' a peak above camp. It was called Khunho (a peak of undetermined height in the Kohima area). If it took Godwin-Austen three hours to ascend, that would have brought him and his party of 20 Nagas to between 1200m and 1500m (c.4-5,000ft) above his campsite at Priphemah, lying at approximately 1600m/c.5,250ft asl. The tough task of peak-clearing on summits was necessary in order to obtain uninterrupted sightings for the survey instruments. He started about 12.30pm and reached the top three hours later after '*a pretty stiff pull*' with '*very bad places up rock faces*'. There was, however, '*a grand view*' from the top.

One even higher peak had been seen in the distance several times from Japvo and the Kohima hills. It was called Saramati (3826m/12,552ft) but lay too far east for Godwin-Austen to be able to tackle it. Instead, he had to settle for Telizo, probably 2485m/8,153ft, on the northern perimeter of Manipur. After Telizo, Godwin-Austen climbed a ridge he named Kopamedja, and satisfied himself that the land beyond should be called 'East Naga'. In creating these definitions the political aspect of this work again shows itself: by defining natural features such as watersheds, and being the first so to do, Godwin-Austen is really declaring whose land is whose. On this expedition (1872-3), however, the always devious Raja of Manipur would not agree to let him and his party proceed to define the watershed further; such unwillingness perhaps explains why the boundary between Burma and Assam, even today, effectively contours parts of the Barrail Range slopes, rather than runs along either the top of the watershed or along a river at the base of any nearby valley.

---

[4] In this chapter, I continue the convention of putting mountain heights first in feet where the measurement is Godwin-Austen's, and in metres first where taken from modern maps.

[5] *JIB*, entry for 5 January 1873.

The remainder of Godwin-Austen's ascents have proved almost impossible to define. On 2 March, he climbed Nunshigung by himself; on 30 March, he is climbing in the Shiroifurar Range, which I have not been able to locate, although from a reference to one river, it must lie to the south of Imphal, the capital of Manipur. He had to start at 6.30am and described the climb as a '*good long ascent of 2¼ hours*' with a rewarding view. He made the second ascent of Pundengroo (his assistant, Lieutenant Beavan, having made the first a few days before), described only as a '*high scarp*' with 600 '*difficult*' feet to the top. Haversham never defines what he means by '*difficult*', but given that 'Difficult' is one of the lowest rock-climbing grades and that Haversham was not a sophisticated rock-climber, the two 'difficults' may well roughly coincide. Marangthang Peak soon follows, and comes in at a lowly 1,389ft (423m). On 20 February he climbed Munghri and finished his work, then returned to camp without the experience having made much impression on him. He ascended various other spurs and ridges, in a similarly rather casual manner, between that day and the end of the *Journal*'s entries in April.

While the views across large tracts of land with considerable topographical variety are usually rewarding from these high points, sometimes Godwin-Austen gets sufficiently carried away not to have finished his work in daylight:

> I ... managed ... by working like a fiend to get through all the angles by dark or rather until I could not see any more to read them and I made the descent by moonlight.[6]

This dangerous activity is repeated several times, although without any accidents.

One important aspect of the surveying work is worth mentioning: the triumph of science over superstition. On the ascent of Kylas (see Ch 15), the locals try to put him off by declaring that they know of no path to the summit, their way of warning Godwin-Austen away from a mountain where ghosts are said to live. Godwin-Austen, who could be quite brusque and sharp when he felt the occasion needed it, was having none of their nonsense:

> I said a path must be made ... but they shook their heads, and said they would not even cut a stick ... Finding our guides of no use, I sent Beni, one of my chuprasies,[7] with our Bengali interpreter on to ... the next village ... to see what they could effect ... After groping about in the forest, and looking here and there through the trees, we tried a likely-looking spur; up, up we went, the ridge mostly clear and open, and leading in the direction I knew could only be the right one; a rocky bit followed, and then the steady ascent again ... and from the height we had got, I felt convinced we had hit the western flank of the peak ... I pushed as far as a fine open spot, on the brink of a southern cliff. This commanded a splendid view of all I wished to see – the rivers and bheels[8] in the plains showing bright as silver in the evening light. It was with feelings of the very greatest delight and intense satisfaction that ... I returned to camp, having put an end to all the wild superstitious reports of both plain and hill people, that Kylas was a peak not to be scaled – an absurdity it was quite impossible to talk them out of.[9]

So science had won, but only because it was well supported by this very determined Englishman.

---

[6] *JIB*, entry for 8 March 1873.
[7] *Chaprasi* (Hindi), literally a 'badge-wearer', a junior clerk or office-boy; here, more usually, a messenger.
[8] Bheel: flood-pond.
[9] *On the Garo Hills*, p15.

FOOD AND DIET

In such ecologically and politically inhospitable country, it was hardly surprising that finding food for the survey party of several dozen was frequently a problem for its officer in charge, even for supplying only two meals a day. Breakfast was taken any time between early morning and early afternoon; dinner was eaten in the early evening or, if it had been a long day in the field, just before retiring. At least twice, the party came within 24 hours of running out of rice, such a shortage leading to bartering or a degree of arm-twisting in getting the locals to replenish supplies. And on another cheerless occasion, breakfast consisted only of '*a biscuit and water*', hardly enough to sustain the considerable physical hard work required of a surveyor. On arrival at Imphal, the capital of Manipur, Haversham's first duty was to head for the bazaar and buy a few *seers* – a *seer* is just under a kilo, in the weighing system which Haversham used – of rice, and some *goor* (as a substitute for sugar) for his men. *Goor* (now more usually spelt '*gur*'), and also known as 'jaggery', is a concentrated product of cane juice without separation of the molasses and crystals; it can vary from golden to dark brown. In taste, it resembles a heavy-duty Kendal Mint Cake. Dinner was often '*cold fowl*', varied by the occasional partridge or other game bird which Haversham had shot or fish which he had caught. As with the Karakoram expeditions, he does not go into much detail about his or his men's diet, partly, no doubt, through lack of interest (glaciers and jungle being no places for gourmets), and partly because the diet was very monotonous. The only food which Haversham describes in any detail, although unfortunately because he finds it so disgusting, is *sukti*:

> Large quantities of fish ... are dried in the sun and packed in bundles for the hill markets; in this state, it is called 'sukti', and is a favourite article of food of the Khasi and Sinteng. Its smell is naturally very strong when stale and old. The village of Ilampur was a very smelly place: near every house was a platform, with posts all round it, on which were drying hundreds of fish, the larger hanging on strings, the smaller placed on bamboo mats. Nets are spread over the whole to keep off the crows and kites.[10]

It was not just the age or the stench of the fish which was objectionable, or that they had been extracted from still water at a relatively low level, quite close to human habitation, but the fact that they were caught by '*indiscriminate poisoning*' (using a form of naturally occurring chemical used to induce paralysis).

Set against such unpleasantness was a convivial, if very modest, dinner together at the end of a long day:

> Against my advice we pushed on over the ridge instead of halting at the last (clean) water. We therefore did not reach the crest until it had got dark and it was nearly 8 before we got into the village of Diumar. Luckily there was good moonlight and all the disagreement of marching in so late was soon forgotten over a warm fire and some hot soup.[11]

To set against the uninteresting food, there was the local alcohol. One variety is called *mot*. It was made from grinding down superfluous grain into a coarse flour, then fermenting it. On one occasion, Haversham was invited into a village house to take a drink of *azu*. His informed opinion was that, '*it is not bad stuff when it is cleanly made, with an acid taste, made out of fermented rice previously pounded.*'

---

[10] *On the Garo Hills*, p3.
[11] *JIB*, entry for 5 February 1873.

MEDICAL MATTERS

With food being fairly scarce and obtainable by less than hygienic methods, it is hardly surprising that illnesses are referred to frequently, if somewhat vaguely. Haversham's best assistant in 1872-3, Captain Butler, was often described as '*seedy*', a comment on his physiological state rather than his moral character. Haversham succumbed in much the same way, as well as acquiring an occasional cough, although he suffered no serious illness on either the Garo or the Naga expeditions. Several men caught 'fever', that catch-all term which is so hard to define medically. 'Jungle-fever' might have been a form of malaria: although there are no descriptions of convulsive malarial attacks, the sudden onset of fever and a feeling that the spleen is being attacked might suggest this disease. The frequent references to diarrhoea are not necessarily connected to the 'fever': these attacks are just as likely to be related to poor hygiene or infected drinking water. Otherwise, in an area where all the mosquitoes had yet to be discovered and classified, further diagnosis is impossible.

Haversham does, however, experience one most unusual moment of upset:

> One of the Nagas caught one of the pretty little green tree frogs and said that it produced great irritation to the mouth or nostrils after handling it. I unconsciously touched my nose & the irritation was very great, causing a copious discharge for about 20 minutes.[12]

The only other specific ailment he refers to is sunstroke.

OTHER HAZARDS

There were any number of other hazards. Rather like the Karakoram, though to nothing like the same degree, rockfall presented an objective danger to any exposed person, '*A coolie carrying bedding got an ugly knock on the head from a stone that came rolling down the hillside.*'[13]

Perhaps the nastiest (if less objective) threat came from the *pongis*, animal traps usually designed to catch deer. These were pits, camouflaged by undergrowth, which contained sharp, upward-pointing wooden or bamboo spikes with poisoned tips on which any unlucky human or animal could be impaled. Haversham narrowly avoided a particularly unpleasant death when he fell into one of these:

> In going after *(some pea-fowl)* I had a very narrow escape from being spiked in a pitfall made for deer, in a run leading to a patch of cultivation; very fortunately, I slipped in near the edge and fell between the side and the sharpened bamboo stakes ...

A local woman was less fortunate. Haversham continues:

> ... the following season a Garo woman met her death from one, under the following unfortunate circumstances: the Survey Party coming suddenly into a patch of cultivation, the Garos weeding it bolted, as they usually do; this woman, in her fright taking the line of the pitfall, received a spike right through the femoral artery ... and bled to death in a few minutes.[14]

Another major danger was fire, in both forest and huts, at any time. Despite it being a bitterly cold

---

[12] *JIB*, entry for 28 March 1873.
[13] *JIB*, entry for 18 January 1873.
[14] *On the Garo Hills*, pp35-6.

January day, for example, a sepoy firing dry moss on a summit let it get out of control. A thick layer of decayed vegetable matter then caught fire, and spread to other areas where '*controlled*' burning was going on. Fanned by the wind, the fire spread '*headlong*' along the summit ridge (Haversham was at first put out because the smoke prevented him taking an angle on a neighbouring peak) and 'it was only by cutting a deep trench and piling it with earth that we saved the Trigonometrical Station.'[15] (No mention of saving the surrounding countryside.) At that point, a large tree above their campsite (it was not unusual to camp on or near a summit) blazed up and '*we had to bolt hastily to another place*'.

In the villages and towns, the biggest hazard was crime. Usually, this involved no more than desertion by the coolies (perhaps under pressure of violence from the Raja's henchmen), but the occasional murder case was referred to. After one memorable crime, a headman was required to produce some suspects. When they failed to appear, the headman was imprisoned, an awkward state of affairs as the headman's son was one of the suspects, '*as fine a specimen of a Garo man as I have seen.*' The police eventually brought in the suspects after they had given themselves up following pressure to do so from their village. The headman was then released. Haversham never discovered the fate of the men themselves, as he left the area shortly after, although he implies that their punishment was singularly unpleasant. Otherwise, crime seemed to be confined to '*impertinence*' or desertion, with one notable exception (see below).

The only other threatening event coming from humans, apart from the shoot-out described below, took place on the visit to the busy Manipuri capital, Imphal. Haversham had been walking around in the bazaar part of the town. (This in itself is a remarkable place, being run, nowadays, exclusively by 3,000 women.) A stampede occurred. Both buyers and sellers suddenly rushed off; Haversham could not understand why. His surprise was understandable as it turned out that *he* was the cause. '*Some blackguard,*' as he splendidly puts it, '*possibly put up to it by the Raja or his minions had put it about that we were up to mischief.*' He concludes that by buying '*without bullying*' they (the Englishmen) will not come across as the devils which some of '*our Asiatic friends*' have labelled them.

THE CACHAR EARTHQUAKE

By far the most complex hazard, which seemed to combine virtually all the other natural dangers which Haversham faced during the Burma-Assam years, was the strong earthquake in Cachar. Quite apart from reminding us about how cool Godwin-Austen could be when in danger, his measured account demonstrates how the dispassionate scientist, the genuine 'natural historian', is emerging.

> Camp Apalos, 14 Jan, 1869
>
> I must now give you a short account of an earthquake we have had here, and such as I do not care ever to experience again ... I had been working in the office-tent during the day, and on its getting chilly, about half-past four, I shut up, and went into the sleeping-tent for my choga,[16] before going outside with the telescope, which required cleaning; while stopping to take the choga from the bedding, I heard call out, 'An earthquake! An earthquake!' I left the choga and ran out; I had not then felt it myself, but by the time I had got outside the ropes, there was no mistaking the fact, for the ground was beginning to rise and fall tremendously, and at last became so bad that it was with difficulty that I could keep my feet. ——'s children were sitting on the ground, all crying, and this, with the shouting of the

[15] *JIB*, entry for 17 January 1873.

[16] In Hindi, a judge's coat; here, therefore, a kind of overcoat.

servants, and the kicking of the ponies, etc, prevented my hearing any particular sound, save the crashing of the large forest trees that are near our camp; these were tossed about in the wildest way, and one very large one close by came down.

The motion, in addition to the waves that passed by, consisted of a jerking or shaking. Everything upon tables or chairs was thrown off, no two-storied, or even one-storied house of brick could have stood it; here where the houses are of wood or bamboo – a mere frame – it would require a terrible earthquake indeed to throw them over.[17]

This account of the earth's convulsions would be graphic enough in itself, but Godwin-Austen goes a step further: the scientist who sees an opportunity for data recording takes over:

After the motion had been going on for about fifty seconds, I went into the office-tent and got out the chronometer. The intensity of the shock had then passed; in fact this must have occurred about twenty seconds from the time of the first shock, and this did not last more than ten seconds. After I had got out the chronometer, and had looked at the time, the last waves were passing, very like those of a gentle swell at sea. It was a curious sight from here (camp) to see the way in which the wave passed over the forest-clad mountain side, as if the trees were bowed by the passage of a mighty wind. The direction of the motion was from west to east, and it went rumbling and roaring away in the distance.

Haversham, with the kind of detachment which comes, oddly enough, from an unusual capacity for concentration, times the shocks, which he records in a short table, then, still unperturbed, notes the aftershocks:

Some fourteen or fifteen happened between this last and 6am next morning. On the 11th we had a shock at 7.20pm and another at 9.25pm, and several during that night.

He goes on to record two days of further shocks, then mentions a further effect:

On the 13th I went up to Mahadeo, a high peak to the south of this, and when observing there with the theodolite, there was a smart shock. It was very curious to see the bubble of the level moving backwards and forwards long after the shock had ceased to be felt by us ... I was taking vertical angles at the time, and had to wait until the earth had composed itself.

It is not just the geologist-seismologist at work, however; the natural historian gets his turn:

It was curious to notice how the hoolocks[18] in the forest were frightened: many of the shocks continued after it had become dark *(on 14 January, four days after the original earthquake)*; these animals kept calling out as they do in the day time, which is what I have never known them to do, and I have wandered a good deal in those forests.

---

[17] *Earthquake in the Cachar Hills. Extract from Letters of Captain Godwin-Austen. Proceedings of the Royal Geographical Society, Vol 13, No 5 (1868-9), pp370-372*. 'Captain' is a mistake: Godwin-Austen was gazetted major on 18 November 1868.

[18] Hoolock gibbons.

Almost three weeks later, on 1 February 1869, Godwin-Austen sent a second letter to the Royal Geographical Society, in which he reflected on and further described his experience:

> The terrible earthquakes, of which I gave you a short account in my last, seem to have ushered in a period of constant disturbance, which is still going on. Not a day has passed without a shock,[19] we have had one whilst at dinner (7pm) only just now; the earth hardly ever seems firm, a constantly recurring tremor is very perceptible, and very disagreeable. Since my last I have been twice up the Mahadeo Peak to observe with the theodolite, and the instrument was several times thrown out of level, the tremor of the ground, even when not perceptible to the senses, was distinctly shown on the level scale.
>
> The river Berak[20] flowed backwards for an hour, and near Sylhet[21] has been so lessened in depth that boats now navigate with difficulty; the earth opened in many places, swallowing up trees and houses, mud and hot water were thrown out of fissures, large areas have sunk, others been raised. Near Cachar, a village has been left on a slope where a long line of low hills has been formed.

Godwin-Austen's report concludes with an uncharacteristically farcical description:

> During the great shock elephants and horses all went over like ninepins, and I found that I and —— were the only two persons in this camp who kept our feet, and we danced a 'balancez'[22] in front of each other, calling out, 'What is going to happen?'

In his final paragraph, Godwin-Austen, sounding now like a geologist, looks to the future:

> It will be most interesting to find out how the levels of the country have been altered. I feel certain that great changes have taken place in the peaks. I am getting together all the data I can, and expect to have a very interesting account to offer, relative to the area disturbed, and the directions of the waves. We seem to be here in a very unstable part of the earth.

At no point does Godwin-Austen mention fear, or indeed much in the way of any other emotion. Instead, we gain an assessment by a scientist whose objective records transcend his personal emotions, and thereby testify to his bravery.

EXTREME WEATHER

The Cachar earthquake was not of course the only seriously unpleasant natural phenomenon to occur in that part of the world, which is also known for experiencing the heaviest rainfall on the planet. Cherrapongee, now renamed Sohra, in the east Khasi Hills, where Godwin-Austen spent his recess (i.e. from April or May to November or December: in 1867), and again in 1869 (see plate 36, 'outside Colonel McCullogh's bungalow') used to be recognised as the wettest place on Earth.

---

[19] The English teacher in me can only deplore Godwin-Austen's habit of separating main clauses by commas, instead of full stops, colons or semi-colons, even if it does enhance a sentence's descriptive pace.

[20] Now more usually the River Barak, which divides Cachar from Nagaland. It flows from north to south, more or less parallel to the Irrawaddy, before debouching into the Bay of Bengal at Sittwe.

[21] Sylhet is located on the banks of the Surma River, now in north-east Bangladesh. It is surrounded by the Jaintia, Khasi, Cachar and Tripura hills. The city's population is about 500,000.

[22] Not, as far as I know, a formal dance, so much as a sustained swaying movement. The whole description of the earthquake draws on nautical metaphors.

(Now, the inhabitants have great difficulty in obtaining clean water because of topsoil erosion causing pollution.) A paper by a fellow officer,[23] Lieutenant EH Steel, written at the same time as Godwin-Austen was there, makes clear the extremeness of the precipitation:

> The mountain-range at Cherrapoongie (which is a British cantonment[24] and a large native village) attains an elevation of some 4000 feet, and as the rise is exceedingly abrupt, the hillside consequently forms a gigantic natural wall, against which the vapour-laden south-west monsoon strikes during the monsoon months from May to October. This produces an immense rainfall, amounting in some years to as much as six hundred inches: in July 1865, two hundred and eighteen inches of rain fell.

British mountaineers and holidaymakers might be interested to know that the average *annual* rainfall in Fort William, one of our wettest towns, between 1861 and 1987, was 76.2 inches (1934.7 mm). Steel continues:

> This immense amount of water falls over a very small tract, some twelve miles broad; for at Silhet, twenty-nine miles from the foot of the hill southward, the rainfall is one hundred and fifty inches; and at Shillong, twenty miles in the interior of the hills, and 5300 feet above the level of the sea, the fall is only eighty inches.

One might be forgiven for thinking that the prevailing weather conditions, outside the rainy season, would be hot and humid. This is not the case. In December, for example, mist often descended. It '*covered all the lower ground and lay like carded wool*'.[25] On another occasion, it was described as '*a pea soup-like atmosphere*'. And of course, the higher the survey parties climbed, the colder it became: on 30 December, a night-time low of 25.5° Fahrenheit was recorded (at a height of approximately 3,000ft/914m above sea-level). On the same day, the temperature had risen by 3.30pm to only 26°, i.e. six (Fahrenheit) degrees of frost. Frost is mentioned surprisingly often, sometimes as a result of a temperature inversion, '*the main range of the Burrail* (the mountains which form the Brahmaputra-Irrawaddy watershed) *westward was quite white with the frost upon the trees*.'[26] Low temperatures were made more painful by the wind being bitterly cold, enough to, '*leave the work two or three times as my hands became so cold I could not turn the clamping screw*.'[27] (This clamping-screw was needed to fix the paper for the ray drawing to the plane-table.) Naturally, high pressure systems sometimes came along: the weather would change dramatically and could raise morale, '*the air felt beautifully fresh & pure as we ascended the hill ... It turned out to be one of the clearest days it was possible to imagine*.'[28] But as the season advanced, thunder could be heard increasingly often over the hills; the weather gradually became less stable, with storms, rain and thunder combined. (Snowstorms never occurred.) As the season approached its conclusion, rain fell more and more often, and more heavily: on 21, 22 and 23 March, no work at all could be done because, '*it continued to rain on and off for three days*.'[29] Still, the heat of the day had been referred to several times by the end of the season, in April. Combined with the rain, such heat was a sure sign that it was time to go into recess and get all the field notes properly written up and published.

---

[23] *On the Khasia Tribe, Transactions of the Ethnological Society of London*, Vol 7 (1869), pp305-312.
[24] A semi-permanent military base. They were very common in British India.
[25] *JIB*, entry for 24 December 1872.
[26] *JIB*, entry for 5 January1873.
[27] *JIB*, entry for 18 January 1873.
[28] *JIB*, entry for 29 January 1873.
[29] *JIB*, entry for 21-3 March 1873.

ART, HUNTIN', SHOOTIN', FISHIN', SCIENCE

Life was not, however, entirely about pain, suffering and danger. When not being attacked by headhunters or carrying out his work with theodolite, heliotrope and plane-table, how did Godwin-Austen spend the little leisure time left to him? The most important activity for him was sketching. There are any number of *Journal* entries where scenes and people are sketched. It is a pleasurable extension of his professional work. (Plane-table rays are best supplemented by sketches of local topographical or artificial features, in order best to fix them: see the Appendix.) For example:

> I left the dug-out and sat down with the villagers, who soon assembled round me. I made a sketch of an old lady with heavy-laden ears of brass rings, and her likeness being a fair one, I made an impression on them, and had numerous offers from other lookers-on to take their likenesses.[30]

A good instance of art getting the better of suspicion.

Related to his sketching is Haversham's interest in Buddhist temples. Virtually everywhere he visited in southern and central Burma as a young man, and on every stop his regiment made across northern India, Haversham made a beeline for the best local temple. His watercolours (see plates 13 & 14) exemplify the detail and artistic sensitivity he brought to his paintings once he looked them over. It is no wonder that he installed a miniature temple in his garden at Nore: it symbolised and summarised the artistic, creative and spiritual sides of his professional career.

Like an English country gentleman, Haversham enjoyed shooting and fishing. He describes fishing with particular enjoyment:

> About 4 o'clock we reached a fine rapid; this looked so tempting I put up my rod and whipped it for about an hour, taking out about half a dozen nice 'mahasir',[31] running from 3 to 5 pounds.

The shooting side is more complex. Paradoxically, in the light of our contemporary standards, birds were often shot for scientific purposes, with a view to being skinned in order to analyse and classify them during the part of the year when fieldwork became impossible. These few years were Haversham's ornithological heyday: he collected, between India and Burma, 3,750 skins. Birds, most frequently duck or partridge, were also shot for the pot, although sometimes the science and the gastronomy became confused:

> I was very fortunate to obtain, on this day's march, a very fine specimen of a rather rare eagle, *Aquila hasta*. When halt was made for breakfast, I took the measurements, while Lieutenant Beavan made a sketch of it to be coloured in camp. It was then handed over to one of the Khasi coolies to carry, but we never saw it again, for on calling for it the next morning, some of them had cooked it the previous evening, not knowing that it was wanted.[32]

A similar misunderstanding occurred on at least one other occasion. Having asked the natives to find a *Cariornis Blythii*,[33] Haversham tells us, '*the Nagas succeeded in finding one, but thinking I wanted the feathers only, they had, to my utter disgust, picked and eaten it.*'

---

[30] *On the Garo Hills*, p23.
[31] *Mahasir*: now more usually *mahseer*, a substantial fish of the carp family found in lakes and rivers, but with some uncertainties in its taxonomy.
[32] *On the Garo Hills*, p5.
[33] There may be a name in English for this bird, but I have not yet found it.

Birds were killed by employing the beater method which is standard practice in Scottish grouse shooting, only this time Haversham was usually seated on an elephant: *Malacocircus (Layardia) rubiginosus* – a kind of babbler – for example, was shot from such a position. On another occasion, Haversham shot a *Chelidon nipalensis*, the little Himalayan martin, and noticed that this was the first time that the Daflas had seen a bird '*knocked over*' on the wing, they being used to killing only by noosed snares. On another occasion, he was obliged to step back from shooting a small, unfrightened bird, in order to avoid blowing it to pieces at close range.

The scientific results were impressive: at least 18 papers on the area's ornithology came out, as well, no doubt, as much gainful discussion among the many bird collectors in the field, chief among whom were Jerdon, Blyth, the Blanfords, MJ Ogle and Viscount Walden (who jointly owned copies of Jerdon's *Birds of India* with Haversham, annotated by both). Three new species of the genera *Spelleornum, Actixura* and *Pomatorhinus* were discovered, and his various papers listed, described and classified around 600 species, not including common birds. In one paper, Haversham also mentioned that a M Oustalet[34] had shown him his collection at the *Jardin des Plantes* (which also houses Paris's zoo), the only reference to Haversham having been in Paris. We do not know when this meeting took place, but on one of his furloughs during the Assam period seems most likely.

Haversham did not seem to take part in much large game shooting. Indeed, he seems, strange though this may appear in the light of his studied attacks on the area's bird population, more concerned about extinction, as he mentions for example the disappearance of the rhinoceros from a particular area, probably, he suggests, because of excessive '*jooming*'. This concern for the disappearance of a species, albeit an inconsistent one and tending to apply more to mammals than any other creature, is one of many of Haversham's ideas which was ahead of its time.

At one point, however, he recorded shooting a large sambar deer (an endemic species now almost extinct) which was not killed cleanly and had to be finished off by his interpreter using a spear, with the wretched animal's cries '*pitiful to hear*'.

Otherwise, he enjoyed excursions, sometimes by elephant, sometimes by dug-out. When going out for the day, Haversham usually had a particular object in mind, such as viewing one of the many monoliths (an important part of Naga burial culture) which they appear to have transported, perhaps like the Druids at Stonehenge, a considerable distance – a granite block four miles into sandstone territory is mentioned – or the dome-like cairns which housed '*celts*', primitive stone, bronze or iron artefacts. These he usually sketched: the diary is full of charming and detailed drawings of country, people, buildings and objects, just as his earlier Burmese diaries were. Short periods of 'leisure' time were also occupied in writing letters, sometimes of course related to work, but more often than not to his family.

There was something of a break from routine while the party negotiated with the Raja in Imphal. Haversham was invited to stay at Commissioner Thompson's bungalow. After the jungle, it must have seemed like the Garden of Eden:

> This place is prettily kept, nice flowers & kitchen garden where are splendid beetroot & peas, & excellent cabbage, onion, lettuce. Neither asparagus or celery will grow, and strawberries not a success.

The point of spending time in this garden was clearly not so much to grow the kind of 'local produce' now so beloved of contemporary chefs and restaurateurs as to try to create some small corner of the Home Counties which was forever England in this distant outpost of the Empire.

---

[34] Jean-Frédéric Émile Oustalet, (1844-1905) was a French zoologist. He was born in Montbéliard, eastern France. Was there something in the air which made Montbéliard, an unexciting little town near the Swiss border, produce two of France's greatest natural historians (the naturalist-zoologist Cuvier being the other)? Having lived in Montbéliard for a year, memorable mostly for what I got up to outside the town, I am as surprised as anyone.

After admiring the garden, Haversham walked to the polo ground to watch a match; polo is said to have originated in Manipur; the world's oldest polo-ground is still there.

Some days were lost through bad weather, but even Christmas Day was spent working until the evening. Then the survey party sat down to dinner. Haversham drank the health of '*all dear friends at home*' and he, '*silently thought of each in turn*'. After dinner, they played whist: the ever-scrupulous Haversham had copied the rules for 'short' and 'long' whist into the back of his diary. (Not a bad idea: for want of the rules concerning 'Scrabble' to hand, and a good dictionary, I could have saved myself an extended argument one night in an otherwise delightful bothy on South Uist.) The party then enjoyed some '*hot grog*' and, with a touch of world-weariness, Haversham records, '*thus passed another Xmas in India.*'

It was hardly a leisure activity, but it is obvious how sustained, intense and systematic Haversham's passion for scientific observation and collecting had become. The *Journal* is a valuable document in showing us not just a record of the flora and fauna of an obscure area, but also an account of Haversham's metamorphosis into the important, indeed leading, natural historian he would become on his return to the big London science institutions after 1877. Molluscs, shells ('*had a hunt for shells in the damp underwood and got a good number*'), birds, such as *spelaeornus chocolatinus godwinaustenii*, the long-tailed wren-babbler, which are listed systematically on p5 of *On the Garo Hills*), butterflies, amphibia and fish are hunted, observed, described, sketched and classified on, at the very least, 32 of the *Journal*'s 120 pages. (Snakes, which abound in this area – the local environmental information centre lists 34 species on its website – are, rather curiously, never mentioned.) Geological observations are to be found frequently, not just as technical descriptions of the topographical formations but with that 'feel' for the nature of the country, whether it be Kashmir, the Aksai Chin or Nagaland, which makes Godwin-Austen outstanding among explorers. (Perhaps his most surprising geological observation in this area was of '*coal in fine thick beds*', an echo of his father's original contribution to 19th century geology in predicting coal reserves in Kent, thanks to the insight that its strata connected with those of northern France.) It was this passion, and what one might call scientific sensitivity, which turned Godwin-Austen into one of the most egregious collectors of natural history objects, anywhere: as one leading Natural History Museum malacologist has remarked, 'his collection represents the basis of all modern science in this area.'

And as befits a good diary, there is also a personal, if largely unconscious, revelation of character. One very odd moment stands out where Haversham made virtually the only unpleasant mention of his father in the whole of his writing, although the entry was subsequently crossed out. It is, however, worth quoting for its curiosity value, rather than because it tells us anything genuinely important about the father-son relationship:

> When the English letters came in I felt ever so much better, but as I walked out of the bungalow (a dak bungalow,[35] not Thompson's) after my solitary dinner, it did make me laugh as I looked at its tumble down appearance that the heir of Shalford should be living in such a place when not one single member of that family is now or has ever gone through such discomfort and put up in such piggish places – & why is this? Perhaps my father would answer, it is Haversham's wish & all he cares for. But the question if I die here is – why did the father take so little interest in his son?[36] Why did he not take that pride which a father should take to shew his son the estates he is to inherit? That he might in his time take that proper interest in them that his ancestors did before him. No quite the reverse. Every matter connected with the estates & family affairs always a sealed letter to his son, not to

---

[35] Dak bungalow: a government-built rest-house for any through-travellers, particularly mail-carriers.

[36] We know how wrong this is, of course, for example from all the strings Robert Alfred Cloyne pulled on Haversham's behalf after General Godwin's death. (See Chapter 7.)

> be brought forward & much kept dark, & when that son asked for a betterment that would bring his annual income up to £600 a year he refused it. A son that had never cost him a fraction or had received for years and years ever since embarking for India in 1852 a single penny. Whose education but £50 a year while at Sandhurst. £620 for 4 years – 415 coming from - -.[37] Yet after 17 years service in India my moderate request of £500 a year was refused out of the income of an estate in Surrey of 5000 acres in extent.

There is no other entry quite like it anywhere else in Haversham's writing. The important thing is that the entry, while still perfectly legible, is crossed out in red crayon. We therefore know that Haversham (and we know it is he because the same censor's crayon is used later in several places to make corrections in his own handwriting) did not consider this a thought worthy of himself or worth keeping for posterity. I presume he must have been tired or depressed (uncharacteristic though that is of him) when he wrote it, as it is so conspicuously wrong and unfair in so many ways. Haversham always preferred truth to rhetoric.

There were other occasions, however, when he felt the force of his isolated circumstances:

> The gloom of these forest camps was quite overpowering; the want of light, the dark sombre green of the foliage, the stillness of the air, and the rank damp smell of decaying vegetation, all intensified the feeling, the only sound that broke the stillness being the murmur of the stream close by and the croaking of frogs. Yet they are not without great beauty; the jagged, denuded surfaces of the grey limestones are set off by the elegant ferns and palms that form the underwood, and by the moss-grown trunks and roots of the trees.[38]

One can see why Thomas Hardy's lyrical yet dramatic novels of man and nature, both subjected to fateful forces which they cannot fully understand, were making such a strong impression on the Victorian public in England at this time. That Haversham derived great satisfaction from his work, where nature was overcome and systematised, cannot be doubted. The last entry in his diary has his happy last word:

> ... It was a lovely spot for a camp & the temperature & all surroundings most enjoyable, & as I sat looking up the valley I could not help being thankful I could still enjoy such scenes & feel contented with my lot.[39]

In any case, back in Calcutta, Haversham had enjoyed a good social life. It is noticeable, for example, from his *Journal* that his social diary when in Calcutta is full.

LEAVING THE SERVICE

It is regrettable that the chapter cannot be left off with Godwin-Austen in such a sanguine state of mind. By 1876, however, he was adamant that his health was '*broken*' (although he appears to have made a speedy-enough recovery on returning to England), and he appears to have provided professional medical opinion for the Army authorities to back this statement up, with advice, moreover, not to return to Assam. Once more, however, he does return, and suffers from '*spleen and fever*' as '*a consequence of waiting for clear days on peaks 7-8000ft exposed to the full force of*

[37] Why the blank? Was the benefactor General Godwin?
[38] *On the Garo Hills*, p22.
[39] *JIB*, entry for 4 April 1873.

monsoon storms.'[40] He therefore sent in his resignation. At first, the Army hesitated, but on reapplication, and with help from his Mentions in Despatches, the Army conceded:

[Government of India embossed stamp top left]

No.4
A 1876

From, Major HH Godwin-Austen
Dy Supt. Topogsurvey of India

To, Col HL Thuillier CSI[41]
Surveyor General of India
At Calcutta. 18th. Decr 1876

Sir,

I have honour to request your kind consideration of the application contained in this letter. I recommend the same for the sanction of the East of India.

For some time I have been led to consider the advisability of continuing to serve in the Survey Department. I am no longer physically so well fitted for the work, and having suffered consid'ly I cannot now without considerable risk to health and strength be toiling in the sun and climbing steep hills all day. Sir Joseph Fayrer[42] when I last appeared before the Medical Board at the India Office said, 'You have had quite enough of Assam. You should not go there again.' I would still very willingly run the risks after dark on working in the malarious district if any credit, the interests of exploring new country has to be obtained, but the boundary work on which I shall be employed will not give this - and it is notably on the most unhealthy part of the Khasi Hills, or even of the whole valley of Assam.

The late application made by the Curator of the Indian Museum for my services arranging and making a Catalogue of the Birds, and which W Medlicott[43] informs me has been supported by all the Trustees affords an opportunity of my being employed usefully, and will lead to the earlier opening of the Museum.

I therefore solicit that I may be transferred from the Survey Depart. And be permitted to do duty in the Museum here, pending the decision of the East of India. I have to ask permission to remain a few days longer in Calcutta before proceeding to the Khasi Hills to secure the changes of No 6 Topogr. Party.

I have the honour to be,

Sir,
Your most obedient servt
HH Godwin-Austen - Major
Dy Sup. TopgSurvey of India[44]

In 1877, Haversham left India, indeed Asia, for the last time, 25 years after first setting foot there. On his retirement, he was promoted to Lieutenant-Colonel (gazetted 2 March 1878) and departed for a second career, as one of Britain's most exacting natural historians. He began his new life, with characteristic application, by subjecting himself to a six-month scientific method instruction course

[40] *Abstract of the Services* of Lt-Col HH Godwin-Austen, unpublished.

[41] Deputy Surveyor-General of India in 1856, when Waugh proposed naming the recently determined highest mountain in the world, Mount Everest, after George Everest. Succeeded Waugh as Surveyor-General from 1861-77. Companion of (the Order of the) Star of India.

[42] Sir Joseph Fayrer, 1st Baronet (1824-1907) was an English physician noted for his writings on medicine in India.

[43] Should be HB Medlicott?

[44] Unpublished private letter.

(mainly in biology) under no less an authority than 'Darwin's Bulldog', TH Huxley (1825-95), the biologist known for his strong early advocacy of Darwin's Theory of Evolution. It was a course of action which led to Haversham being awarded great honours, some of the highest in their field, but which would be set against much more mixed personal fortunes.

PART FOUR

SCIENTIST, ARTIST, LANDOWNER

CHAPTER SEVENTEEN

JESSIE AND HER SQUIRE

Love, Faith, Death

WHEN GODWIN-AUSTEN RETIRED FROM THE army, more than half of his long life remained. This period of 46 years set much private suffering against public success. It is a moving story: the great satisfaction and honours which his scientific studies brought to him were often undermined by worldly difficulties, mainly to do with finance and health, or emotional troubles, originating either within his immediate family or from the sudden reappearance of ghosts from distant parts of his personal history. The outcome of this struggle was a new spiritual direction, a turning towards Buddhism.

JESSIE ROBINSON

Godwin-Austen married Jessie in 1881 when she was 20 and he was 47. She undoubtedly possessed striking good looks (see plate 41), but gossip, which often goes with marriages reflecting age and class differences, and jealousy of her looks, were to become reasons for an unflattering and misleading later characterisation of Jessie as a spendthrift. In fact, this badmouthing of Jessie was simply the result of muddled thinking – the 'spendthrift' wife was undoubtedly Pauline (see Chapter 14).

The professional life of Jessie's father, John Harding Robinson (1833-90), reached its high point in 1869 at Westminster, when he was appointed Examiner of Standing Orders in the House of Lords, Examiner of Private Bills in the House of Commons, Taxing Master, Head of the Private Bills Office and Clerk to the Chairman of Private Committees. (Two Examiners of Petitions for Private Bills were created by the House of Commons in 1846 to investigate whether petitions for leave to introduce private bills had complied with standing orders relating to private legislation. From 1854 the Examiners were appointed to investigate those petitions for private bills originating in the House of Lords.

One Examiner is appointed by the Speaker of the House of Commons while the other is appointed by the House of Lords. Each Examiner sits separately and usually investigates half of the petitions. Once an Examiner has investigated a petition for a private bill he makes a report to the House of Commons on whether the standing orders have been complied with. The House of Lords receives a certificate. If the Examiner finds that a petition for a private bill has complied with standing orders the bill begins its passage through Parliament. If the Examiner finds that the petition has not complied with standing orders he must report his reasons for this conclusion. The petition is then referred to the Standing Orders Committee in each House, which will decide if the standing orders should be dispensed with and whether the bill can proceed. If an Examiner is uncertain about how a standing order should be interpreted he can make a special report which is referred to the Standing Orders Committees of both Houses for a decision.)

John Harding Robinson had moved to The Square, Godalming, in 1866 when Jessie was probably five; her birth date is uncertain, although the year was 1861. He then moved to Guernsey in 1880, where she and Haversham were married in 1881. (Like the young Haversham, Jessie spent her formative years in the far south-west, in her case at Poughill, just outside Bude, in Cornwall, whereas Haversham had lived for his first four years at Ogwell, near Newton Abbot, Devon.)

Jessie's mother, in Gertrude Jekyll's and the Margaret Watts's ('MSW') *Memoir* is described as 'devoted' and if an anecdote in it, in the present tense, is to be believed, she outlived her daughter. It seems likely that Godwin-Austen would have met the daughter of his Surrey neighbour shortly after his return from India in 1877, since he would have had the leisure to be sociable as he recovered his health, broken by the extreme working conditions he experienced in Burma and Assam.

It is not known how Haversham came to know John Harding Robinson, but being neighbours must have played a part, and Robinson shared some of Godwin-Austen's interests – he was an artist of considerable ability, having had two paintings accepted for the exhibition of the Society of British Artists in 1875-6. He might well have met Haversham at the Surrey Exhibition, where Haversham was on the selection committee. He was also an accomplished and noted rifle shot, having developed this talent through earlier military experience. If his background rise from a 'mean house' in unfashionable Camberwell (south London) to Lincolnshire, Twickenham and Bude sounds rather suburban and provincial, it is worth remembering that even the aristocratic Godwin-Austen, on a number of London visits, wrote from an Ealing address and made not infrequent trips, mainly for health reasons, to St Ives. The two worlds evidently merged at 'Deepdale' in Reigate, where Haversham and Jessie lived out their first three years together after their marriage on Guernsey. Godwin-Austen shortly after inherited the estate, on the death of Robert Alfred Cloyne in 1884.

Given the personal histories of her father and her husband, Jessie would certainly have had to get used to the ideas and practices of living a country life with only occasional trips up to town, and living out the role of a squire's wife. Even allowing for its near-hagiographical and sentimental tone, the memoir published by Gertrude Jekyll (though spelt 'Jekyl' in what appears to be upper-case handwriting) and 'M.S.W.' (Margaret Watts), just after Jessie's death suggests very strongly that despite their age and social differences, Haversham and Jessie's marriage was indeed a happy one. The early part of their relationship resembles Othello's defence of his love for Desdemona:

> She loved me for the dangers I had pass'd
> And I loved her that she did pity them.

The *Memoir* makes this relationship clear, at some length:

> Early in their married life he would tell her about his adventures on the great survey, when, day after day, in deep snow at immense altitudes, or among treacherous hill tribes or open enemies, he would carry his life in his hand – only regarding it as simple and necessary duty. She would write out the story of these adventures, and they remain in her writing – a series of narratives of splendid hardihood and devotion such as should be in the hands of every boy and young man who loves his country and has pride in his countrymen's gallant deeds.

The marriage began with a romantic excursion. A sign of this can be deduced from some mature watercolours at Papplewick Hall showing that rather Romantic view of Lake Geneva near the Château de Chillon, looking east, and another of the Matterhorn from the north, the artist looking over what is probably Zinal (see plates 39 & 40); when combined with Godwin-Austen's 1884 geological paper, *Observations on certain Tertiary Formations at the south base of the Alps, in*

North Italy, originating from a talk first given in 1883 in which he refers to his visit to various places on the Italian side of the Alps '*the previous year*', it seems reasonable to suppose that the Godwin-Austens' honeymoon, most likely in the spring or summer of 1882, was an Alpine grand tour. The Jekyll and 'MSW' *Memoir*, indeed, mentions two Alpine tours but does not date them. The second, however, we do know followed the '*lower Rhone valley below Geneva; then from Turin to Ivrea on the south side of the Alps to Lake Maggiore and Varese*'. When on Lake Maggiore, they stayed with the then well-known bon vivant, explorer and natural historian, Sir George Macleay (1809-91). Given these dates, and the date of the south Alpine geology article, it seems most likely that both trips were undertaken in the early or mid-1880s.

Godwin-Austen's final marriage, unlike his youthful passionate folly of an Islamic marriage to Kudikji, or the continuous emotional pains and financial strains of his marriage to Pauline, was securely and maturely founded on common interests. In Jessie's obituary in the *Surrey Advertiser* she is described (admittedly in the rather overstated style of the time) as 'an ardent lover of nature', with 'acute powers of observation developed by habit', a description vindicated by her article published in *Nature* 45, 416-7 (3 March 1892):

> A Swan's Secret
>
> Now that the breeding-season for birds is coming near, it would be interesting to note if the following sight I saw last spring is common to swans. A pair of swans built on an island on the River Wey, which runs through our grounds, and I stood on the bank opposite their nest, and watched for a view of the cygnets, which were just hatched out. The male bird presently picked up an empty half egg-shell lying beside the nest, and carefully carried it to the edge of the water, some 20 feet from where the nest was built, and proceeded to fill it with mud, and then pushed it into the river, where it sank to the bottom. He then fetched the only other remaining piece of shell, and did the same. On returning to his nest the last time, he placed a few sticks across the small track he had made, as if to conceal his actions. Evidently this process had been done to each piece of shell, as no other pieces were to be seen, although five cygnets were hatched out.

A rather cute incident in the Jekyll *Memoir* confirms Jessie's sharpness of eye, a characteristic she shared with the hare-eyed Haversham:

> Our neighbour, Mr Archibald Thorburn when engaged in the drawings for the coloured plates of a book on British Mammals by Mr JG Millais, wanted a live specimen of the short-tailed field vole ... Riding one day alone, her quick eye caught sight of one of these tiny animals. She flung herself off her horse and contrived to catch it and to tie it up in her handkerchief. All women who are riders know that it is no easy task to mount on a side-saddle without help, and it was something of a further feat ... to do this, and at the same time to take care of a small wild animal that must be kept alive and unhurt.

Jessie's interests in nature extended to horticulture. The garden of their country house at Nore was designed or more correctly, planted, by Gertrude Jekyll, (1843-1932), who had design-planted several gardens in the Hascombe area and who also, with her brother Sir Herbert Jekyll, attended Jessie's funeral at Hascombe. (The Jekyll family were neighbours, having moved to nearby Bramley from London in 1848. Another minor literary Godwin-Austen connection was thereby established – Gertrude Jekyll's younger brother Walter was a friend of Robert Louis Stevenson, who used the unusual surname in the title of his famous novella.) Jessie succeeded at Nore, moreover, in breeding both a pink-tinted primrose and a dark red pansy named 'Red Velvet Nore', thus demonstrating a sense of the kind of '*stickability*' her husband brought to his work as well as an acquaintance with some of Mendel's laws of inheritance.

While at Lake Maggiore, she drew a magnolia which was hung at Pendell Court, the English home of their political ally and fellow natural historian, Sir George Macleay, after which the room where it hung was named 'the Magnolia Room'. (Some of Haversham's watercolours depict shooting scenes on a Scottish landscape which resembles parts of Caithness, where the Macleays came from.) What was, however, most significant about Jessie's love of nature was her ability to 'assist her husband in his scientific work' through her art; she used a microscope and *camera lucida*, an optical device used as a drawing aid by artists. (It performs an optical superimposition of the subject being viewed upon the surface upon which the artist is drawing. The artist sees both scene and drawing surface simultaneously, as in a photographic double exposure, thus allowing the artist to duplicate key points of the scene on the drawing surface as an aid to reproducing perspective accurately. At times, the artist can even trace the outlines of objects.) Jessie became an accomplished painter of both the bloom and seed of wild flowers, a talent probably inherited from her father: some of her work found its way into illustrating Godwin-Austen's books, particularly *The Land and Freshwater Mollusca of India*, although exactly at which points is now impossible to determine. Her training in natural observation also helped Haversham in the sorting of his vast collections.

If we take Jessie's obituary and the *Memoir* at face value, we see that she also demonstrates an ethical sense in her campaign of systematically tackling local butchers in order to promote the humane slaughtering of animals. Her intellectual life extended to membership of the Primrose League, an organisation for spreading Conservative principles in Great Britain. It was founded in 1883 and remained active until being wound up in December 2004. Jessie was an 'earnest and zealous worker for the Conservative cause', a political commitment which required sustained analytical and organisational effort. Not just intellectual effort, either. The *Memoir* describes the following political incident:

> One evening in the Summer of 1909, they *(Haversham and Jessie)* were walking between two local stations in a Western London suburb when they came on a large crowd of nearly all men, who were being addressed by a man mounted on a chair. When they heard, in answer to a question to a policeman, that the speaker was a Socialist, and when Jessie caught the drift of the address, she quickly made her way through the crowd, and, compelling the orator's attention by laying a hand upon his arm, told him he ought to be ashamed of himself!

One wonders if this oddly impulsive and rather rash behaviour gave a sign of the nervous troubles which beset Jessie in her dying months four years later.

The obituary also mentions her interest in local schools, at Shalford and the villages of Grafham and Hascombe; it included her provision of annual treats for the children. This activity should not be elevated to an interest in education but should rather be seen when combined with her other activities as correctly fulfilling her role as mistress of the Godwin-Austen estate and as the wife of the distinguished old scientist. (They were a very prominent family: any articles in the *Surrey Advertiser* about the Godwin-Austens were often headlined, no doubt in recognition of their local distinction, with an unusually large font.)

Those who attended Jessie's funeral (all listed by the newspaper, as was then customary) overlapped substantially with those who attended Godwin-Austen's ten years later, a recognition of the general acceptance of her standing in the local 'squirearchy' despite her lower middle-class origins; she was evidently sufficiently hard-working, charming and upright to have won over local people of the Godwin-Austens' social standing.

It is, moreover, the range of her interests which her obituary describes which perhaps most of all demonstrates her compatibility with her extraordinary, polymath husband. This, then, was a marriage of common and sustained intellectual interests – it lasted 32 years – and was doubtless satisfactorily kept going for these reasons. Three months after Jessie died Godwin-Austen cannot contain his sorrow, as we see in a letter to his friend J Scott Keltie:

> I had to catch a train at Victoria, for home – a desolate house now, & my visit to the Society yesterday was a most sad one – the day I saw you all there last was the Antarctic Night to which I and my wife went, it was the last day she ever went out with me.

No doubt Godwin-Austen would also have been recalling the loss in January that year of Captain Scott and his fellow-explorers – Godwin-Austen knew Scott well (see page 179).

Jessie's physical troubles, which hint at the beginnings of mental problems, are mentioned in a letter which also betrays Godwin-Austen's financial anxieties as early as 1890. Whether her later depression was related to continued financial worry, combined with household removal to Nore and some shuttling between it and Shalford Park, or whether the causes were more purely physical, is not clear. There were, however, happier family moments where the couple's togetherness is casually referred to:

> ... my wife and I went yesterday to see her young brother get his Commission into the R Artillery. He has I am glad to say been posted to a battery at Kurrachee after a 2 months course at Shoeburyness – just what we wanted. He has done well.

That Godwin-Austen was still very fond of Jessie as her death approached is evident from a number of letters describing her final illness, which in Dr Richard Godwin-Austen's opinion might well have been stomach cancer. Haversham's devotion to Jessie is conspicuous. But by May 1913, he is anxiously describing Jessie's condition to one of his scientific friends:

> I found my wife very unwell, indeed quite a breakdown, so I am taking her down to Looe in Cornwall for perfect quiet and rest as she has not been able to sleep for some time.

The plan, however, does not succeed – Jessie is described with increasing concern:

> I was up in the Garden Room at work at 5am when I looked in on my wife and found her asleep ... She is very weak but far better than she was all Saturday. She was regularly near down. The week in town, the horticultural show, a concert, Quo Vadis,[1] and finally the Geographical at the Albert Hall on Wednesday was more than she ought to have attempted. She felt nothing but sea air in Cornwall to set her right and would make the attempt to go. She only got to Cranleigh Station. We were actually in the train, 1 min to start, when she said she couldn't do it. Fortunately our Cart had not left the station. I got her onto it and back here and to bed. She had seen the doctor the previous day. He and I and my sister did our best to persuade her not to attempt the journey, but she would. It was an escape from far greater trouble had she broken down en route. Here she has many friends and perfect quiet.

The part-incoherence of Godwin-Austen's grammar and ideas ('many friends' hardly being compatible with 'quiet', for example) reveals his anxiety. Despite, once more, the strong evidence of shared interests shown by the events 'in town' mentioned at the beginning of this extract, it is now obvious that Jessie's illness is mental as much as physical.

By June, Godwin-Austen tersely observes:

> My wife has not made the progress I should wish. On Saturday she was much

---

[1] Still a fashionable London restaurant.

> better and got into the garden in the afternoon but yesterday there was a turn for the worse. It is so difficult to get her to eat.

Matters have taken a grave and disturbing turn by late June:

> I am sorry to say my wife makes very little progress and is still in a very sad state with all kinds of strange imaginings. She slept better last night than as she expressed it, the last 3 weeks, and was very bright about 9-10 am but it has not lasted and she is again in a depressed state. At times, it is difficult to get her to eat, and great patience is required. I have 2 excellent nurses and I do hope before long to be able to take her somewhere for a change of scene and air.

Jessie died on 22 July 1913, aged 53, exactly four weeks after the observation about her '*strange imaginings*'. It is pointless to speculate on the causes of her depression, particularly given the complete lack of evidence for any direct cause or indeed any precise description of her physical illness. Jessie's depressive and physical decline was a heart-rending ending to what seems to have been an unusually happy marriage; despite the protracted financial problems, one of mutual interest and dependence. In a letter written a week after her death, Godwin-Austen's unhappiness and his devotion to his wife become painfully clear in what is essentially Jessie's obituary:

> I have been going through 10 weeks of dreadful anxiety such that does not come to most men. To witness a dear devoted clever wife gradually passing away and suffering terribly, so that it was a merciful thing when life ended in a sleep – no more worry or the cares of the world to face.
>
> ... She always had a great liking for you *(Sir Aurel Stein)*. To me, her loss is irreparable, she was part and parcel of all I had done in the past & since we were married it stimulated me to work. She was so clever and well-read & beloved of high and low around here, & the letters I am getting astonish me. She was never idle, in fact she overdid it – her body was not strong enough for her active mind – Nore is desolate – it will not be the same to you when you may come here again.

THE EVERYDAY LIFE OF A SURREY SQUIRE

From what we know of the many papers delivered to his scientific societies and to the many meetings Godwin-Austen presided over, with other references to concerts, dining out at the smart Quo Vadis restaurant and so on, we readily form a picture of the rhythms of his life during the period of his return from India until the outbreak of the First World War. The core activity is his daily routine at Nore:

> Here I can begin at 6 & work till 9am, then Breakfast and again in the afternoon. In this way, I have lately ... catalogued and mounted 2123 species and put them in the glass-topped Boxes, and I am taking with me all my specimens of Diplomminatina to Cornwall. While I am there, for there is sure to be many a bad day ... the results will be published. For this sort of work I have a much better room and light than at the Museum.

At Shalford Park, and later at Nore, Haversham led a very busy life. Apart from the scientific papers, offset by walks, he was quite heavily engaged in local political activity: Haversham was a Conservative local councillor; the *Surrey Advertiser* describes him, in its unmistakably partial fashion, as 'adopting a broad-minded view of the national and imperial responsibilities of England.'

Much of his time was taken up by attending meetings of bodies on which Robert Alfred Cloyne had sat: for example, he represented his ailing father on the Shalford School Board and at an agricultural meeting of the Royal Counties' Association held in Shalford Park. He attended parish vestry meetings. These were mainly concerned with church and local land-management business. He was, like his father, a member of the Burial Board. In 1893, he was, 'one of three members representing Shalford on the Joint Committee considering sewerage for Shalford, Bramley and Wonersh.' In the election of 1894, Haversham indicated his unwillingness to stand. He did stand in 1896, however, and came sixth out of nine, having been substantially outvoted by his own head gardener, James Scobie. (The winners were prominent builders, tradesmen and employers, such as George Unwin of the Chilworth printing works.) In the School Board elections of 1887, he came fifth out of five, and fourth in 1890, not, one suspects because he was particularly unpopular, but because the electorate simply did not know him very well.

He took 'a leading part' in the Surrey Art Loan Exhibition – possibly where he met his future father-in-law (see above) – and was 'a valued member of the committee of selection, where his ripe judgement and cultural knowledge were of material service'. He also made frequent visits 'to town', where he usually spent the night. While in Surrey, he evidently enjoyed his role as the grandee among local scientists and landowners: the frequency of his correspondence and prominence at local events with his wife, e.g. supporting his MP from the platform during general election campaigns as well as Primrose League meetings, testify to that.

He was in a position of authority to pronounce on a whole range of subjects. He makes his views known about local access and property rights, for example on the area abutting Chantry Wood (just outside Guildford) which Sir Henry Austen had first planted in the 18th century. He gave a paper in 1884, *On the New Railway-Cutting at Guildford:—The Pleistocene Sands and Drift-Gravels* in the *Quarterly Journal of the Geological Society*, and, in the same year, an archaeological paper *On the Discovery of some Worked Flints, Cores, and Flakes from Blackheath, near Chilworth and Bramley, Surrey*. He regularly attended meetings of the Surrey Archaeological Society, in Guildford; it still meets there. I had the pleasure of addressing the Society on the subject of Godwin-Austen's life in September 2010, and was given an enthusiastic welcome from an audience which had three members of the present Godwin-Austen family in it. He also left to Guildford Museum some Roman funeral urns, still containing incinerated bones, and a statuette of an 18th century Austen – although who is unclear, because of mislabelling – and some bones and shards from Tyting Farm tumulus (on Godwin-Austen land until its sale in 1887): if not a very large collection, at least a testimony to the breadth of Godwin-Austen's archaeological curiosity.

His main preoccupation remained, however, natural history. On this subject, his time was often occupied in correspondence, sometimes on whimsical subjects, as in this case with *The Field* magazine when Godwin-Austen quotes a letter he has received from a Mr Tweedie, probably a member of the family of his solicitors, describing how:

> BIRD CAPTURED BY FRESHWATER MUSSEL
>
> My sons ... were able to get some freshwater mussels in one of which they found the accompanying bird's toe. The mussel was alive, and the tendon attached to the toe was outside.

Godwin-Austen then comments:

> There is no doubt that this is of no infrequent occurrence to wading birds ... The mussel lying on the mud or sand with its valves slightly apart, would close them instantly on anything being introduced between them ... they would effectively hold down many species of birds and the young of others. Nor would there be much chance of escape: for as long as the bird struggled, the mussel, with its strong retractor muscles, keeps its valves closed ... The great interest that surrounds this well-authenticated observation is its connection with the distribution of species ...

> The genus Onio and Anodonta are ... constant over enormous areas ... and here we find an admirable example of how they must ... be carried from one piece of water or one river-system to another ... Not only might the freshwater mussels be thus conveyed but many other genera living with them, such as Limnes, Pianorbis, etc, the species of which ... deposit the ova in a jelly-like mass or in strings upon stones ... I am unable to identify from the toe the species of bird which was so unhappily made prisoner.

It is not just the detail of the comment but the ability correctly to derive general principles from a particular instance which makes the article so remarkable.

Haversham's technical expertise extended to his role as Chairman of the parochial local Sanitary Committee. He was, consequently, obliged to inspect the parish privies, and wrote (for example in 1896 to the *Surrey Advertiser*) on the Shalford drainage question. He remained opposed to water-closets and sewers, and preferred a system of filter-beds on the Common, although this scheme was dropped on financial grounds and because the Common could not be interfered with. These views appear to have been based on a mixture of good science and reluctance to see rate values rise at a time when he was feeling considerable financial pressure. While there is undoubtedly an element of anti-climax in one of England's greatest scientists being reduced to the banalities of lavatory and drain inspection, nonetheless these were serious matters; one only needs to read Ibsen's *An Enemy of the People* (1882), to realise what passions at that time were provoked by the clash between the imperatives of public health and economic necessity. All the relevant correspondence points to the earnestness with which Godwin-Austen carried out his inspector's and other public duties.

While much of the 30-year period leading up to the death of his wife and the outbreak of the First World War a year later was therefore a happy, busy and constructive time, it was overshadowed by increasing concerns about money. How Godwin-Austen tried to stave off bankruptcy, and how he recovered from it, now needs to be looked at in order to get a proper perspective on his private life, before we turn to his amazing range of public honours and distinctions.

CHAPTER EIGHTEEN

MONEY (I)

Financial Strain

IN THE EARLY 1890s, ALTHOUGH at the height of his scientific and intellectual powers, Godwin-Austen's strengths did not extend to a mastery of his finances. These progressively deteriorated: late in 1898, at the age of 64, he was declared bankrupt, although he succeeded in discharging himself from it quickly, by 1902. It is clear that by the time of the bankruptcy he had been feeling the pinch for almost a decade, and equally evident that from then on, financial strain remained, without a break, until very shortly before his death. There are complex reasons for this momentous loss of much of the Austen estate(s).

In an obituary,[1] an 'unfortunately worded will' which caused 'financial embarrassment' in Haversham's later years is referred to, but nothing to the present Godwin-Austen family's knowledge or mine has survived to support this claim. It may refer to the entail resulting from Sir Henry Edmund's youthful marriage in 1805 when only 20 and therefore below the legal age when he could act without the estate trustees' consent. This entail appears to have limited the properties which could then be sold in order to relieve financial pressures, and could therefore be a contributory factor in the attempts to stave off the bankruptcy.

The least significant 'cause' of the bankruptcy originates in an anti-feminist stereotype.

THE SISTERS

This 'cause' is known simply in the family as 'the sisters'. This Chekhovian quartet - Helena, Beatrice, Alice and Elfrida - remained spinsters. (Their combined ages totalled 359 years.) They produced no earned income, although they would have received some (it is not known how much) from investment dividends. To have to maintain four dependants would certainly have put pressure on the estate but it was hardly sufficient - there are no tales of their extravagance, though there are some of their generosity, not least Beatrice's looking after Haversham during the last decade of his life - to cause its ruin.

THE CAUSES

In considering the genuine causes of Godwin-Austen's bankruptcy, it has been asserted that he did

[1] *Nature 122*, 946, 29 December 1923.

not care about money;[2] it might be more accurate to say that Godwin-Austen was not much interested in money but that he was quite careful about where it came from and where it went. The implication is that financial negligence through too much immersion in scientific research caused the bankruptcy by his failing to keep track of the estate accounts. In fact, Godwin-Austen's Burmese *Journal* and late letters both show a keen interest in precise account-keeping. Letters to Robert Alfred Cloyne sometimes cover financial matters, also. During his early years in India, particularly after his marriage to Pauline in 1861, moreover, Godwin-Austen writes quite insistent letters (somewhat greedily recorded by Phillimore[3]) to the Indian Army authorities concerning his pay being inadequately suited to his household and marriage expenses (i.e. the marriage to Pauline, and, presumably, funds to the Milners for Edward's upbringing). There is no reason to suppose that during his third marriage he would have neglected his interest in his finances, acquired perhaps during his second one, although with a young bride to spoil, and a newly-inherited estate in 1884, for a time he might have become more indulgent.

After about 1893, the overwhelming majority of Godwin-Austen's surviving letters expresses not only one anxiety or another about his financial situation but also shows him thinking ahead and trying to resolve the problems. The only argument which might counterbalance this evidence is that Godwin-Austen was known for his kindly and generous disposition towards his dependants: he looked after his four unmarried sisters apparently without question and he is said to have gone on paying servants and keeping his tenants' rents down when he could no longer afford to do so, a character trait consistent with his treatment of his 'coolies' in Baltistan.

And we know that some expenditure must have gone on maintaining his position as local squire. In 1897, for example, he was chairman of the committee for organising Queen Victoria's diamond jubilee celebrations and, 'bore part of the expenses for a dinner held at the village hall.'[4] Afterwards, he accompanied a crowd to Chinthurst Hill, just north of Bramley and Wonersh, where one of hundreds of bonfires across the nation was lit.

The most powerful cause, historically, of the bankruptcy was the Great Agricultural Depression. After unprecedented profits in the 1870s, agricultural bankruptcies began to increase rapidly across the chalk-land and clay farms of southern England. While Surrey resisted the trend in the 1880s, hardship increased from the late 1880s into the 1890s through raised rents and static income for the tenants and farm workers, who left their homes and posts to find better circumstances in the towns or abroad, the effect of this being to decrease income for the landowners.

This recessionary spiral occurred because of improved global food production and technological improvements in its distribution. For example in 1876-7, the French engineer Charles Tellier and the steamship *Frigorifique* completed the first overseas shipment of meat under artificial refrigeration. Three methyl-ether refrigerating machines kept the cargo in a chilled state during the 12,000-kilometre (7,456-mile) voyage from France to Argentina, and indeed the return trip. In 1879-80 the *Strathleven*, equipped with a Bell-Coleman air machine and loaded with beef, mutton, butter and kegs, sailed from Melbourne to London and arrived with the frozen cargo in good condition after a nine-week voyage of about 24,000 kilometres (14,913 miles): the era of refrigerated transport of otherwise perishable food had begun. Argentinian and Antipodean prices could undercut British prices very substantially; to the farmers' loss of rental income was therefore added loss of income from sales. English memories of arable land cereal production were revived, until the completion of the trans-continental American railways undercut that hope when cheaper American cereals came to be imported from the mid-West. The hardship caused was unsentimentally and accurately recorded by the Surrey watercolourist WH Allen (1863-1943).

[2] Richard Godwin-Austen: *Seizing Opportunities* p15.
[3] Phillimore, *Survey of India*, Vol V.
[4] Letter from Margaret Dierden to the author, 26 April 2010. I am much indebted to Margaret Dierden for all the detail she has supplied concerning Haversham's role as local squire during this period (and in particular in Chapter 17).

ATTEMPTS TO AVOID BANKRUPTCY

Godwin-Austen was a practical, resolute man. When his financial affairs worsened and bankruptcy threatened, he took action. With a nod to Julius Caesar's tactics of 'divide and conquer' he decided as early as 1884, as soon as he inherited the estate, to sell off some presumably unentailed agricultural land around the Common, possibly around 80 acres. (Ten acres were bought by local builder Edward Mitchell for £1210.) Tyting Farm, just outside Guildford, a source of Roman pottery burial urns and their charred, bony contents now in Guildford Museum, was sold in 1887 to the Duke of Northumberland for £12,600.[5]

By 1890 (by which time, perversely enough, it might already have been too late to stave off the bankruptcy eight years hence) Godwin-Austen was hoping to try to sell or rent Shalford Park and remove to Nore, the very substantial country house south-east of Godalming which had been an Austen property since the 16th century. (The beams in the upper part of the house were said, with good reason, to pre-date the Armada.) Next, he would try to sell his vast collection of bird skins, molluscs and shells. Some insects were also added. Finally, he would earn extra money by doing the classificatory work for his collection and for those of others. Obtaining due payment for this work, originally intended for the ever-expanding British Museum, would engage Godwin-Austen in 25 years of niggling correspondence with the British Museum's new branch, the Natural History Museum in South Kensington. It was opened in 1881, although the Natural History collection's move was not completed until 1883. This shambolic event, a worthy descendant of so many previous eccentricities and follies in the museum's history, provoked an exasperated outburst from Godwin-Austen, '*What a disgraceful state for a new building to be in, to require repairs … before occupation*'.[6]

Although his foresight was not enough to save him, as early as 1890 Godwin-Austen could see his likely financial difficulties,[7] '*The farming prospect is so bad that I must look ahead to check* (?) *the heavy outlay it entails for I do not see how it is to come back*'.

In the same letter, he discusses, with some pain, the sale of his beloved collections:

> It is a great wrench to part with collections, every specimen of which has its own pleasant memories of days gone by in the mountains & forests of India and of the men one has worked with, but I must get rid of my collection of Bird ... My collection contains about 4000 skins, with many unique specimens and types obtained from very inaccessible parts of the Eastern Frontier ... I cannot afford to give it to the BM Trustees, if I were well off it would be disposed of in this way. If they can purchase it, I make them the offer of it ... Failing the BM, do you know any other Museum where it might be wanted?

This last question betrays an already acute, if understated, anxiety about his financial situation.

---

[5] I am, as ever, most grateful to Margaret Dierden for supplying such detailed and helpful information relating to these land-sales.

[6] Letter to Günther, 8 January 1883. Albrecht Karl Ludwig Gotthilf Günther FRS (also *Albert Charles Lewis Gotthilf Gunther*), 1830-1914, was a German-born British zoologist, born in Esslingen in Swabia (Württemberg). He studied theology in Bonn and Berlin and later studied medicine in Tübingen. He went to the British Museum in 1857, where he worked on ichthyology. In 1875, Günther was appointed Keeper of Zoology at the Natural History Museum, a position he held until 1895. The major work of his life was his eight volume *Catalogue of Fishes* (1859-70). He also worked on the reptiles and amphibians in the Museum collection. Günther was elected Fellow of the Royal Society in 1867 and served as Vice-President 1875-6. He served on the Council of the Zoological Society for nearly forty years (1868-1905). He was elected Fellow of the Linnaean Society in 1877, and was President 1896-1900. He became a naturalized British citizen in 1874. Günther died at Kew Gardens (then in Surrey) on 1 February 1914.

[7] Letter to Flower, 8 July 1890. Sir William Henry Flower KCB, FRCS, FRS, (1831-99) was an English comparative anatomist and surgeon. Flower became a leading authority on mammals. He eventually became Director of the Natural History Museum.

The terse reply on the letter's front page, written in official red in a crabbed hand, is, '*Declined. No money for purchasing birds at present*'.

Godwin-Austen's second tactic was being thought about by 1894:[8]

> I am obliged to let my house at Shalford Park near Guildford and may have to leave it very shortly. I wish to remove into some safe quarter my collection of Indian Land Mollusca. It is my desire that this collection should eventually be incorporated in the National Collection ...

It was not just the letting of Shalford Park, perhaps even its eventual sale, which concerned him, but that he could not, under the present circumstances, work there properly:

> I have found that when Shalford House has been let even for a short time that it is most inconvenient to get access to the collection and work becomes stopped, while there is always a danger of its being meddled with.

Godwin-Austen must have been disappointed by Günther's rather starchy reply:[9]

> ... it is not our usual practice to receive private property on deposit, the available space being very limited and becoming more so from year to year ...

... although he then grudgingly added, '*I am glad to be able to assist you in finding a place of safety for it*'.

Günther went on to specify four characteristically tight conditions for his acceptance: the collection would remain Godwin-Austen's private property; no-one, other than Godwin-Austen, the Director of the Museum, Günther himself or Mr Edgar Smith[10] would have access to it; the BM and its Trustees were to be absolved from all responsibility for the safety of the collection; the Keeper of the Zoological Department might require Godwin-Austen at any time to remove it. This response to the offer of the greatest collection of *Mollusca* hitherto assembled, anywhere, is, to say the least, rather ungenerous.

THE SALE OF THE BIRD-SKINS

In the following year (1895), Godwin-Austen tried a different tack and reverted to his original plan of selling his extraordinary collection of bird skins. By that time, he had catalogued them carefully. There were 3,731 birds, representing 592 species (although, for unexplained, perhaps sentimental reasons, a collection of skins from the Mishmi Hills, a south-easterly but sub-tropical branch of the Himalaya (in terms of height and vegetation cover) in Arunachal Pradesh, north-east India, was not included).[11] He described the basis for the collection:

> When I was doing duty in the Indian Museum, Calcutta[12] ... all birds (which were) drying at Jamracks[13] place were brought to me every morning and anything

[8] Letter to Günther (addressed only as 'Sir', but the reply is from him), 14 November 1894.
[9] Letter from Günther to Godwin-Austen, 16 November 1894.
[10] Edgar Smith, 1847-1916, Curator of Molluscs at the Natural History Museum.
[11] Letter from Godwin-Austen to Günther, 9 January 1895.
[12] Godwin-Austen assisted Dr John Anderson, FRS, (see n20 below) in cataloguing the bird collection for the new Indian National Museum in Calcutta, in 1876-7, just before leaving the Indian army to return to England.
[13] Jamrack appears to have been a well-known sea captain in Calcutta who traded informally in a number of exotic objects.

> interesting I had purchased, & was not wanted by the Museum, I retained for myself. Also by visiting the Calcutta market every morning I got a good many species brought in by the birdcatchers from the Sunderbunds.[14]

Financial difficulties had become so pressing, however, that even later the same month he was obliged to write again on the same subject to Günther:[15]

> I should myself like to know with more certainty what is to become of the collection and what I may rely on getting for it within the next year or two. With a diminishing income from rents and farm produce I must try and make it up somehow.

Günther's cautious reply suggests that an arrangement was being discussed. Late in February 1895, Godwin-Austen despatched '*16 cases of Birds*' from Shalford Station by goods train. While packing the collection he came across '*a lot of insects collected in the Khasi Hills:*[16] *11 insect cases full, principally* Hymenoptera.'[17] It is a mark of his utterly exceptional collecting record, as well as his absentmindedness, that such a substantial collection by the first western explorer of northern Burma was mentioned in not much more than an aside.

Some wrangling over payment must have followed. Günther and the museum's trustees were perhaps intent on striking a hard bargain. Godwin-Austen's final arrangement in this phase of his life, to the discredit of the trustees who, whatever their financial constraints, seem singularly to have failed to recognise the outstanding importance of what was being offered to them, saw him reduced to not much more than genteel beggary. In a moving and pitiful letter, Godwin-Austen attempted to extract any little dignity he could from a hopeless bargaining position:[18]

> The price you offer me is so much below what I anticipated I would get. You said, if I remember rightly, the Trustees had paid 3/- a skin all round to Shelley[19] ... Walking back from the Museum with Anderson,[20] the conversation turned to the value of skins all round. I said I thought 2/- was a fair price, but he said he thought that low and he has had considerable experience ... There are more than 4000 specimens in my collection but assuming that is the number, the price you offer is only 1/- per skin. Is not this putting it as a minimum one? ... I think if you could see your way to giving me 1/6 a skin or £300,[21] assuming the number to be 4000, it would be nearer the value of the collection, which although it gave me a great deal of pleasure to bring together, cost me a good deal more than £300 and a considerable amount of time and hard work, in a trying climate.

'Trying' is a magnificent understatement, even by Godwin-Austen's tight-lipped standards. The exceptionally high humidity in the Burmese and Assamese jungles helped to produce the infections which 'broke' Godwin-Austen's health in the 1870s.

---

[14] The Sunderbunds were the great marshy areas lying just outside Calcutta (now Kolkata).
[15] Letter to Günther, 22 January 1895.
[16] Jungly hill range on the border between northern Burma and Assam, first explored and surveyed by Godwin-Austen in the 1870s.
[17] Letter to Günther, 27 February 1895.
[18] Letter to Günther, 14 May 1895. Günther is now being addressed as 'Dear' rather than 'My Dear' as in all previous correspondence.
[19] Captain George Ernest Shelley (1840-1910), English geologist and ornithologist and nephew of the poet.
[20] John Anderson FRS (1833-1900) was born in Edinburgh and graduated in medicine from the University of Edinburgh in 1861. From 1864, during his time in Calcutta, he supported the creation of its zoological garden. He become the first curator of the Indian Museum at Calcutta in 1865 and held that position until 1887.
[21] Between £25,500 and £32,500 at 2013 values.

THE BANKRUPTCY

Like many farmers during that period of the Great Agricultural Depression, Godwin-Austen was held in the grip of historic forces larger than he could properly understand. He filed for bankruptcy on 28 October 1898 'on the debtor's own petition', with unsecured debts of between £296,000 and £348,000 at 2013 values.[22] Debts to fully secured creditors were ten times that amount. Godwin-Austen's assets amounted to £3,100 (only a little less than the unsecured debts) in farming stock and growing crops of £950; household furniture and effects, £3,030; a 'collection of shells' worth £1,481 18s 6d; 'book debts' expected to realize £154 4s 10d and retired pay from the Indian civil service of £185 4s 10d (between £16,400 and £20,000 at 2013 values). Trustees were appointed and further creditor meetings arranged, all reported, no doubt to the distress of the bankrupt and his wife, in the *Surrey Advertiser* under the pathetic heading, 'Failure of Col Godwin-Austen.'[23] It was agreed that Godwin-Austen's pay would not be touched and that he would be allowed 'four guineas a week' until his debts were discharged; this was successfully achieved on 3 January 1902.[24]

A twist in the story occurs in 1899, however, with the intervention by letter at a creditors' meeting of Arthur, Godwin-Austen's son and heir, with the startling claim that his father owed him '£30,000'. At a meeting on 7 October 1899, Arthur is said to have 'met with a serious accident quite recently' and was therefore not able to pursue his claim in person; the hearing was postponed for three months (and not subsequently reported). Given the general integrity of Godwin-Austen's character, it stretches credulity to believe that this was some sort of device plotted by father and son in order to distance themselves from the creditors. That it should be seen as an outrageous attempt at profiteering by Arthur, while less unlikely, is nonetheless not very plausible as the sum is so unrealistically large, despite the recent sale of Shalford Park's contents. Some kind of vengeful attack by Arthur also appears unlikely: relations between father and son were undoubtedly sometimes very uncomfortable, but discomfort hardly warrants this kind of revenge. The remaining – and to my mind most likely possibility – was that this was a muddled, indeed befuddled attempt by Arthur to protect his father from his creditors, a loyal but entirely unethical action. In any case, it produced no result: there is no further mention of Arthur's strange claim which, once the bankruptcy had been discharged, became irrelevant.

THE DEMISE OF SHALFORD PARK

Shalford Park's magnificent contents had to be sold. The house had several halls and landings, 27 bedrooms, a boudoir, an armoury, several dressing-rooms, several drawing-rooms, a music room, all with many splendid contents. The library, for example, had a fine collection, although the librarian-collector's taste was fairly unadventurous, running through Metastasio, Dante, Milton, Butler, Byron, Scott, Johnson, Evelyn, Goldsmith, Gibbon, Burke (in their order in the catalogue) as well as a number of geological, zoological, natural history and geographical treatises. Works on drama were generally lacking: Shakespeare is not mentioned. The other rooms were a pantry, a harness room, a stable, a servants' hall and servants' bedrooms.

The sale took place from 21 to 27 April 1899. Some items from the auction catalogue (see plate 9) hold the attention even on the title page, where expensive pieces such as 'grand pianoforte', or 'painted Sheraton satinwood suite' or 'about 750 ounces of old silver plate' are given due prominence. Nine hundred and fifty individual items are listed, such as a 'capital 4ft 3in mahogany tray-top washstand' or 'half-length portrait of a Venetian lady in striped dress and pearl ornaments, by Veronese' or, its excellence drawn attention to by its being printed in gothic script, 'very

---

[22] Sources: www.thisismoney.co.uk and www.measuringworth.com

[23] *Surrey Advertiser*, 29 October 1898, p5. 'Failure' was the newspaper's euphemism for 'bankruptcy'.

[24] Source: *London Gazette*.

handsome 6ft inlaid Louis XV writing table, beautifully decorated with elegantly inlaid brass mountings and fitted with 3 drawers'.[25] Certain items were as a condition of sale to be preserved, such as the dining-room fireplace, the ceiling and doors in the drawing-room, the panelling and armorial fireplace in the oak room, the fireplace in the Tudor Room and the armorial stained glass throughout the house. When Shalford Park was demolished in 1967-8, some of the items were preserved by the Godwin-Austens who were by then in a position to buy back some of what had originally been sold off.

The press were presumably excluded from the proceedings as no local paper carries an account of the event. What is much more interesting, however, is Haversham's detached attitude to the sale – or perhaps he could not bear to be present. On the first day of the sale, he spent his time writing from Nore, several miles away, to the editor of the local newspaper:[26]

> Sir, All lovers of nature will, I am sure, deplore with me the recent destruction of animal and bird life witnessed as constantly going on around us and which is surely and rapidly wiping out what is left of the old fauna.

He then goes on to describe various examples of wanton destruction by hunters and poachers of the local fauna, particularly in relation to mother birds and their young, and to badgers, before concluding:

> One of the saddest things after a visit to the Zoological Gardens at the Regent's Park is the thought that in 50 to 100 years so many of the beautiful animals and birds there will have ceased to exist.

While this is a characteristically perceptive remark, and a shrewd assessment of future trends, it is a strange subject on which to dwell during the opening day of the sale of the contents of the ancestral home. It was not, however, an isolated action. While some sources say it happened during the bankruptcy hearing and others claim it occurred during the Shalford auction, it is agreed that at some point, Haversham reached into his back pocket for a matchbox and took out a shell (of, one assumes, one of the molluscs on which he had been working), then declared, '*this shell is worth more to me than all the money I owe*', a remark which attests to his commitment to scientific truth while hardly endearing himself to his creditors.

After the sale, Shalford Park remained empty for a while (Haversham and Jessie having already removed to Nore). In 1907-8, it was rented out to a group of clergymen and in 1910 it became a hotel (not unlike the history of Conan Doyle's Undershaw which was not far away). It became the property of Guildford Borough Council some time before 1938 and from then until 1955 was rented to the Cornhill Insurance Company as their wartime headquarters. In 1955, it was leased as a furniture store, although by then it was said to be damp and in need of repair. It was almost completely demolished in 1967 and this sad part of its long history completed by its having to make way for the River Wey Water Abstraction Scheme. A small plaque behind St Mary's church in Shalford marks its former site.

With such splendid objects having been sold, it was not surprising that the bankruptcy was discharged so quickly. Its effects, however, were to haunt Godwin-Austen for the remaining 20 years of his life.,

---

[25] Shalford Park Sale Catalogue, published by Alfred Mellersh, Guildford; printed by Stedman of Godalming.
[26] Letter to the editor of the *Surrey Advertiser*, 27 April 1899, published 1 May 1899.

CHAPTER NINETEEN

MONEY (II)

The Aftermath

BY 1905 THE PRESIDENT OF the Alpine Club, George Forrest Browne, Bishop of Bristol, was bemoaning the fact that Godwin-Austen was 'less often seen in town'; the self-disciplined discharging of his bankruptcy undoubtedly imposed considerable travel restrictions, as did the likely decline in his wife's health. And the First World War, of course, necessarily restricted Godwin-Austen to more local than national or international travel.

Godwin-Austen's next tactic was to restructure his property assets. After the sale of Shalford Park's contents, forced on him after the bankruptcy hearings got under way in 1899, it would have seemed sensible to attempt to let Shalford Park indefinitely because it was much too large, and by all accounts too cold, for a couple, Arthur having joined the army, and of course too expensive to maintain its large grounds and the numbers of servants required for its upkeep. Like Sir Walter Elliot in *Persuasion*, Godwin-Austen was forced into acknowledging that he must 'retrench', and consider a move to a less expensive property still commensurate with his social standing.

NORE

The first time that Godwin-Austen had used a letterhead from Nore was 1890, such use becoming more frequent after 1894. The local newspaper suggests that he finally moved there around 1900-1. It was probably a good move. Nore was the oldest of the surviving Godwin-Austen properties. It lies about eight kilometres (5 miles) south-south-east of Godalming, on the quite steep scarp of one of the Surrey Hills. It possesses the advantages of remoteness and some of the drawbacks of isolation. A later, not very accurate, article in the *Daily Telegraph*[1] described it as:

> Set in ten acres, it was more of an estate, with two cottages and a studio in addition to the house itself, which boasted ten bedrooms[2] and eight bathrooms. Nore was certainly conducive to inspiration. The peg-tiled house is set in eight acres of grounds on southwards-sloping land looking towards the South Downs. It also has an outdoor swimming pool and a rather moth-eaten tennis court.
>
> Although parts of the house are a little tired now, its chief glory is its dining hall, parts of which are believed to date from the 12th century. Nore has the

---

[1] I have edited out the inaccuracies. *Daily Telegraph*, 25 July 2004.
[2] The house is now two separate dwellings.

> distinction of being where Captain Scott spent his last night in England before boarding ship for his fatal mission to the South Pole: the house was then owned by friends of his family.

(The 'rather tired' remark is no longer relevant: under its present owners, Mr and Mrs David Kemp, both house and garden have been elegantly modernised.)

One of the pleasures of life at Nore was working in the downstairs Garden Room, which had space for microscopes, collecting boxes and other paraphernalia. It looks south over the most beautifully designed part of the garden then across the leafy woods, rolling hills and open fields of Surrey and Sussex – an ideal workplace for a great natural historian. Time and again, Godwin-Austen refers to some thought having occurred to him, or some significant piece of work having been achieved in the Garden Room.

As regards the Captain Scott story, there is no independent confirmation of it, but there are two very strong pieces of circumstantial evidence. Scott and Godwin-Austen knew each other well enough, through the Royal Geographical Society. Just before Scott's main Polar Expedition left, Godwin-Austen recorded his regret at missing 'the Luncheon to Capt Scott and his officers', adding that, 'I could not get away: I have business in Guildford to attend to before I go off to St Ives, Cornwall, on Thursday.'[3] (Scott's expedition, on the *Terra Nova*, left Cardiff on 15 June 1910. Scott, having stayed on in England to complete fundraising and other business before taking a faster ship to overtake the main party, followed them a couple of months later to rejoin his expedition once it had reached Cape Town.) And in the *Memoir* of Jessie, published by Gertrude Jekyll and 'MSW'[4] in 1913, the following rather lengthy comment is well worth reproducing:

> An opportunity for an act of such kind consideration (*as a general trait in Jessie's character*) occurred in connection with the Scott Antarctic Expedition. Colonel Godwin-Austen and his wife were both interested in Mr Brooke,[5] a promising and clever young fellow, who went out to explore far Western China. He had been to Nore to talk over his plans, and they were in correspondence with him when the news came of his murder by the Lolos.[6] When his travelling companion, Mr Meares,[7] came home, he too came to Nore to tell the sad story. Meares's love of adventure led him to volunteer for the Antarctic, and he became one of that noble crew, under Captain Scott, of the "Terra Nova". Just before they sailed Jessie packed up a box of things to cheer their Christmas and adorn their table. Some months afterwards, to her surprise and pleasure, the thanks of the ship's officers and crew came to Nore in the shape of a post-card, bearing on one side a picture of the "Terra Nova" with Captain Scott's portrait inset, and on the other the whole of their signatures. To some of the names – of those who so nobly gave their lives, after terrible suffering, Jessie afterwards put a little cross. It so happened that the last effort of her life ... was going with her husband to hear Commander Evans, at the Albert Hall, relate the story of the voyage and journey to the South Pole. At the conclusion they went upon the platform to welcome Mr Meares and invite him to come again to Nore. It was all too much for Jessie; it was her last look at the outer world ... She returned home with shattered nerves, never to go away again.[8]

---

[3] Letter to Fagan, 31 May 1910.

[4] MSW was Margaret Watts, the wife of George Watts, an ancestor of the present owner of Watts Gallery, near Guildford. I am indebted to Ann Laver for this piece of sleuthing.

[5] John Weston Brooke FRGS (1880-1908) was a military officer and explorer born at Fenay Hall near Huddersfield, Yorkshire.

[6] Lolos: the Yi people, the older name 'Lolo' or 'Luoluo' now being considered derogatory in China, though used officially in Vietnam as *Lô Lô* and in Thailand as Lolo, are an ethnic group in China, Vietnam, and Thailand. They live mainly in Yunnan, Guizhou, and Guangxi, usually in mountainous regions. (Adapted from Wikipedia.)

[7] Cecil Henry Meares (1877-1937): lieutenant-colonel, linguist and Antarctic explorer.

[8] *Memoir*, pp11-12.

The *Memoir* contains an image of the moving postcard sent by the whole expedition to Haversham, postmarked 29 November 1910, from Dunedin. On it Jessie has marked crosses against the signatures of Scott, Wilson, Bowers and Oates (see plate 42).

The most famous owner of Nore after it passed from Godwin-Austen hands was Dirk Bogarde, who bought the place in 1962 for what seems the unbelievable sum of £38,000.[9] His guests at Nore included Ingrid Bergman and Judy Garland, who in 1965 was acting in 'A Month in the Country' at the newly opened and still flourishing Yvonne Arnaud Theatre in Guildford. It was Bogarde who rediscovered the strange but important little stone pagoda in a patch of brambles while clearing the garden. His ribald remark about it, ' ... a rather curious, and very detailed phallic symbol standing erect in the very center! *(sic)* So I am not absolutely certain that it was only spirits who went there to worship ... '[10] could hardly have been further from its original intended purpose. The little pagoda,[11] – which contains several phallic parts – is undoubtedly in the Burmese style, though not necessarily of Burmese origin, and is a central piece of evidence, discussed in Chapter 24, for the unusual spiritual journey Godwin-Austen was to set out on during his final years.

Still, he enjoyed other sources of satisfaction: in 1906, for example, he seems to have made a trip, perhaps one of several, to Scotland, to the Orkney Islands, as suggested by a dated watercolour at Papplewick of the mighty Standing Stones of Stenness. Other watercolours, regrettably undated, suggest that he might have taken in parts of west or possibly the north-eastern corner of Scotland on this journey or others.

SCIENCE FOR SALE

In 1910, with financial strain still showing, Godwin-Austen tried a new tactic. Having made some headway in getting the British Museum to accept his collections, he now attempted to negotiate an annual grant from the museum in return for him classifying and cataloguing his (and others') collections. His idea seemed to be to negotiate a fixed amount per annum, then per month, out of which he would bill the museum for his expenses and fees for work done. He acknowledged to Harmer '*the promptness with which you acted on my letter to Fagan.*' He followed this up with a recognition of, '*what you propose for work at the Museum. This I think is very liberal, and would meet travelling expenses to town and occasionally staying there.*'[12] The second aspect of this plan was then revealed: '*By this means the Blanford*[13] *collection of Indian shells could be put into order.*'

He was taking on the work of classifying others' collections: in the Victorian (or by now, Edwardian) world of obsessive collecting this was a never-ending task, a kind of milch-cow producing rather watery milk, though also one eminently suited to Godwin-Austen's vast scholarship and capacity for hard work. This was a shrewd move, as a letter the following year made clear:[14]

> I am glad to hear from you that the Treasury have sanctioned the expenditure of £75[15] for taking over sketching and arranging of shells in the late Col Beddome's[16] collection. That is a good thing settled.

However, the obvious hitches produced by an inefficient or reluctant bureaucracy soon appeared:

---

[9] Current value £713,640. Source: www.thisismoney.co.uk.
[10] Private letter by Dirk Bogarde to Jean Lion, 23 March 1980. Source: typed MS at Nore.
[11] See plate 44.
[12] Letter to Harmer, from 14, The Terrace, St Ives, 7 June 1910.
[13] WT Blanford (1832-1905), geologist and naturalist. He and Godwin-Austen collaborated on Godwin-Austen's chef d'oeuvre, *Mollusca*, in the *Fauna of British India* (1908). Blanford's collection had not been properly arranged and catalogued by the time he died.
[14] Letter to Fagan, 30 March 1911.
[15] Between £6,262 and £7,518 at 2013 values.
[16] Colonel Richard Henry Beddome (1830-1911) was a British military officer in India and a naturalist. While an important collector, little is known about him.

> I am sorry to trouble you. Bill for April & May - £14.12.3 sent in on 23rd ult. ie 15 days ago. It may have been overlooked.[17]

Godwin-Austen must have been glad of the income: his correspondence with the British Museum between the mid-1890s and his death is peppered with financial details of this sort. Nor is it just the financial aspects of the sale – Godwin-Austen was characteristically concerned that descriptive details should be correct before any classified object was made publicly accessible; in doing this, he earned a rather unjustified reputation for tetchiness – which is how his insistence on demanding standards of accuracy must have seemed to his fellow naturalists. For example:

> I am anxious while able to attend to it, to assist in the catalogue and registry. I wish during my life-time to see the collection in safe keeping, arranged at the Museum for future workers. A catalogue containing hundreds of Indian local names, unknown to the writers, often very young clerks, cannot be made from labels, without innumerable mistakes such as 'Tajendra' for 'Rajendra' which I have lately had before me and have constantly to note. It appears to me that all difficulties, and I think my very reasonable desire, can be met by the Trustees permitting me to make a catalogue of the 3 Parts of the collection commencing a series with a prefix letter, A1 – X, on loose sheets similar to those in the Register Books, these sheets when complete can very easily be bound up with the regular series.
>
> As I am unable to reside in London and can only carry on work here, I hope the Trustees will as a special case, allow me to take out as required any specimens entered on a Register compiled as above. Also that during my life-time only those specimens I have fully done with, are to be accessible to workers in Malacology and Conchology, and these will be finally worked out.[18]

Attempts at further sales went rather quiet during the decade surrounding the bankruptcy, perhaps as a result of the meanness of the British Museum. Certainly, Godwin-Austen's 'collection of shells',[19] valued at almost five times the amount of the bird collection, was listed as an unsold asset during the bankruptcy process. By 1910, however, in order to boost his income, Godwin-Austen had embarked on new attempts to sell off his assets:[20]

> ... some time back I offered the Museum the collection of watercolor[21] and other drawings of Indian Birds formed by Dr Jerdon[22] illustrating his work on the 'Birds of India', including his MS notes on the Birds of Assam & Burma with drawings, also a copy of the Birds of India interleaved with numerous drawings of head and feet of many species. I gave a list of them all at the time to Bowdler Sharpe who declined to take them.[23] I still have these books and drawings here. They are no use to me as I shall never touch ornithology again & want to get rid of all books on the subject.

---

[17] Letter to Harmer 8 July 1914.

[18] Letter to the Trustees of the British Museum, July 1904.

[19] Shells, of course, but also the greatest collection of freshwater molluscs yet assembled.

[20] Letter to Fagan, 24 April 1910. Louis Alexander Fagan (1845-1903), etcher and writer, Dept of Paintings and Drawings, British Museum.

[21] Godwin-Austen always spells 'watercolour' this way.

[22] Thomas Caverhill Jerdon (1811-72) was a British physician, zoologist and botanist. He is best remembered for his pioneering works on the ornithology of India. Several species of bird are named after him.

[23] Richard Bowdler Sharpe (1847-1909) in 1867 became the librarian of the Zoological Society. In 1872, he joined the British Museum as a Senior Assistant in the Department of Zoology, to take charge of the bird collection. He became Assistant Keeper in 1895. Sharpe founded the British Ornithologists' Club in 1892 and edited its bulletin. He wrote half of all 27 volumes of the *Catalogue of the Birds in the British Museum*.

There is no explanation for Godwin-Austen's rather violent abjuration of ornithology, unless various physical ailments which were beginning to accrue around this time (see Chapter 26) had made him aware that the long hours of outdoor observation required for successful ornithology were no longer possible for him. This is a reasonably plausible explanation, as these illnesses were mostly muscular or neurological in character and likely to be exacerbated by damp or cold. It hardly explains the abruptness of his renunciation, however. In the same letter, Godwin-Austen provides a tally of the watercolours and drawings, in a characteristically neat, precise table:

Drawings* made for Jerdon's 'Birds of India'

	From Nature	Copied from Various Works & Other Drawings
Birds of India in 6 cases	756	596
Assam & Burma	17	33
Malaysia	71	23
China & Japan	–	29
	844	681 = 1525

* a large number are by native artists

These various attempts at alleviating his financial circumstances might have provided a steady income at the level Godwin-Austen was hoping for. But then the disaster which overtook Europe occurred and, very quickly, all prospects of an improved income disappeared.

THE FIRST WORLD WAR AND AFTER

After the war broke out, Godwin-Austen's income from the museum stayed the same. When offered the same grant for 1915-6 as he had received in previous years, Godwin-Austen declared, '*I shall be glad to work it out for the war has very considerably reduced my income.*'[24] The difficulties induced by the war are repeated the following year, '*The war has whittled me down to little more than my Retired Pay*'.[25]

This sense of near-despair and anxiety about financial prospects unsurprisingly worsens as the war grinds on:

> A line or two to ask how the Treasury have met you this year as to a Grant for work on the Land Mollusca. I hope it will be the same as before. It will make every difference to me as to coming to town for this, as regards expense is now far more difficult to do than a year ago. It has been a miserable Spring, I don't know when I have seen things looking so backward, & farming prospects so bad. To keep poultry is no longer profitable and large numbers will be killed for food when they ought not to be.[26]

Godwin-Austen's difficult circumstances were evidently known to the family. By the end of the war, his nephew, Annesley, decided to act. He wrote to the President of the Royal Geographical Society, Sir Thomas Holdich, an old friend and comrade-in-arms from Burma and Bhutan days, and probably Godwin-Austen's finest obituarist, and to the Secretary of State for India.[27] From the former, he requested a 'memorial', most probably a lecture or series of lectures the proceeds of which would benefit his uncle; from the latter, he requested an increase in his uncle's pension as

24 Letter to Harmer, 29 September 1914.
25 Letter to Harmer, 13 March 1915.
26 Letter to Harmer, 22 April 1917.
27 Edwin Samuel Montagu PC (1879-1924) served as Secretary of State for India between 1917 and 1922.

it had dropped in value by half since being first paid it in 1877, and which in any case amounted to only £185 per annum.[28] The Royal Geographical Society Council cunningly absolved itself of responsibility under the cloak of generosity by recommending that the government of India paid an increase to:

> This most distinguished servant of India ... now, at an advanced age (*he was* 84) in straitened circumstances; they *(the Council of the RGS)* most earnestly hope that the Government of India may be willing to grant some substantial increase in his small pension to so eminent a man.[29]

The government of India then did its bureaucratic duty, i.e. to be self-serving rather than meet the needs of those whose lives it was originally designed to improve, and asked for various bits of proof of circumstances, for example a medical certificate. This process naturally resulted in considerable delay as the letters had to journey between Surrey and Calcutta. It was only by 1923 that a pension increase of £100[30] was squeezed out of a mean-minded India, for Godwin-Austen belatedly to enjoy during the last few months of his life.

'BAD' ARTHUR AND THE ESTATE

Family difficulties made further demands on Godwin-Austen's naturally tough spirit and buoyancy. His son and heir, Arthur, brought dishonour on the family, his wife Betty (they had no children, which is why the main Godwin-Austen inheritance passed through Haversham's next-youngest brother, Godwin, and his estranged son,[31] Annesley), his regiment and himself through being cashiered for running up debts. He was also said to be a womaniser and a drunk. Unsurprisingly, little detail of such habits can be confirmed although they are attested to by separate branches of the family. A lamentable ethical character is, however, to be set against Robin Godwin-Austen's memory of Arthur when, as a 16-year-old, he plucked up the courage to visit Arthur one day, and found him 'good-humoured and charming'. Any uncertainty about Arthur's being the black sheep turned into hard evidence, however, after Haversham's death in 1923. Instead of inheriting the £90 7s. 11d (probate, 29 October 1924) left by Haversham – somewhere between £4,217 and £4,606 at 2013 values – the estate was put into trust; Arthur was allocated a not too self-destructive monthly allowance which continued until his death in 1948. Only one clue tells us that the relationship was not a wholly bad one: the copy of the *Memoir* published by Gertrude Jekyll and 'MSW' in 1913 after Jessie's death which is lodged in the V&A is inscribed 'from HHGA to Arthur and Betty with much love'. (Arthur's wife, Elizabeth Augusta Mary née Hamilton died on 29 June 1945, aged 63. Family reports and obituaries suggest that she was very well liked and got through her difficult marriage cheerfully.) To the best of my knowledge, no letters, or indeed any other forms of substantial evidence of the nature of their relationship survive between Haversham and Arthur, although Arthur and his wife are recorded attending both Jessie's and Haversham's funerals. Their relationship can otherwise only be guessed at. Arthur would not be the first son to have gone to pieces under the influence of a psychologically powerful and successful father; Haversham would not be the first squire to have been bitterly disappointed (one assumes, from what is known about the character of both) by his son and heir.

The sale of the Shalford contents and the move to Nore are the bridge between Haversham the private man and his public position as a Tory squire and scientist of international reputation. If

[28] Between £7,370 and £10,805 at 2013 values.
[29] Letter to the Government of India from the Council of the RGS, 26 November 1918.
[30] Between £4,665 and £4,786 at 2013 values.
[31] See Chapter 3.

his private affairs caused him protracted difficulty, his public role balanced that with one success and honour after another.

CHAPTER TWENTY

MEDALS AND HONOURS

The RGS Founder's Medal

POSSIBLY HAVERSHAM'S GREATEST HONOUR, BEING awarded the Founder's Gold Medal of the Royal Geographical Society, did not arrive without a long wait. That it was blocked largely through the malign influence of Sir Clements Markham (1830-1916), secretary between 1863 and 1888 and the Society's president from 1893 to 1905, seems certain: Montgomerie had, after all, received his Gold Medal as early as 1865 for considerable but lesser achievements than Godwin-Austen in a similar area of geographical expertise and innovation. Godwin-Austen's geographical work was well known and had been mostly completed, moreover, by the time of his departure from the Indian Army in 1877, a long 33 years before being awarded the medal.

History has not been kind to Markham, an overbearing, priggish, violently prejudiced moraliser whose somewhat amateur naval enthusiasms rather than an interest in serious science led to his support for the disastrous Scott Antarctic expedition (he was godfather to Peter Scott, the explorer's naturalist son) as well as working against good human relations, for example in his embittered attitude towards Shackleton. His moralising seems to have extended to not awarding the Gold Medal to someone with an illegitimate son, or even a bankrupt. Perhaps, as the descendant of an Archbishop of Canterbury who was a tutor to William IV – the 'Founder' who donated 50 guineas for the medal in question – he felt himself to be some sort of guardian of the moral purity of the Gold Medal winners. An example of Markham's spitefulness occurred in 1894: he had been asked to comment on the suitability for publication of a paper by Godwin-Austen, *On Bhutan, the Himalayan ranges East of Darjeeling*. Godwin-Austen probably knew more about this subject than any other Westerner, and had been delivering papers to the RGS, both groundbreaking and well-received, since the 1860s. Markham's 150-word review spent itself mostly on summarising the contents, then snappishly concluded:

> A report on the country was written in 1866 by Colonel McGregor. I am, therefore, unable to recommend that Colonel Godwin Austen's *(sic)* paper should be printed in the Journal. It is out of date and the ground is already fully occupied.[1]

These statements are absurd: Godwin-Austen's unsurpassed map of western Bhutan, and his work in the Duars (the low-lying Indo-Bhutanese border area) were completed in 1865. McGregor's work was, at best, supplementary to Godwin-Austen's achievements; and as I know from my own

[1] Clements R Markham, CB, FRS, 15 October 1894. RGS archive.

expedition in 2008 to near the base of Gangkhar Puensum (7541m/27,741ft), lying on the Tibet-Bhutan border and still the highest unclimbed peak in the world, large swathes of Bhutan are still unknown to Westerners (particularly in the north-east and east) – the ground is and was anything but 'fully occupied'.

Within five years of Markham's retirement, and with Sir Aurel Stein as the worthiest of Gold Medal predecessors, Godwin-Austen's award in 1910 was cited as follows:

> For geographical discoveries and surveys along the North-eastern frontier of India, especially his pioneer exploring in the Karakoram.

The Society's *Monthly Record* provided rather better detail:

> Colonel Godwin-Austen receives the Founder's Medal for his geographical discoveries and surveys along the north-eastern frontier of India. He was the pioneer explorer of the Karakoram and Kashmir borderland, and his name will always be associated with the first acquisition of accurate knowledge respecting Mount Godwin-Austen or K2. With his surveys, he completed geological researches, and in both respects his work was of the highest order. He was the first to open up the palaeontological records of the Siwaliks, and, as a member of the Eden Mission, the first to reduce to order the topography of Western Bhutan.

In his letter of acceptance,[2] Godwin-Austen's genuine pride showed through:

> It is hardly necessary to tell you how much I value the honour bestowed upon me, by the award of the Founder's medal for the work I have been able to do, & that it comes from the Royal Geographical Society.
>
> There is one thing I prize almost as much, viz the reference in the Statement (*i.e. the citation*) on which the Award was made to the work of Sir Joseph Hooker,[3] and comparing his work in the N.G.J.[4] with mine on the (*illegible*). – no greater compliment could be paid to anyone, I always tried to follow in his footsteps, observing and noting the physical features of the country I was sent to survey.

THE NAMING OF K2

The Royal Geographical Society had also figured significantly much earlier in this period of Godwin-Austen's life, by raising the question of naming K2 after him. The mountain had first been sighted on 10 September 1856 by Montgomerie and a civilian, Mr Beverley, from Haramukh, a 5143m-/16,873ft-peak lying 35 kilometres (22 miles) north of Srinagar (not to be confused with Haramosh, 7409m/24,308ft, which is located in the Karakoram about 65km (40 miles) east of Gilgit).

But it was of course Haversham's 1861 expedition which had explored the arduous way to K2 and defined its position and height. In 1888,[5] at the Royal Geographical Society, after the reading of a paper by Sir Francis Younghusband, General JT Walker, (1826-96), Superintendent of the Great Trigonometrical Survey 1861-77, then Surveyor-General of India 1878-83, an old friend of Godwin-Austen since Sandhurst days, proposed that K2 should be renamed 'Mount Godwin

---

[2] Letter to J Scott Keltie, 23 March 1910.

[3] Sir Joseph Hooker (1817–1911), botanist, biogeographer and traveller. He was educated at Glasgow High School and later at Glasgow University, where his father was Regius Professor of Botany.

[4] *New Geographical Journal.*

[5] See *Proceedings of the RGS*, 1888.

Austen'; the proposal was seconded by Sir Henry Rawlinson and carried unanimously by acclaim. In a letter to the *Morning Post*,[6] Walker added that:

> In the absence of a name known to the people of the country, this name may be accepted as more appropriate than any other; it has been accepted by German and other geographers.

The RGS Council rejected it on the principle, established ironically enough by Sir George Everest, that only local names should be used for 'new' peaks. (The exception for the highest peak came from Everest's successor as Surveyor-General, Sir Andrew Waugh, who, because Tibet and Nepal were closed at the time of the mountain's height's determination, had found difficulties in establishing a local name.) Godwin-Austen was untroubled by the rejection. Having thought about it for a while, in a letter to the editor of the *Geographical Journal*[7] he made his views quite clear. Having first remarked that, '*I never succeeded in obtaining a* (local) *name for it, or I should have at once put it down and verified it*'.

He went on to say:

> I should myself have preferred, long ago, to see the peak called 'Mount Waugh', after the Surveyor-General of India, during whose tenure of office it was first observed, and following the precedent of naming the highest point in the Himalayas 'Mount Everest' some years before. I am at the same time not in favour of surnames of persons being given to unknown peaks, still less am I in favour of such English compounds as 'The Watch-Tower of India', 'The White Lily', 'the Pioneer Peak' which Mr Conway delights in, because eventually according to the taste of the traveller we shall be presented with all sorts of fanciful, incongruous or vulgar titles, perhaps to the exclusion of the native names.

This matter, however, has never completely gone quiet. In correspondence two years later, Godwin-Austen himself called the mountain '*K2 (Godwin-Austen)*'. Much directionless correspondence then ensued about whether 'Chiring' or 'Muztagh' or the anaemic 'Chogori' (Chinese: 'Qogir', and still used by them) might be suitable. The *Times Atlas* used the 'Godwin-Austen' name until very recently and the Google and *Encyclopaedia Britannica* maps still give it as an alternative. But it has also been asserted that, 'when Mrs Godwin-Austen wrote to Sir Francis Younghusband "in great state of mind" asking him why he insisted on referring to the peak as "K2", he did not bother to answer her letter.'[8]

The debate broke out again in the 1950s, when the great mountain literature translator, Hugh Merrick, somewhat wildly asserted the following in a letter to the editor of *The Times* on 23 October 1956:

> Sir, - There can be no gainsaying the contention of the Surveyor-General of Pakistan that K2 is the proper name for the second-highest peak in the world ... It has definitely been accepted as a geographical name even in Tibetan. So much so that the Balti porters, having heard its sound so often from their sahibs, have long referred to it as 'Ketu'.
>
> It must, however, be remembered that many of the older generation had 'Godwin-Austen' drummed into them at school with the connivance of the erring atlases ... Colonel Innes Wilson's contention that no name is known for the peak

---

[6] 18 February 1892.
[7] 16 March 1893.
[8] Patrick French, *Younghusband*, HarperCollins, London 1994, p63.

> locally is perhaps more open to challenge. In his well-founded book on the great Himalayan peaks, Zum Dritten Pol, Professor GO Dyrenfurth[9] authoritatively gives a Kashmiri name 'Lamba Pahar', signifying 'The Great Mountain'. And he makes out an even stronger case for 'Chogori', the Balti name, bearing the same significance, so often heard in Askole, the nearest village to the mountain, which after all rises in Baltistan. That even greater authority, Professor Longstaff,[10] also acknowledges this name, but rejects its narrower application to the peak itself, preferring to relate it to the huge tract of the Baltoro mountains generally.

It is hard to know where to begin in unravelling Merrick's illogicalities and inaccuracies – I mention this letter for its historical value, to illustrate how the debate had evolved, rather than because it throws any light on the subject. It is disingenuous or misleading on several counts: K2 cannot be seen from Askole; it is about eight days' trek from there to Concordia, where the mountain is properly first seen; 'Baltistan' is a rather vague term to apply to a precise place being included in it. The problem with all the local names – 'Great Mountain', 'Ice Mountain' etc – is that they are unspecific: K2 or Godwin-Austen both have the value of being clear-cut.

Still in the 1950s, moreover, I certainly recall 'Mount Godwin-Austen' being used in my primary school atlas, and being puzzled as to why a familiar-looking yet peculiar name should be stuck on a mountain in such a distant place. In the 1960s, the pendulum swung back towards 'K2', for its shortness and geographical accuracy. (It was so named, of course, as the second Karakoram peak to be surveyed. For example K1 is now usually known as Masherbrum (7821m/25,659ft), K3 (8051m/26,414ft) as Faichan Kangri or Broad Peak.) Moving up to the present day, in *The Times* of April 19, 2010, the mountain is loyally if cumbersomely named 'K2 (Godwin-Austen)'.

Convenience has won out, however, and K2 looks set to stay. Godwin-Austen's descendants (and biographer) will have to remain reluctantly content with occasional and haphazard 'Godwin-Austen' references, for example on the websites of many Pakistani tour companies and some maps, along with the permanent consolation prize of the Godwin-Austen southern access glacier. But if one follows through the 'convenience' argument, I cannot help feeling that if Haversham had been surnamed merely 'Austen', that name might well have stuck. And there is a good case for 'Godwin-Austen', because there is no genuine, specific local name. The last word must go to *The Times*, perhaps the most consistent of the supporters of 'Godwin-Austen': in its crossword, number 11,349, of 1 October 1966, the '1 Across' clue is, "Freethinker and Novelist meet by mountain" (6-6)', not only giving the answer 'Godwin-Austen', but presupposing that *The Times*'s readership will know that that is the great peak's correct name.

GEOGRAPHICAL PUBLIC BODIES

The references above to the high reputation which Godwin-Austen enjoyed at the Royal Geographical Society (he was also on the Council for a time) provide a useful link to the other distinctions, mainly to do with London scientific societies, which Godwin-Austen received from about 1890 onwards. As early as 1883, indeed, he had been made President of Section E (Geography) of the British Association.

In his *Presidential Address*[11] to the Association, Godwin-Austen's theme is how tertiary geology can inform geography. He begins by defining the extent of the Himalayas and how the 'Central Asian Chain' of mountains, from the Kun-Lun to the 'Sub-Himalaya' should be divided; he goes

---

[9] Gunther Oskar Dyrenfurth, Himalayan climber and film-maker. Merrick translated his *Zum Dritten Pol* into English.

[10] Tom Longstaff (1875-1964) was an English doctor, explorer and mountaineer, most famous for being the first person to climb to a summit of over 7000m/22,966ft..

[11] *Proceedings of the Royal Geographical Society, New Monthly Series*, Vol 5, No 10, pp 610-25, October 1883.

on to explain how he arrives at his opinion on the different divisions. These subdivisions are described in quite fine detail, all the more impressive as one realises that in much of what he describes, Godwin-Austen is talking from first-hand experience. About two-thirds of the way through the *Address*, however, he turns to the Alps, particularly their southern face as far as the River Po and the plains of Lombardy, and compares their geological formation with parts of the Indian Himalaya. The *Address* concludes with some remarks on comparative glaciation. In concluding, he praises the greatness of Lyell, a British lawyer and the foremost geologist of his day. Lyell is best known as the author of *Principles of Geology*, which popularised the idea that the Earth was shaped by slow-moving forces still in operation today. Lyell was a close and influential friend of Charles Darwin who is also praised and of course in so doing, Godwin-Austen thereby nailed his colours to the evolutionary mast, a position he had occupied, as a researcher, for decades.[12] He did, however, with perhaps unintentional wit but undoubted deliberateness add '*The study of these changes must create in our minds humble admiration of the great Creator's sublime work*' where not only is a periphrasis ('*the great Creator*') drawn in to protect him, as in earlier days ('*a follower of the Prophet*') from directly naming his god, but rather suggests, in a very modern way, that our sense of creation is simply what our minds are capable of constructing about it.

On 27 October 1898 Godwin-Austen acknowledged the honour of life membership of the RGS. Sadly, this honour was conferred probably in recognition of the need to alleviate his deteriorating financial position. On the same day, he wrote to J Scott Keltie at the RGS to resign his Fellowship as he could no longer afford to pay it:

> The stress of the Agricult. Depression has at last been brought home to me, & I have had to seek the protection of the Bankruptcy Law – this has rendered it my duty to send in my resignation of the Royal Society; the Zoolog.l.; the Malacological as its President, & of many other public bodies ... Members of the Council may not know all the circumstances under which I had to take so serious a step, but they are known to my friends Dr Sclater[13] & Mr WT Blanford.[14]

But there was a happy outcome: a note at the top of the letter says, 'does not require to resign. President says tell him to go to Council.' The other societies to which Godwin-Austen belonged appear to have taken a similar line, a testimony to the esteem in which this extraordinary scientist was held.

---

[12] See the reference in Chapter 5 to his observations in the Calcutta Museum in 1853.

[13] Ornithologist, possibly malacologist who worked with Godwin-Austen on the natural history of Socotra (archipelago 80km/50 miles east of the Horn of Africa).

[14] Letter from Godwin-Austen to J Scott Keltie, 27 October 1898. William Thomas Blanford (1832–1905) was a geologist and naturalist. In 1851 he was induced to enter the newly established Royal School of Mines (now part of Imperial College London), which his younger brother Henry F Blanford (1834-93), afterwards head of the Indian Meteorological Department, had already joined. He studied under Henry de la Beche. Towards the end of 1854 both he and his brother obtained posts on the Geological Survey of India, remaining for 27 years until retiring in 1882. After his retirement he took up editorship of the *Fauna of British India* series.

CHAPTER TWENTY-ONE

SCIENTIST

Introducing the Expert

TO ASSESS GODWIN-AUSTEN'S CONTRIBUTION TO the history of science, I have turned to the expert. (When I was at school, biology was not offered as a subject; chemistry and I parted company at the earliest occasion possible; and my physics teacher generally treated me with well-deserved derision.) John Mathew is not only well-qualified through his academic experience, and as a writer of a well-received BBC Radio 4 play, but is also the recent recipient of a Wellcome Foundation grant to carry out research, recently completed, at Godwin-Austen's old haunt, the Natural History Museum in South Kensington. John's essay extends our understanding of Godwin-Austen as a human being, by guiding us through the outburst of academic spleen which accompanied the publication of Godwin-Austen's chef d'oeuvre, *The Fauna of British India*, but also demonstrates just what an important contributor Godwin-Austen was to this and several other branches of Natural History.

GODWIN-AUSTEN AND *THE FAUNA OF BRITISH INDIA* SERIES

by

John Mathew, Ph.D, Department of the History of Science, Harvard University

The genus (*Glossula*) has been treated by Mr GK Gude in the *Fauna of British India*. He approached it with a great knowledge of conchology, bibliography, and especially synonymy – the last most useful to workers, but unattractive. They have to thank Mr Gude for undertaking such labour. It shows, like so much work of its kind of the series to which it belongs, that he had never been a collector in India and knew little of its physical features and all that that comprises. There is an absence of original matter, such as Dr Jerdon, the Blanfords, Lydekker, Oates, Day and others brought to bear on and embellished the history of the Mammals, Birds, and Fishes of India which they had collected and which had passed through their hands.

It is easy to find fault, and it may appear I do so with Gude's work. I am only animated by the desire and striving to make the record of Geographical Distribution as correct as possible; thus under G. *tenuispira*, p.379, I notice all the errors of determination which Blanford, Theobald, Nevill, Beddome, and myself have perpetuated. I have to point out that these determinations were made 40 to 60 years ago, much too long ago for such data to be reliable. I am able to say they

> were often without sufficient material at hand, or on shells erroneously named in the first instance. I take, for example, G. *baculina*, p.379. Khasi Hills (Godwin-Austen), evidently on the authority of Nevill in the 'Hand-list,' p.176. It is a distinct species, which he did not notice; I have named it *subbaculina*, for I cannot find in my collection from the Khasi Hills any *Glossula* that matches in the Henry Blanford collection.[1]

This veritable broadside launched by HH Godwin-Austen (as he usually signed himself) against GK Gude in the lines above was not limited to the pages of the first part of the third volume of the *Land and Freshwater Mollusca of India*. In a review in the journal *Nature*, published in 1921, Godwin-Austen did not desist from showing the author of the third volume of the malacological section of *The Fauna of British India* series in a light that was frankly unflattering:

> Among the Indian land mollusca the family *Cyclophoridae* especially attracts attention by the beauty of form and variety in the shells. The range is extensive, and their study began so far back as 1849, when WH Benson, of the Bengal Civil Service, and Capt T Hutton were writing on the animal of *Diplommatina*. After seventy years the above families are still very imperfectly understood, so little is known of the animal which constructs the shell. I think I am correct in saying the time had not arrived for publishing a volume on these molluscs, for sufficient material had not been examined; in truth, much has yet to be collected. It would be interesting to know for this reason why Mr Gude was selected to do it, and the work then left to all intents and purposes in his hands; why malacologists with knowledge of the subject were not consulted; and why the present writer, with more than forty years' connection with these land operculates, both in the field and in collections, was in complete ignorance that such work was under compilation. It leads me to think of past workers in this field of natural history. The foremost among them was the late Dr William Blanford, the editor and founder of *The Fauna of British India*. No one would have known better how much preliminary work there was to do, or what material to obtain; and had he lived he would have prepared the way for it, as he did for Vol I for the Mollusca Series (the *Testacellidae* and *Zonitidae*). It recalls the type of paper on the animals of *Raphaulus*, *Spiraculum*, and other tube-bearing *Cyclostomacea* so long ago as 1863 in *The Annals and Magazine of Natural History*. There is even now food for thought in this paper, while it is an indication of how the history of the land operculates should be approached, which is to be looked for in vain in the publication under notice.
>
> ... Space does not admit of quoting many and various errors, but looking at the contents I begin with *The Systematic Index* and naturally turn to the genera I am best acquainted with. On p.xii I have to notice, very fully, an unfortunate, inaccurate record of species under the sub-genus *Raptomphalus* of *Alycoeus*; eight species are put into it, whereas it is represented by one, *R. magnificus*, described by me in *Land and Freshwater Mollusca of India*, vol. 2, part 12, p. 366, plate 156, figs. 1, 1 a, 1 b.
>
> ... Such a record does not help zoology; it is gone forth to the world, striking directly at the accuracy of geographical distribution, and before it can be disproved it will be used, trusted, and quoted by workers, both at home and abroad, for years to come ...

---

[1] HH Godwin-Austen: *Land and Freshwater Mollusca of India, Including South Arabia, Baluchistan, Afghanistan, Kashmir, Nepal, Burmah, Pegu, Tenasserim, Malay Peninsula, Ceylon, and Other Islands of the Indian Ocean*. Supplementary to Messrs. Theobald and Hanley's *Conchologia Indica*, Vol III, part I: November 1920. London: Taylor and Francis, 1920, pp5-6.

> ... The volume adds very little to our previous knowledge of the land operculates of India. An opportunity has been lost of bringing that knowledge up to date. The money spent on it might have had a better result. There is an absence of editorial supervision. Mr Guy AK Marshall is not mentioned in the preface, all seemingly devolved on Mr GK Gude. It is purely conchological and was brought out in a hurry. It should not, in fact, have been commenced until it was first ascertained who were to be engaged upon it and what collections were available. This is most important to those who, at great expense of time and toil, collect material and desire to see it in the hands of the best malacologists and so far have a voice in the publication.[2]

What might have occasioned such searing vitriol on the part of Godwin-Austen? After all, the obituary devoted to Gude upon his passing, in the *Proceedings of the Malacological Society of London*, had this to say of him:

> Modest, unassuming, and generous to an extraordinary degree, he was ever ready to help others from his store of knowledge, while one who knew him in his early days has described him as the best of colleagues and companions that a man could hope to meet with during a long life-time. His loss to Malacological Science is indeed great.[3]

This latter comment did not exist in isolation: Gude's standing was such that he had served as Secretary to the Malacological Society from 1909 to 1918 and as its President from 1919 to 1922. By denigrating one of its stalwarts in print, Godwin-Austen was essentially challenging the establishment of which he was, both nominally yet also historically, a part.

And yet we might profit by going back to the kernel of the matter. Godwin-Austen was not alone in calling into question the credentials of a taxonomic expert who had never laid eyes upon the geographical region whose molluscan fauna he had been invited to describe. Similar concerns had been voiced as early as 1891, when an anonymous reviewer in the *Journal of the Bombay Natural History Society* metaphorically eviscerated GA Boulenger for his book on Indian reptiles, itself, incidentally, part of the same series (*The Fauna of British India*). The tensions were legion between the systematic worker functioning in the metropolis and the field-worker in the 'colony' whose knowledge of local ecology was second to none. Dorinda Outram points to the work of Georges Cuvier, himself a somewhat controversial figure in this context, in this regard, given his own extensive work in the field, when he wrote:

> ... if the sedentary naturalist does not see nature in action, he can yet survey all her products spread before him. He can compare them with each other as often as is necessary to reach reliable conclusions. He defines his own problems ... The travelling observer can only travel one road. One can only roam freely through the universe, by staying in one's study. For that, a different sort of courage is needed ... courage which does not allow its possessor to leave a subject, until by observation ... and connected thought, he has illuminated it with every ray of light possible in a given epoch of knowledge.[4]

---

[2] HH Godwin-Austen: Review of *The Fauna of British India, including Ceylon and Burma: Mollusca. – III. Land Operculates (Cyclophoridae, Truncatellidae, Assiminedae; Helicinidae.)* by GK Gude, ppXIV+386. London: Taylor and Francis; Calcutta: Thacker, Spink, and Co; Bombay: Thacker and Co, Ltd., 1921; *Nature*, Vol 108, No 2708, pp106-8.

[3] BB Woodward: *Obituary Notice: GK Gude; Proceedings of the Malacological Society of London*, Vol XVI, pp 205-6.

[4] Georges Cuvier: *Analyse d'un ouvrage, de M. Humboldt, intitulé Tableaux de la Nature ou considérations sur les déserts, sur la physionomie des végétaux, et sur les cataractes de l'Orénoque*. Institut de France, Fonds Cuvier,

There is a key difference however, between the travelling naturalist (or voyageur-naturalist) of whom Cuvier speaks in contradistinction to the European metropolitan worker (be that worker in London or Paris, Amsterdam or Berlin), and the colonial worker, whose knowledge of place is predicated precisely upon a long sojourn. The people of whom Godwin-Austen spoke all fell within this category: Thomas Caverhill Jerdon, William and Henry Blanford, Richard Lydekker, Eugene Oates and Francis Day all worked assiduously on different aspects of Indian fauna through an intimate knowledge of the subcontinent. The main narrative of this book has already shown how much a part of this specialist group Godwin-Austen was. And the nub of the matter was this: Godwin-Austen, who had completed the first volume of the *Mollusca* for *The Fauna of British India* series upon the untimely death of William Thomas Blanford (and consequently named co-author with the deceased) in 1908, who had written the monumental *Land and Freshwater Mollusca of India* in three volumes between 1887 and 1920, had not been invited to become the author of the second and/or the third volume of the *Fauna* series, nor, by his admission, even been consulted. The response might have appeared curmudgeonly, but Godwin-Austen seemed at least in part justifiably put out by what seemed to be his exclusion by the votaries of the very discipline for which he had laboured incessantly over at least four decades.

It is fascinating to leaf through the pages of Godwin-Austen's annotated copy of Gude's books in the *Fauna*. As early as Volume II, the first salvo was fired in the preface. Gude had this to write in the second paragraph:

> It was at first anticipated that the second volume would be the joint production of Lieut-Colonel Godwin-Austen and the present writer. Unexpected circumstances have, however, deprived me of the advantage of the former's cooperation. This is the more to be regretted as it involves the loss of a considerable quantity of additional anatomical details, the result of his recent researches, which would have greatly increased the value of the present volume; and further, that his own collection and that of the late Dr Blanford – both very rich in Indian Mollusca and both in the custody of the British Museum – have in consequence been inaccessible to me.

The last phrase, 'have in consequence been inaccessible to me' was underlined in red by Godwin-Austen and attended by the indignant comment, '*This is not true vide my letter 4th August 1914 to Mr. Robson and my meeting with Gude on 31st July.*'

There were substantial criticisms in the text. Godwin-Austen pertinently pointed out that in the attribution of particular species to authors, Gude merely wrote 'Blanford' without distinguishing which of the brothers made the designation, i.e. Henry F or William T (in some cases, both). There were discrepancies noted in terms of range, habit, morphology, all of which appear in the marginalia.

But the appearance of Volume III[6] excited Godwin-Austen's splenetic fury even further. As he wrote on the title page, '*Received 19th March 1921 from Mr. William Foster, India Office, London. Applied for by me, on hearing of the publication from G.C. Robson. Natl. History Museum. I was never*

3159. Quoted in D Outram. *On Being Perseus: New Knowledge, Dislocation, and Enlightenment Exploration.* Max Planck Institute for the History of Science; Berlin, 1997. Jean Léopold Nicolas Frédéric Cuvier (1769-1832), known as Georges Cuvier, born in Montbéliard, Eastern France, was a French naturalist and zoologist. Cuvier was a major figure in natural sciences research in the early 19th century, and was instrumental in establishing the fields of comparative anatomy and paleontology through his work in comparing living animals with fossils.

[5] GK Gude: *The Fauna of British India including Ceylon and Burma. Mollusca - II.* (Trochomorphidae-Janellidae); Taylor and Francis, London, 1914; Preface. The annotated copies are held in the Mollusca Department of the Natural History Museum, London. Special thanks to Dr Fred Naggs for bringing them to my attention.

[6] GK Gude: *The Fauna of British India including Ceylon and Burma. Mollusca – III. Land Operculates* (Cyclophoridae, Truncatellidae, Assimeinidae, Helicinidae); Taylor and Francis, London, 1921.

consulted from first to last and did not know it had been placed (?) in hand. 25. iii. 21.' The result was the review in *Nature* that we saw earlier in this article. It is of note that Gude himself, in the preface to the Volume III, made specific mention of all the revered workers in the field, ending with the only living one at the time, Lieutenant-Colonel Godwin-Austen, of whom he wrote, 'The last named has an intimate and extensive knowledge of the Indian Mollusca, both conchological and anatomical, equalled by no man living.'[7]

And yet the question remains as to whether Godwin-Austen could have been the popular choice for the task. There is no doubt that his later volumes of the *Land and Freshwater Mollusca of India*, virtually coincident with those of Gude for the *Fauna*, were meticulous, exhaustive and painstaking, mirroring the efforts that had marked his first works. A review of the tenth part of the series placed in Volume I in 1907 in *Nature* indicated just how difficult the task was, not least since time and opportunity had clearly been an issue:

> It is just nineteen years since the first six parts (1882-1888) of this work were noticed in these pages (Nature, July 5, 1888, Pp. 217-218), and the author is now the sole active survivor of the celebrated band of Indian malacologists that included Theobald, the Nevills, Stoliczka, Benson, Beddome, and the two Blanfords. One by one they have nearly all gone from us (Dr WT Blanford's demise is still fresh in our memories), leaving, alas, no successors in the field of their labours.
>
> This but adds to our hope that Col. Godwin-Austen may long be spared to carry on the work he has begun, but, so wide is the field, can never expect to complete, especially since of late it has been only now and then he has been able to find time for the examination of material that has come to hand. As a consequence, seven years have elapsed since the appearance of the preceding part of the work before us.[8]

Clearly, the time taken by Godwin-Austen in producing the volume was of marked concern. Furthermore, a review in the *Madras Times* relating to Volume I of the *Fauna* (under the co-authorship of Godwin-Austen and the late WT Blanford) by R Lydekker (whom Godwin-Austen had quoted in the group of reputed subcontinent-based workers on fauna) suggested another area of worry: age. By 1908, Godwin-Austen was already 74.

> How the work is to be completed is not stated in the volume before us. Of course, Colonel Godwin-Austen is the one man fully qualified for the task; but then he is rapidly advancing in years, and may feel that his powers are unequal to the strain involved in such serious labour, or may consider himself entitled to pass his remaining years in well-earned leisure.[9]

We are now in a position to reflect on these goings-on. By 1908, it already seemed that Godwin-Austen's powers were on the wane, despite his unparalleled knowledge of Indian Mollusca, always duly acknowledged, and, as a consequence, work on the subject would perforce need to be advanced by others.

The one person, however, who refused to be put out to pasture was Godwin-Austen himself. The final parts of *Land and Freshwater Mollusca of India*, concluded a mere three years before his passing in 1923, are their own impressive testament to his tenacity. Indian Malacology remains in his debt.

John Mathew, Harvard University, September, 2010

---

[7] *Ibid*, Preface, iii.

[8] BB Woodward: *Review of Indian Malacology. Land and Freshwater Mollusca of India &c. by Lieut.-Col. H.H. Godwin-Austen. Supplementary to Messrs. Theobald and Hanley's "Conchologia Indica."*; Vol ii, part x, pp147-238; plates ci-cxvii; Taylor and Francis, London, 1907. *Nature*, No 1967, Vol 76, July 11 1907, pp244-5.

[9] R Lydekker, *Review of The Fauna of British India: Mollusca, Testacillidae and Zonitidae by the late WT Blanford and Lieut. Col. H.H. Godwin-Austen, London, 1908. The Madras Times*, 19 September 1908.

NATURAL HISTORIAN

Just before his Presidency of the British Association Geographical Section, a different branch of natural science had conferred its own honour in 1895-7 on Godwin-Austen, when he became Vice-President of the Zoological Society of London, then was placed on the list of Extraordinary Members of the British Ornithological Union.

But what for Godwin-Austen would probably have been his most satisfactory honour was his presidency of the Malacological Society of London from 1897-9. This small but dedicated society was founded on 27 February 1893; it never seemed to contain more than a few dozen members. (It is, however, still flourishing, and their attractive website at www.malacsoc.org.uk promises an enjoyable visit.) The members met in the 'apartments' of the Linnean Society, Burlington House, London: founded in 1788, it remains the world's premier society for the study of natural history, taking its name from the Swedish naturalist Carolus Linnaeus. The society Council, of which Godwin-Austen was a member for well over a decade, often met at the Natural History Museum, South Kensington. Their commitment and the quality of their science shine through in their records. (These now lie in a couple of wooden chests at what was Wye College, just outside Ashford, Kent, but which is soon to become part of the University of Kent at Canterbury.) Godwin-Austen appears to have attended from the earliest meetings; the last meeting he seems to have attended was on 7 April 1911. The Society's archives are helpfully clear: the accounts, for example, show that in 1907, Godwin-Austen paid his subscription of 10/6d (between £45 and £51 at 2013 values) on 10 January (a notably prompt, early payment). A paper of his, entitled *Further description of the Animal of Damayantia carinata, Collinge, showing its similarity to D smithi Collge & GA, with remarks on this genus of Issel, Collingea of Smiroth and Isselentia of Collinge* was read out on 13 March 1903, suggesting that weather, illness or economy had prevented him from giving the paper in person. The records also tell us something of the range of Godwin-Austen's malacological interests, for example in his paper given on 16 June 1905, *On the extension of the genus Macrochamys* (sic) *to the Island of Mauritius*, his previous work having been confined to the sub-Continent, Burma and Malaysia.

Otherwise, the archives record the Society's humdrum business, for example the Council meeting of 12 March 1897, presided over by Godwin-Austen:

> The minutes of the last meeting were read and confirmed.
> The Secretary reported that Mr A. T. Daniel had remitted his subscription for 1896 and that he stated he desired to continue his membership; the Secretary was instructed to inform Mr Daniel that his resignation in the summer of 1896 having been accepted, he had ceased to be a member of the Society.

The high point of his presidential tenure for Godwin-Austen would have been his valedictory *Presidential Address* on 10 February 1899. He began by defining his own specialist area, the land and freshwater molluscs of India, then acknowledged the contributions made by the two previous presidents, as well as '*the Indian workers*', all of whom turn out to be English or German scientists! He traced the history of his research area in the sub-continent, and Afghanistan, with particular praise for Benson,[10] Hutton[11] and Stoliczka.[12] He then described the development of his own interest, from his first visit to 'Little Tibet' (Ladakh and Zanskar) from1862 onwards, through the rich research area of the 'eastern frontier' i.e. Burma and Assam, and some of the outlying Himalayan foothills, such as the Siwaliks. Godwin-Austen then rightly described the work which

[10] William Henry Benson (1803-70) was a civil servant in British India and an amateur malacologist. He made large collections of molluscs and described numerous species
[11] Thomas Hutton, fl. 1840s–60s; captain in the Bengal Army. Author of works on natural history and scriptural geology. Fellow of the Geological Society.
[12] See Chapter 16, n1.

remained to be done. As part of this work, the president suggested, in a regrettably rare anecdote, that the deep Pliocene beds of mud in Kashmir should be investigated:

> Mammalian remains might also possibly be found, but the teeth, or bones, of a rakhshus, or demon, were never reported to me in Kashmir, though such relics are well known to the natives in the neighbourhood of Jammu and Aknur, for there they say was a field of a great battle between demons. Rhunbeer Singh,[13] the Maharajah of Kashmir, had one brought to show me when I paid him a visit, and we had a long and amusing discussion. I tried to show him it was a molar of an elephant, but he would not be convinced.[14]

From this point onwards, the discussion becomes somewhat arcane, although it pursued a clear line of review of the work which still needed to be done in classifying the various molluscan *genera*. The President passed from the north-west frontier through 'peninsular' India to Burma and 'Ceylon' and eventually to Malaysia; China, particularly Yunnan Province, 'Siam' and the Andaman Islands were also mentioned; there was then much discussion about the significance of shell-lobes. Many other malacologists and conchologists were referred to: the *Address* was altogether magisterial in its range, both geographical and historical, and its depth of detail. His conclusion was simple:

> A great deal more has to be done, and I hope this Society will be the means of helping to do it. It will never be completed as it should be, until some malacologist goes out to work in the country, and on fresh material ... Other species would be made known, for which only a trained naturalist knows how and where to look.[15]

Given that much of his paper is taken up with discussion on the nature of molluscs and their shells it was not at all surprising that Godwin-Austen became President of the Conchological Society from 1908 to 1909. He was asked for his *Who's Who* entry in 1916.

Godwin-Austen's public honours were, therefore, a fitting recognition of his remarkable scientific abilities. Although of a much more private nature, his abilities as an artist, the results of which have provided an equally enduring legacy, must now be considered.

---

[13] General Shriman Rajrajeshwar Maharajadhiraj Sri Sir Ranbir Singhji, Indar Mahindar Bahadur, Sipar-i-Sultanat, Mushir-i-Khas-i-Kaiser-i-Hind GCSI, CIE, Kaiser-i-Hind (1830-85) was the son of Maharaja Gulab Singh, Maharaja of Jammu and Kashmir and head of the Jamwal Rajput clan.
[14] HH Godwin-Austen, *Address of the President*, p246, 10 February 1899; Malacological Society archives.
[15] *Ibid*, p256.

CHAPTER TWENTY-TWO

ARTIST

An Evaluation

HAVING CONSIDERED GODWIN-AUSTEN'S PUBLIC DISTINCTIONS as a scientist, we should now look at his character as an artist. From his early training at Sandhurst, where his distinctive natural talent for drawing was enhanced, he became a good amateur watercolourist and exceptional draughtsman. Again, I have turned to the experts (my own artistic qualifications including the narrowest of 'Pass' marks at 'O' Level, a student summer job painting registration numbers on trawlers, and a Sixth Form History of Art Course taught by someone who seemed unaware that 'Vienna' and 'Wien' were the same place). They offer the following assessment of Haversham's character as an artist, his principal achievements, and his significance in nineteenth-century Anglo-Indian art.

GODWIN-AUSTEN THE ARTIST

by

Peter Mallalieu, Keeper of Pictures at the Alpine Club

with additional notes by

Jerry Lovatt, past Vice-President and Honorary Librarian, the Alpine Club

THE MILITARY ARTIST AND DOCUMENTARY ART RECORD

Most military action results in the creation of public or private records: reports, letters, diaries and memoranda, and works of documentary art in the form of sketches, watercolours, oil-paintings and prints.

A 'military artist' is trained, often at a military academy, to create visual documents. The artist's acquired skills could be applied to military assignments or duties: such training results in easily understood visual records. Artistic education at British military academies usually consisted of producing watercolours and drawings mostly in notebooks or sketchbooks, generally of small size. Redrawn work, done later, tended to be of a larger size which could subsequently be made into prints.

Once trained, the artist could respond to specific military requests, for example the need to record some aspects of a country for navigation, spying, colonisation, survey, or to record a construction used for purely military purposes.

ARTISTIC TRAINING AT SANDHURST

The Woolwich Academy had appointed a Drawing Master by 1744; Sandhurst would follow later. Haversham Austen (as he then was) entered the Royal Military College, Sandhurst, in January 1848, as a Gentleman Cadet aged 13 years six months and standing just under five feet tall. (He includes himself in some of his paintings.) While there, he was taught topographical drawing and mathematics by Captain Robert Petley. He was to be commissioned, with a College Recommendation, into the 24th Regiment of Foot in December 1851.

Haversham's youthful notebooks are full of detailed mathematical theorems, equations and solutions written in an immaculate hand as well as numerous associated drawings relevant to the classwork, e.g. capstans and pulley systems, together with amusing and lively sketches done when his attention might have wandered. He obviously possessed an ingrained ability to draw, which was undoubtedly developed by his tutor.

Haversham's larger, more formal, sketchbooks reveal topographical drawings and plans of the training areas close to Sandhurst, done with great clarity often reflected in the outlining, hatching and washes which echo Petley's own work. The Royal Military College, Sandhurst, has preserved one of these sketchbooks in its History Room.

ROBERT PETLEY (1812-69)

Robert Petley had himself joined Sandhurst as a 14-year-old Gentleman Cadet in 1827, some 21 years before he was to teach Haversham; both followed a similar curriculum, which included military and landscape drawing in pen and brushwork.

Having graduated in 1830, he soon joined the Rifle Brigade and was posted to their Canadian HQ in Fredericton, New Brunswick in 1831. He served there, and in Nova Scotia, from 1837 to 1838 as part of the military response to the threat of an American invasion, as well as to subdue internal rebellions.

While in Canada, Petley produced a quantity of watercolours and drawings of local topography; these sometimes related to military matters, and were much admired. Many of his watercolours were later turned into hand-coloured lithographs, which were then coming into fashion, with three suites being published in 1837, although Petley published somewhat reluctantly, only after being encouraged by his fellow officers. His pen and pencil drawings suggest that his training had included stress on outlining and cross-hatching, similar to that used by engravers. He was later to impart these ideas to the young and receptive Austen.

An exhibition of Petley's works was held in Nova Scotia in 1999. The exhibition highlighted aspects of his technique: *Immigrants at Bedford Basin near Halifax* (see plate 10) is a Romantic watercolour executed with clarity and liveliness. *Windsor, Nova Scotia, from the Barracks*, 1837, (see plate 11) is a hand-coloured lithograph showing cross-hatching, outlining and shading, together with the precise architectural details which would later be ingrained in Haversham.

After travelling to India and briefly to Burma for the Second Anglo-Burmese War, the newly promoted Ensign Godwin-Austen found himself in the Punjab in 1854. There, in Sialkot, he met and worked with the professional artist William Carpenter (1818-99).

WILLIAM CARPENTER (1818-99)

Since he is the major influence, with Petley, on Haversham's artistic development, Carpenter's Indian career is worth a brief description; it has been compiled from my own researches and combined with some adapted comments from *Indian Life and Landscapes by Western Artists* by Pauline Rohatgi *et al.*

William Carpenter painted and travelled widely in Kashmir, the Punjab, Afghanistan and

Rajasthan and other parts of Northern India between 1850 and 1857. He accumulated a wealth of images, 280 of which were exhibited at the Victoria & Albert Museum, South Kensington, in 1881. They were subsequently bought by the V&A and are still held in its collection.

Carpenter was essentially a watercolour artist, though he painted in oils early in his career. It is possible that *The Illustrated London News*, which was publishing articles about India, commissioned him to record people and scenes in that country. Illustrations in colour and black-and-white from his Indian watercolours appeared in that journal between September 1857 and October 1859. He did not exhibit many Indian scenes publicly except for five pictures, four of which portrayed scenes in Kashmir, at the Royal Academy in 1857, 1858 and 1866. Among the mid-19th century professional artists who went to India, however, William Carpenter remains relatively unknown.

WILLIAM CARPENTER IN KASHMIR

It was only after the annexation of the Punjab in 1846 that Kashmir became more accessible to European travellers. Carpenter's first of at least three annual trips to Kashmir was probably in 1853. Vigne (see Chapter 10) was probably among the first observers to be captivated by the romantic vision of Kashmir: besides general views of the Vale and its lakes, or Shah Hamadan's Mosque, romantically attractive views included the quaint wooden houses and streets of Srinagar, bridges across the Jhelum River and Mar canal, Kashmiri women and *nautch* (dancing) girls, and the Temple of the Sun at Martund, to which Carpenter made a special excursion. He also met the Governor of Kashmir, Nawab Shaikh Imam-ud Din.

Carpenter's many watercolours convey the impression of snapshot-like images, captured in an instant with the versatility of an artist highly skilled in the watercolour technique. He selected the minute rather than the broad aspect of a scene. His penetrating vision, as revealed in these small-scale watercolours, evokes a gentle romanticism. Brilliantly executed in a range of sparkling colours that were dominantly warm, Carpenter captured vivid chromatic effects, comparable with George Chinnery's graphic landscapes in pencil, pen and ink, and his delicately subtle tones in oil and watercolour.

Connections with Haversham, in time, place, technique and subject matter are not difficult to discern. Godwin-Austen's watercolour of *Skirdiwala, Sealkote (sic)* dates from this mid-1850s period and begins to show Carpenter's influence: neat pen/pencil outlines and light, clear washes. He is at his best dealing with the details of buildings, which would later culminate in the strong and convincing portrayal of mountain structures. He was never as secure as Carpenter when painting figures, although invariably accurate with all details and colours of (ethnic) garments. His accurate depiction of trees also compares with Carpenter. And a glimpse of the Himalayas, for example, is caught in *View of the Wood Bazaar, Simla*, the mountains which Godwin-Austen would later explore, survey and paint so extensively.

In 1855 Haversham moved north-west to Peshawar. There, he painted aspects of that city such as: *View from the British camp at Nowsherah towards the Khyber Pass*, 1855 (see plate 16) at the same time as William Carpenter painted *View of houses adjoining a stream, Peshawar*, 1855 (see plate 12). Here the influences are clearer: glowing effects of sunlight on cityscapes and architectural monuments; his handling of colour shows a warm, earthy range which evokes this characteristic 'gentle Romanticism'. There is a resonance between the two watercolours which cannot be entirely by chance. The seemingly chaotic scene is handled, in both cases, with a selective eye and a distinctive harmony so that the overall impression has a convincing reality. Haversham appears to be following in the footsteps of his sub-Continental tutor.

THE GREAT TRIGONOMETRICAL SURVEY

From 1856 until his retirement in 1877, Godwin-Austen was attached to the Great Trigonometrical Survey of India. He ranged widely and into previously inaccessible high places in a series of arduous surveying expeditions; he appears always to have kept a careful visual record of his work and travels.

Godwin-Austen comes into his own when painting mountains, valleys and glaciers, where figures are of lesser significance within this massive scale. *K2 from above Urdukas* (see plate 29) is of historical importance in that it shows Haversham in position to fix the height of K2, the snowy peak to the left at the rear, behind the heights of Biange, and with Broad Peak and Gasherbrum IV to the right. This viewpoint of K2, showing Godwin-Austen's 'coolies' carrying his plane-table and other pieces of equipment, is situated high above the Baltoro Glacier on the flanks of Masherbrum. Although Godwin-Austen did not, strictly speaking, discover K2, he did uncover the Baltoro approach to the mountain and he was, of course, the first to sketch and define its topography, during the trip on which this painting was made. This impressive watercolour, one of the most important in 19th-century mountaineering, since there is no equivalent for Everest, was evidently painted by someone with a fundamental and scientific understanding of geology which makes it thoroughly convincing yet without sentiment or prettiness. The painting of the flowing glacier and its lateral moraines, and the rocky ribs of these Karakoram peaks has exactly the right 'feel'.

Haversham's knowledge of geology is supported by a comment he made, later in his life, to the eminent Himalayan historian (and sometime Superintendent of the Survey of India and Professor of Geography at Oxford) Kenneth Mason. Mason himself describes Godwin-Austen's work in the region as a beautiful example of the surveyor's art and goes on to say that the then elderly Haversham described how he had learnt from studying the topographical drawings of the 'old French pictorial school' and that he drew topography as close to nature as possible. '*I was always mapping with a geological eye and that makes a man look well at the country.*' (Mason: *Abode of Snow*, pp77-8.)

But Godwin-Austen's lasting artistic legacy lies in the wealth of chronological documentary material he compiled throughout his extensive travels and surveying work on his Karakoram, Central Asian and Burmese expeditions. What he has left behind constitutes a treasure of national importance.

This last phrase is, however, relevant not just to Godwin-Austen's watercolours, drawings and sketches. What he gave, through his research, collections and publications to natural history and geology is not just a national treasure, but a global one.

Godwin-Austen's artwork extended beyond his time in India. Various watercolours survive of Buddhist mandalas, Alpine and Scottish landscapes, and at least two drawings of what was presumably a trip to Anglesey, all painted or drawn from his return in 1877 until at least 1909, as well as an immense number of molluscan drawings. Painting and drawing continued, therefore, to be a satisfying outlet both for scientific rigour and private emotions, generated by a happy marriage with a grim end, the loss of friends and family, many of whom Haversham outlived, bankruptcy and unending straitened financial circumstances, a dreadful war, increasing physical pain, a discreet religious conversion and a haunting by a ghost or two from his early years while he struggled with academic recognition and 'grand old man' national status – Godwin-Austen's final quarter-century was to pass through heavy turbulence.

PART FIVE

THE SHADOWS LENGTHEN

CHAPTER TWENTY-THREE

LIFE IN WARTIME

War and Politics

GODWIN-AUSTEN REMAINED BY NO MEANS lost in the world of natural history during the war. His other social observations are varied. Some are predictable in referring to his farming prospects:

> I am finishing my haymaking today, quite a stupendous rick. I have not made one so large since 1886. Not a drop of rain on it which can seldom be said.[1]

In terms of more mentally demanding distractions, Godwin-Austen kept up a fairly regular political commentary. Just before the outbreak of the First World War, his sense of unease is betrayed by the repetition in his first sentence:

> We are living in curious & anxious times & I am anxiously looking out for the M.P. The Govt. have found out too late that the Regular Army is not raised to carry out the dirty work of a Political Intrigue, the worst kind the country has seen for many years. Lloyd George's speeches in Huddersfield are nice reading; he wants a serious *(?)* working up for revolution.[2]

A similar theme is developed, and similar emotions expressed, not long after war breaks out:

> War news occupies a lot of time, and there is a disturbing element all around one. Now it is the billeting, in the villages here. We have started a Rifle Club in Hascombe for men & boys miniature rifle ... As of last Sunday we had 40. The best shot was a boy of 12 ... 25 years to begin with ... It should be going on everywhere.[3]

His observation a year later was of more unusual local interest:[4]

> A Zeppelin passed over here at 10.30 & I heard dropped bombs on St

---

[1] Letter to Harmer, 1 August 1916.
[2] Letter to Harmer, 24 March 1914.
[3] Letter to Harmer, 14 November 1914.
[4] Letter to Harmer, 14 October 1915.

> Catherine's, Guildford with what result is unknown. My coachman was up, & it apparently came from the side of Red Hill.[5]

But the war years were not periods of complete isolation: visits from scientist or explorer friends occurred quite often. Godwin-Austen was a hospitable person; invitations to others flowed often enough through his letters of this period. In 1916, for example, he enticed Sir Aurel Stein, his only serious rival as the greatest explorer of his age, apart from the eventually disgraced Sven Hedin, down to Nore. After a talk about Kashmir, on which Stein was to deliver a paper to the RGS, and which Godwin-Austen had been the first to map, he offered to send on a '30 inches long' engraving of a panoramic view of the Kashmir mountains from Sealkot to help to illustrate the paper, along with the name of his engraver, and some of his original notes.[6] These apparently helpful gestures caused Stein some difficulties, as the notes did not fit into the material which his paper was to cover. As Stein rather crisply remarked:[7]

> They have no bearing from the point of view of critical scholarship upon the questions discussed by me.

He then moderated his tone:

> ... they have a certain personal interest as they embody recollections of a veteran Survey Officer for whom I feel sincere respect reaching back half a century.

Stein tactfully asked Hinks, the infamous but durable RGS Secretary, to print his abstract, only, in the *Geographical Journal*, and to insert Godwin-Austen's note as an addendum when the full paper was published. There the matter rested, and is included simply to illustrate both the kind of exchanges which Godwin-Austen dealt with, and also to suggest that by referring rather patronisingly to Godwin-Austen as a 'veteran Survey Officer' Stein was implying that Godwin-Austen's scholarship might by this time be being superseded by a new generation of geographical and anthropological scholars, of whom Aurel Stein was a leader.

By late in the war, however, objective social observation has given way to a kind of rather blimpish though still not imperceptive irritability:[8]

> Thank you for a batch of war literature ... those German posters, they are brutal, but I really wish our people had a little of the German hard character & decision. The Irish[9] require it, sadly. From the Govt. downwards this country is full of old women. They seem afraid to act for fear of something from some side or other. There is no-one to act as was done in the Indian Mutiny, when fortunately for India it was not in close contact with the Home Govt. No general is safe from attack now & it does not strengthen one line on any front.

Godwin-Austen's frustrations, shared by millions, about the inconclusive nature of trench warfare, were not just political or military. In the same letter, as his uncharacteristically careless self-expression makes clear, he added, in a painful contrast to his optimistic comment earlier in 1914:

> The war plays the devil with everything *(sic)* away and a lack of labour. 16 acres

---

[5] Now Redhill, an important railway junction town about 30km/19 miles east of Guildford and just north of Gatwick.

[6] Letter to Hinks (RGS Secretary), 30 November 1916.

[7] Letter from Sir Aurel Stein to Arthur Hinks, 11 January 1917.

[8] Letter to Fagan, received 24 April 1918.

[9] Possibly a reference to the Easter Rising in Dublin, 1916, and the ensuing troubles.

> of hay now lies *(sic)* rotting near this house. I could not set it out until until *(sic)* the 10th Aug. & there it is still & can only go into the yard. But this is nothing to the loss of neighbouring farmers with wheat are in a dreadful state. Weather is an enemy both at the front and here.

By the end of the war, 11 months after the Armistice, he returned to a pre-war concern,[10] the threat of revolution, a fear echoed by many who had seen Britain rise to imperial greatness only to see its economic and political power broken by the war:

> London letters read this morning, so it seems safe to post letters again. After seeing all my life this country rising to greatness year by year, it is a dreadful sight to see what its rulers are bringing it to. There is much hope yet for this attempt at revolution must have its effect on the mass of the people.

As the war was ending, Godwin-Austen's political and scientific interests fused in a justifiably angry attempt to resolve an immediate ethical problem: the future treatment of German scientists. Godwin-Austen's ideas were expressed with characteristic vigour:[11]

> I have just read Lord Walsingham's excellent letter in NATURE of Sept 5, and agree with all he says ... It is impossible we can meet them *(i.e. German scientists)* as if nothing had happened since 1914 ... It is impossible to dissociate the mental attitude of the population of that country, by no means excepting the highly educated and scientific classes, from the world-conquering aspirations of their rulers, or from the barbarous atrocities committed by them in pursuit of that national ideal.
>
> I have not heard of a single letter from the very large number of the above scientific classes in Germany to acquaintances in this country in which such acts have been denounced, nor have I seen any protest or condemnation of German methods coming from the Germans in our midst ... who have enjoyed in this country friendship, hospitality, and even protection, such as no British subject could hope to receive in Germany ...
>
> Instead of protecting objects of science and art by leaving them intact for the benefit of other nations and the world in general, the German has raised looting and destruction into a devilish art. Soldiers are trained ... to commit useless destruction, combined with every conceivable atrocity on man, woman and child. It is lamentable to think of the geological and natural-history collections which have been destroyed in Belgium alone – a country famous for its scientific men, the work of their lives gone for ever. I trust all this will not be forgotten when the war is fought out to a proper issue, and that all Lord Walsingham suggests will come about and just punishment thus be meted out.
>
> The German scientific man has been spoilt by success in the past; he was first in the field in many countries, particularly our own. I knew him in days gone by in India, when he filled the best appointments in the Geological Survey, the Forestry Department, etc. Many were friends of my own. In these days they were quite different men in every way from those of today, so complete a change has come over the whole German population. It is sincerely to be hoped they will never be employed again in any capacity.

---

[10] Letter to Harmer, 2 October 1919.
[11] Article in *Nature*, 26 September 1918.

This unforgiving anger, rather like 'Tiger' Clemenceau's when the Treaty of Versailles negotiations got under way the following year, is understandable (if not necessarily tactically sound should it have been acted upon). Unlike Clemenceau, however, Godwin-Austen seems to have taken up a more forgiving, peaceful point of view once his immediate anger had settled. After the war, indeed, a much more positively exciting distraction hove into view: the first ascent of Mount Everest (see Chapter 24).

LATE MILITARY THOUGHTS

Although he only spent about six years on active military service, and the rest on the Trigonometrical Survey, Godwin-Austen certainly considered himself to be a military man. In the early 1900s, shocked by British reverses in the Boer War, he was moved to write a longish essay which appears to have been designed for private circulation and discussion among his military friends:

(The Education of an Army Officer)[12]

> To what extent the education of an officer should be carried, & what he should know before obtaining his Commission are two questions very difficult to determine and it is certain any attempts to make the attainments of all officers alike can never be hoped for. In practise *(sic)* we find every type of English Society, from the man who has excellent brains, had all the advantages of first rate education, yet never putting his talents to any use, as soon as men enter the Army so many seem wanting in energy of mind, have no ambition and no mental pursuit, dropping at once into that easy going life of amusement the Service offers them. On the other hand there are some less gifted as to brain power who are learning all their lives, have some account of moral courage and are not led by the opinion and remarks of the idler section of their brother officers. All the schooling in the world will not give a man that tact and strength of mind so necessary to command and lead others. A thorough knowledge of professional duties and his own language undoubtedly assists an officer as greatly when in a position of responsibility, but it is not increased by any extra mathematical or linguistic attainments. The ordinary duties which the greater number of officers have to perform both in Peace and War time are not of that order requiring a high class education, we know many most able and distinguished officers have had little or no knowledge of a single foreign language or of Latin & Greek, but they had the power of application of working themselves and thus lead by example or command others to do the same.
>
> As in individuals so there is a remarkable difference between Regiments in our service, a difference I have no doubt is to be found in the armies of other countries. It has been noticed on many a campaign, and in quarters. It is difficult at first sight to understand why the difference should be so pronounced in corps recruited very much from the same source; but there is no doubt this variation depends very much on the Commanding Officer, and the Senior Officers, and may very often be traced to those who have long left the service, the impression of their work and example still remaining. The education of the individual officer has only an indirect bearing on Regimental efficiency. There is no reason why the Regimental unit should not be more uniform in tone and efficiency throughout the Army ...

---

[12] No title in the handwritten original.

Having completed his introduction, Godwin-Austen then went on to review, with increasing sharpness, the present situation concerning the Second Boer War, 1899-1902, rather than the Zulu War, where Freddie died. Arthur Godwin-Austen's father-in-law (i.e. Betty's father) fought in it and no doubt discussed the British defeats and eventual victories with Haversham:

> The South African War has taught us many lessons one that neither extreme youth nor extreme age when armed with a rifle is a bar to usefulness & ability in the field. The Boers were not a military people in the European sense they had no military organisation, no system of examination yet from the farm and from the desk came new *(?)* with all the qualities to command all the natural inborn knowledge of tactics, & how to meet & how to avoid an enemy, no people have shown better how the essence of success and holding out is to inflict the greatest amount of loss on your enemy with little or none to yourself. Every male who cd ride and shoot joined the commandos all imbued with the spirit to do their best and fight, prepared to rough it in a way unknown to the majority of men & officers in our Army. Their efficiency to a great extent was due to their rough primitive life which teaches men to be helpful and self preserving, a Boer could never feel himself lost even when alone on the veldt.[13] These are qualities which the state of comfortable even luxurious existence in Great Britain and most of Europe does not call forth – we may try to teach these qualities but for the majority they can only be purchased by experience in the field. Take even our mounted branch how many horses have been ruined and their full period of usefulness cut short, not by intentional cruelty but simply by the lack of care and ignorance of the rider who has never handled a horse previous to enlistment.
>
> I think we may safely say by insisting on a high standard of education the Service will lose a number of young men of moderate ability, who having been born and bred in country parts, have knowledge of animals, can ride across country & shoot – who besides mix a good deal with the labouring classes & understand them, & many other very practical things, not to be learnt at school or in a town. Many such young men would be considered ignorant by a University Professor & are in comparison with others ignorant, yet they make excellent officers, & when the time comes for real work to be done are quite useful, in many respects more so, than the highly gifted better educated men. The fact is we cannot afford to lose any good & willing material we have got. We want to remove blemishes to the Army System which in some way or other have been conspicuous in South Africa; the short-comings recorded by such witnesses as Lord Roberts[14] and General Ian Hamilton[15] are of a very serious nature. Some of them have been present in our Army for a long time in one form or another. The writer can recall the feeling and tone in some Regiments of the Line in India just after the Mutiny when a number fresh from Home were in the country. The tone was markedly different to what one had been accustomed to. The officers, especially the younger ones, were rather proud to say none of us go in for 'black classics'. Not 'good form' in fact a form dead against efficiency. How valuable such efficiency and a knowledge of the native languages proved only a few months before when the Mutiny burst out. In my own a Queen's Regiment, then ten years in the country, so many officers had the P

---

[13] South African grassy plains, where the uncamouflaged British at first suffered heavy casualties.
[14] Field Marshal Frederick Sleigh Roberts, 1st Earl Roberts, VC, KG, KP, GCB, OM, GCSI, GCIE, PC (1832-1914) was a distinguished Anglo-Irish soldier and one of the most successful commanders of the Victorian era.
[15] General Sir Ian Standish Monteith Hamilton GCB, GCMG, DSO, TD (1853-1947) was a general in the British Army and is best known for commanding the Mediterranean Expeditionary Force during the Battle of Gallipoli.

> 'Persian' or P H 'passed Hindustani' standing after their names in the Army list ... At the same time other officers were on staff employment of different kinds, assisting in the advancement of the country & our knowledge of it. This state of efficiency for anything that might come about, was due solely to the example and advice of the senior officers.
>
> Esprit de corps and pride in the past services of a Regiment is one thing & one to be fostered, but the 'not the form to do this' & 'not the form to do that', & exclusiveness, is quite another thing. It begins with the Seniors and goes down to the last joined subaltern, becomes meaningless & silly & should be stamped out. It may shew itself in extravagance, against which the young officer, with small means, can make no resistance, he gets into debt; is considered useless in the Regiment, becomes a burden on his relations, or has to leave the Army. So few young men have the moral courage to battle with such a system.

The 'debt' idea may be a nod towards Arthur, although the exact history of Arthur's waywardness is unclear. Having, he believed, dealt with some of the existing problems, Godwin-Austen turned towards making some practical recommendations:

> Are young men at the most impressionable period of life undergoing instruction at Crammer's Establishment, properly supervised, & if not are they improved? The easy going life led at a Crammer's led by some young men away from the Public Schools where they are under regular discipline, they often pass to the Crammer where they would become quite as efficient under the masters they leave.
>
> Joining the service young has been a success in the Navy, engendering promptness of action and independent thought at an early age, why should it not be applied to the Army? Would it not be better at the age of 17 or 18 at the latest for a young officer to be learning his work in a Regiment, mixing with men other than himself, & who would exert considerable influence over him – it is too late for this at 20, when a good many imagine they know as much as their seniors. Taking it for granted that methods of Cram cannot compete with steady slow instruction, as this is imparted the future officer should be examined by a Board of Officers (assisted by Professors as Examiners) in each subject viva voce as much as possible, and judge whether the candidate came up to a certain standard of qualification, the board would judge for themselves on his physique and appearance, and weigh his General Knowledge of one subject against another before recommending him for a Commission, this is surely a much more practical test than that of marks.
>
> What is more to be desired for our Army is a liberal sprinkling of well educated officers in every Regiment rather than a high standard of mutual ability for all. Such is only to be obtained by a course of study at a Military College for not less than 3 years – a certain time devoted to reach subject some of them compulsory and passed in at half yearly examinations. As Commissions would be obtainable at 17 or 18 at the latest, admission to College would be at the earliest 15.
>
> We want to remove blemishes in the Army System ... the shortcomings recorded by such witnesses as Lord Roberts & General Sir Ian Hamilton are of a very serious nature.

The essay may not be the most original of its kind, nor indeed the most concisely expressed, but it does at least show evidence of the mature understanding of someone who has been through the military system and reflected on it. Godwin-Austen's military concerns were therefore geographically quite wide-ranging. During his final decade – he turned 80 when the First World War was ending – his interests again turned towards the East where, I suspect, he still lived in spirit.

CHAPTER TWENTY-FOUR

LAST CONTACTS WITH THE EAST

Mount Everest

ALTHOUGH K2 WAS MENTALLY REVISITED through papers and discussions during Godwin-Austen's final decade, as a grand old man of mountaineering – and a very conservative one at that – his opinion was publicly sought, and where it was not sought it was sometimes still given, on that great British enterprise, the first ascent of Mount Everest. If Godwin-Austen's way of presenting his advice indicated that his mental powers were in decline, it is nonetheless true that the shrewdness and experience behind some of the advice would almost certainly have led to an earlier British success on the mountain, and would also have saved lives, possibly sparing us the endless and largely fatuous controversy about the deaths of Mallory and Irvine.

THE FIRST EVEREST EXPEDITION

Although Mount Everest originally had its height calculated in 1852, no serious discussion about exploring or climbing it seems to have occurred until about 1885. Godwin-Austen had connections with the enterprise from the start: a member of Sir Martin Conway's historic Karakoram expedition in 1892, for which Godwin-Austen had been one of the principal advisers, was Charlie, later General Bruce, one of Himalayan exploration's most memorable characters; this amiable, ebullient walrus, who never failed to inspire devotion in his men and who had a grasp of the ethnic complexities of Himalayan culture like few others, was proposed as the leader of the first Everest Expedition in 1921. Because of army commitments, he was unable to go, but he did lead the 1922 expedition. And much of the RGS support for the project had been pushed along by Sir Thomas Holdich, Godwin-Austen's friend since Burma days and one of his eventual obituarists. Permission from the Tibetans (Nepal remained closed to incursions by Westerners throughout this period) only arrived in 1920: the proposed expedition, which after much debate was defined primarily as a reconnaissance, since at that time no Westerner had been within 40 miles of the mountain, had to be organised rather speedily, despite the British, in a typical bout of last-minuteness, having had about 70 years to think over the matter since Everest's first sighting.

Godwin-Austen, despite his 86 years, outlined and argued for his views in a widely circulated article early in 1921.[1] He began by acknowledging the groundwork covered by the Survey of India,

[1] Article in the *Surrey Advertiser*, 22 January 1921, and circulated to the principal London scientific bodies thereafter.

and then recognised the influential role of Younghusband and Bruce in arguing the case for an expedition. He recommended that the expedition make an early start, as '*old and tried methods are the best*' and as much time would be required to complete a survey. He recommended that Colonel Ryder, then Surveyor-General, should choose the surveyors, '*for I do not think they are to be found in England*'. Although the article was designed for general consumption, Godwin-Austen's scientific enthusiasm carried him away at times, so that when he made the comment following, only a tiny proportion of Godwin-Austen's readers would have understood what he was talking about:

> The geology alone is of great interest, for Hooker reported finding a fossiliferous limestone on the Donkia La containing doubtful nummulites, which, if proved to be that genus would indicate the Eocene age of the formation, with all that would mean.

He then turned quickly, however, to the nub of the matter – who should the expedition members be? Should there be a majority of climbers, or surveyors?

> At a time when we should be sending able observers into this country, sending a lot of mere climbers was suggested by Captain Longstaff ... he said, 'In the first year, when the reconnaissance party is to go out, this party should include as many experienced climbers as possible, not with a view to climbing the mountain but with a view of finding the route.' Those climbers, in my opinion, with their impedimenta and servants, would be an encumbrance to any single surveyor, eat up the limited amount of food, and increase the number of porters and their cost. On the other hand ... what Captain Longstaff[2] put forward was practical, he has a good record. I have closely followed his career in the Himalaya, and he says truly about the coolies, 'Feed them well, blanket them well.' And, I would add, get to know them personally. He also says that climbers should not remain long at high altitude, losing strength.

His summarising pronouncement is both clear, and incisively foresees, 50 years hence, the development of relatively speedy Alpine-style ascents in the Himalayas after the much more cumbersome 'siege'-style expeditions, such as the Duke of Abruzzi's 1908 attempt on K2 or Brigadier (later Lord) Hunt's triumph on Everest in 1953, have had their day. '*I am convinced, the smaller the party and the less fuss made about it, the better.*'

He developed this belief with a personal recollection of Everest which, given the fame which is attached to him through K2, makes for interesting reading as a counterpoint to his first sighting in 1861 of 'the great peak on the watershed of Asia', and because of the K2 man's memory of Everest having been overlooked until now:

> I first saw Everest from Senchal, a trigonometrical station near Darjiling, where I camped three days in 1863, observing to the snowy peaks, and starting my plane-table work for Bhutan. When at the Deyra Dun office, not knowing exactly what was before me, and in order to be prepared for surveying in any direction, I projected two plane tables on the 4 mile to an inch scale, one to take all the country on the Nepal frontier, the other all to the eastward Bhutan ... The peaks of the Kinchenjunga *(sic)* mass came into both plane tables. That for Bhutan eventually came into use, and on it the country as far as Punakha ...[3]

---

[2] Tom George Longstaff (1875-1964): one of my mountaineering heroes. See biographical note in Chapter 10.
[3] Punakha: the former winter seat of the Bhutanese government: a low-lying *dzong* (fort, monastery and law-courts) lying north-east of Thimphu, the Bhutanese capital. Godwin-Austen's furthest point in Bhutan in 1864.

> On the other hand, had I gone to the westward of Sikkim, I should have been in a position to secure the greater part of the Arun[4] drainage, and the approaches to Everest in the same time and with much the same native establishment. This is what is just at present required, before any attempt on an ascent of Everest is made. Finding the route is not, therefore, such a difficult matter as it is supposed to be.

This recommendation for a small, Alpine-style ascent party going in via Nepal would have saved the British the expense, trouble and deaths of both sherpas and Westerners over six Everest expeditions on the Tibetan side if Godwin-Austen's advice had been heeded (and if the Maharajah of Nepal had been persuaded to permit access). The 1953 expedition travelled to Everest through Nepal, along branches of the Arun river system – the route was in those days not at all easy, indeed more or less unexplored, but was always less extended and arduous than the pre-war route through Tibet via Kampa Dzong and Rongbuk.

Godwin-Austen returned to his sighting of Everest:

> The three days spent *(at Senchal)* were cloudless, and I had an unrivalled opportunity of observing the great snowy range from Everest to Chumalarhi[5] (*sic*) in Bhutan, of which I was soon to see more.

He concluded his article with a lyrical, almost mystical and rather unhelpful (in its vagueness) statement of his main theme.

> The peaks and ranges still stand as they were in my time; the rivers still flow to the sea, carrying the waste of the Himalaya to build up the delta of the Ganges and fill the Bay of Bengal, as they have done for countless ages. They await the surveyor to lay them down and fill a gap in the map of that wonderful and diverse country, India.

A very unfunny and uninventive satire of the RGS meeting where the discussion about the nature of the Everest expedition took place appeared in *Punch*[6] shortly after. For example, Sir Martin Conway is called Sir Marcon Tinway; Younghusband is evidently 'Sir Francis Oldmead', and so on. The whole piece is far too laboured to quote in full, but the leaden section satirising Godwin-Austen's 'contribution' is worth quoting, partly to show how Godwin-Austen was perceived in journalistic circles at that time, and partly to show why *Punch* mercifully closed rather too many decades later.

> Professor Parbatt expressed his keen appreciation of the vivid descriptions of Himalayan scenery given by the lecturer, and the admirably-selected photographs which had enlivened his address. He wished, however, that he could have furnished more details as to his camp equipment. Had he, for example, used Nummulitic beds for his party? Then there was the question of geoidal deformation, on which he had remained unaccountably silent. As for the vital problem of approaching Mount Amaranth,[7] he ventured to differ from all previous speakers. The Northern, South-western and Eastern routes were all equally impracticable, as he would

---

[4] The River Arun, which forces its way through the Himalaya next to Everest.

[5] Now more usually, Chomolhari, approx 7317m/24,0006ft, on the border of Bhutan and Sikkim. First climbed by F Spencer Chapman in 1937.

[6] Quoted by JBL Noel, *Through Tibet To Everest*, Hodder and Stoughton, London, 1989 (first edition, 1927), pp95-7.

[7] As 'Amaranth' means 'everlasting', there is a play on 'Ever-rest'; this is not untypical of the rather strained wit which pervades the article.

> conclusively demonstrate from the photographs he had brought with him. But there were at least fourteen routes from the West, of which he would confine himself to four. (1) starting from Yeh, the party might cross the Tablung-La pass to Gorkpa Nor, and thence follow the Yombo to Chilgat, where they would be only 25 miles from the foot of the western face of Amaranth. (2) They could follow the old Buriat pack-road to Amdo, diverge by the narrow defile of Koko-Pir-Panjal to Tumbung, and thence make for Chapchu-Strong and Chyang-Chub-Gyultshan. (3) They might start from Pongrot and cross the Tok-Tok pass to Pilgatse. (4) They might construct a tube from Darjiling to Grogma-Nop, and thence proceed by aeroplane to the saddle of Makalu, or, better still, to the summit of Amaranth itself. The last route was far shortest and quickest, but it involved a certain amount of preliminary expense.

(The lunatic idea of laying a pipeline, through which climbers could crawl, protected from the weather, as far as the Lhotse Face was seriously suggested to Lord Hunt during his Everest expedition's preparatory stage as one of the possible ways of facilitating progress during the ascent. See *The Ascent of Everest*, Hodder & Stoughton, London, 1954, p45.)

Godwin-Austen's advice on the primacy of surveying was not heeded. The 1921 expedition consisted of both surveyors and climbers, almost all of whom could manage something of both. There were some difficult clashes of personality, but a thorough exploration and photographic reconnaissance were carried out on the Tibetan side and a feasible route to the summit eventually found. Assuming that the weather had been no worse than on the 1921, 1922 or 1924 expeditions, however, I think it could be quite confidently asserted that, led by Bruce and with Norton's drive and Mallory's ability, Godwin-Austen's advice of an approach through Nepal would have led to a much earlier British first ascent and much less loss of human life as well as an entirely different economic development history for Nepal.

CONVERSION TO BUDDHISM

It was not just Everest which drew Godwin-Austen's attention, however. The long endgame to his life was, in sum, a struggle between public recognition of his great work and a successful marriage on the one side, and his private financial difficulties and increasing physical infirmity on the other. The outcome of this struggle appears to have been a movement towards Buddhism, a philosophy which, more than any other, concentrates on detachment from the pain of the everyday world, although, that having been achieved, contains the option of returning to help others to achieve a similar goal. The Buddhist principle of not killing any sentient being must have been particularly attractive for such a committed natural historian as Godwin-Austen (although not entirely compatible with Godwin-Austen the ornithological collector), and something he must have learnt about after he reached Burma in 1852 as an impressionable young man, perhaps reinforced through his contacts with Tibetan Buddhists in the Lake Pangong and Ladakh area in the 1860s.

There are four pieces of circumstantial evidence to suggest his conversion. It is a long leap from having been 'to all intents and purposes a follower of the Prophet' back in the middle to late 1850s and is not entirely compatible with several references in his letters, between 1890 and 1920 to 'going off to church' (although Godwin-Austen was certainly sensible enough to realise that Buddhism might have seemed a rather alien philosophy to the inhabitants of early 20th-century Surrey, quite apart from the necessity of keeping up his apparent public position as the local squire).

Haversham had been involved in local church affairs for some time. From 1885 until 1898, he was a regular attender at parish vestry meetings, as was Robert Alfred Cloyne before him. When he stood in 1896 for the parish council elections, however, he came sixth, the places being more popularly occupied by local builders and tradesmen. The result perhaps suggests that he was not particularly well known in the parish, despite the general familiarity (which continues to the present) which the family enjoys in the neighbourhood.

The deliberateness of Godwin-Austen's various 'Buddhist' actions, however, does suggest at least a private leaning towards the philosophy he had come across in Burma on his first posting abroad.

In considering this final philosophical meander, the first evidence comes from both Lieutenant-Colonel Robin Godwin-Austen and his brother, Dr Richard, recalling that their mother mentioned, on several occasions, that Haversham had become a Buddhist. The idea never seems to have been developed by her beyond that assertion, however, and must therefore be considered unreliable.

The origin of the idea nonetheless does have definite roots. At Nore, one can still see to this day, a miniature Burmese pagoda,[8] about the size of a largish beehive, which was installed by Godwin-Austen some time after 1901. It appears, furthermore, that it was installed in some out-of-the-way part of Nore's extensive garden, so that casual visitors could not see it or, therefore, discuss it. This secrecy is underlined because a subsequent owner, the film star Dirk Bogarde, discovered it, as he recalls in a private letter, 'among some brambles' where it appears to have lain neglected for almost 30 years. The 'pagoda' is made of weathered pinkish sandstone. Its features are distinctly Burmese: the guardian lions, the four arched doors, the phallic pillars, the stepped style of the *stupa* (the dome-like central section usually used to enshrine a saint's relics). What is less clear is whether this strange edifice was imported from Asia, or made to Godwin-Austen's specifications out of British stone. A Cranleigh (i.e. local) stonemason has examined a piece of the 'pagoda' and affirmed that it is not Surrey stone; whether it is British (e.g. Nottinghamshire sandstone), or northern Indian pink sandstone is harder to say. Certainly, Calcutta, from where Godwin-Austen left Asia for the last time in 1877, possesses a number of Burmese pagodas, no doubt because there were suitable craftsmen there, close to Burma, who could replicate a pagoda's features. (The 'pagoda' now occupies a position of some prominence at Nore: it can be clearly seen from the kitchen: it had acquired an attractive mossy cover which has recently been cleaned, and is surrounded and well set off by paving, flowers and shrubs.)

The existence of the 'pagoda' ties in with one of the oddest events in the last few years of Godwin-Austen's life, the presentation of 'an old Burmese bell'[9] to the Shwesandaw Pagoda, in Bagan, Burma, in 1920. A letter of thanks from the Burmese Chief Secretary's Office, in English, is followed by a description of the installation ceremony, first in a beautiful gold and turquoise embossed statement, written in Burmese (which Godwin-Austen certainly had a grasp of), then in English. The document describing the installation ceremony talks of the 'returning' of the bell which suggests that, if linked to the acquisition of the Nore 'pagoda', is an act of Buddhist piety, but might also have been the closing of an old wound from the Second Burmese War where the British, then and before, were known to have removed pagoda bells, e.g. the Great Bell of the Shwedagon Pagoda in Rangoon (Yangon), in order to demoralise the people they had conquered by displacing an object of significant religious veneration. The description of the ceremony tells us that British troops did in fact remove the bell in 1852 and present it to General Godwin, which explains how the bell fell into Haversham's possession.

At any rate, the installation ceremony was sufficiently grand for it clearly to have been a momentous event. Given that the *stupa* is said to contain some of the Buddha's hair, this elaborateness is hardly surprising.

The restoration ceremony began at 7.30 in the morning of '*the full moon day of Kason*' (1 May 1920). The British District Commissioner was met by the pagoda's trustees under a golden umbrella (canopy). The commissioner then addressed the trustees (in Burmese) on the subject of the history of the pagoda and its bell. The bell's history is a charming and fanciful one.

---

[8] See plate 44.

[9] Letter to Godwin-Austen from the Chief Secretary's Office, Maymyo (signature unintelligible, possibly 'Newnham'), 2 June 1920.

> The power of the bell was such that its sound penetrated far and wide and all who heard it were bound to do homage to the great King. At this time the Peikthano Princess ruled over the far off Mahamyaing country between Shwebo and the Upper Chindwin. She too possessed a magic drum given her by the Nats, the sound of which reached far and wide and compelled all who heard it to pay tribute to the Princess. Her country being filled with the sound of this drum, the sound of the Thayekettaya bell could not penetrate it. So King Dwittabaung's crafty Minister went up in disguise and after various adventures succeeded by stratagem in damaging the drum. Then the sound of the Thayekettaya bell penetrated the Mahamyaing country and the Princess was bound to come and do homage to King Dwittabaung, and become his Queen.

Having diplomatically if casuistically suggested that the removal of the bell by the British was so that Western nations could admire it as a work of Burmese Buddhist art, the Commissioner returned to immediate reality by stating that Lieutenant-Colonel Godwin-Austen, the general's grandson, 'who himself had pleasant memories of Prome in 1853',[10] has now sent back the bell to be restored and 'thereby earned the thanks of the Buddhists of Prome'.

The Commissioner then moved on to less diplomatic and more political territory, by suggesting that Burmese youth disregard the religious message sent out by pagoda bells, because:

> They no longer pay heed to their parents, teachers and elders; they drink liquor, eat opium, gamble, and indulge in various forms of vice and crime.

Presumably the British-directed local police forces were undergoing difficulties in law enforcement at that time. The Commissioner piously continued by mentioning that temperance societies had been formed and that the taking of the pledge against alcohol and opium consumption was to be applauded.

Perhaps more tactfully, Mr Graham (the Commissioner) then proceeded to cite recent examples of Burmese-British co-operation, for example in British Government credits to help the Burmese through recent crop failures.

When Mr Graham had (eventually) concluded, the Chairman of the Trustees, U Aung Nyein, read out his address, beginning with expressions of continued loyalty to the King Emperor (George V) and the Queen Empress (Queen Mary), followed by declaring the joy which the Buddhist community felt at the return of the bell, and the community's gratitude to Colonel Godwin-Austen. The choice of date was said to be particularly happy, 'the full moon of Kason being Buddha's birthday and also the day on which he obtained omniscience'.

> The bell was then carried in procession along the strand and up the Pagoda road to the northern gate of the shrine accompanied by music and dancing, the people applauding from the sides of the street, and was finally replaced on its former site on the platform. The installation of the bell was followed two or three days later by heavy rain after a long season of drought, an event which caused general rejoicing.

This act of considerable generosity, at a time when he could hardly have spared the expense of shipping a heavy and bulky object from England to Burma, can be linked to the Papplewick mandalas. A mandala (from Sanskrit: *mandala* 'essence' and 'having' or 'containing', also translates as 'circle-circumference' or 'completion'), is a concentric diagram containing spiritual and ritual

---

[10] See Chapter 6.

significance in Buddhism and Hinduism. In the Tibetan branch of Vajrayana Buddhism, mandalas have been developed into sandpaintings, often destroyed after long and painstaking completion in order to symbolise the ephemerality of earthly existence. The Papplewick mandalas take the form of two watercolours (see plate 43), unattributed and undated, in an artist's portfolio containing other works both by and about Godwin-Austen and his time in Asia. Although the paintings show fine detail, they are not especially detailed by the standard of the most complex and sophisticated mandalas, perhaps suggesting a 'learner' who knew his limitations; that they are watercolours, and the style of the circumference decoration on one being strongly reminiscent of the Burmese lettering practised by Godwin-Austen in the flyleaves of his 1852 Burmese diary combine to suggest strongly that he must be the artist.

Although all these pieces of evidence of Haversham's 'conversion' are entirely circumstantial, it is hard, seeing them together, not to conclude that at the very least Godwin-Austen's interest in Buddhism was a serious and sustained, if private one. That this interest had endured, probably from early days, comes across when Haversham is in a reminiscing mood late in his life, during the First World War (i.e. an historical event standing for everything which goes against the tenets of Buddhism). He tells Sir Aurel Stein that one of his scientific papers:

> ... has brought back to the foreground of my memory when I was really a lad, deeply interested in everything Indian, & reading with a Moonshie[11] every day ... gave me just that knowledge which renders travel & intercourse with the people so delightful.[12]

[11] Hindu term used here for a teacher or interpreter.
[12] Letter to Sir Aurel Stein, 14 December 1916.

CHAPTER TWENTY-FIVE

A SURPRISING CORRESPONDENCE

Kudikji's Ghost

JUST A FEW MONTHS BEFORE his death in 1923, Godwin-Austen suddenly received a letter from Mrs Helen Barclay, one of his granddaughters by his son with Kudikji.[1] For three months he underwent what must have been a severe emotional test; he appears at first to have handled this sudden reminder of his indiscreet youthful passion with instinctive kindness and skill but, as the implications of the correspondence dawned on him, he began to use rather less diplomacy.

The marriage of Edward Henry Hastings Milner (Haversham's son by Kudikji) with Emma Smith produced 15 children. (The first six were given acrostic initials to spell 'Milner', the surname of Eddy's adoptive parents.) The fourth child, Naomi Florence (who married a Mr Barclay and, confusingly, called herself 'Helen', despite there being a younger sister called Muriel Helen) wrote to Godwin-Austen in May 1923, five years after Eddy's death on 19 February 1917 of a mastoid abscess. (Eddy's parentage and its status were a publicly unacknowledged family secret.) A month elapsed before Godwin-Austen replied. The original of his reply has been lost, and a number of badly copied (and photocopied) versions remain. The Royal Geographical Society, however, possesses an attested true copy of the letter, which is worth quoting in full (for its intrinsic interest, and to establish the following as the definitive version):[2]

> Dear Mrs Barclay,
>
> Your letter of 3.5.1923 has reached me. I can quite understand your wishing it and wanting to know in what way we are connected. I may say at once you are my grand-daughter and I will answer all the many questions you ask. I have nothing to conceal, nothing to be ashamed of under strange conditions and very unusual conditions. I always tried to do the best.
>
> When Mr and Mrs Milner adopted Edward they acted splendidly. I supplied the means and when he went to Roorki[3] they saw him through, he passed out well and got into the Public Works Department and did good work in it. What could be better than this: it makes me proud of him and I am glad in my old age – for I am 89 on the 6th July – to be thus writing to you. I began life early. I was in the

[1] See Chapter 9.

[2] Godwin-Austen to Mrs H Barclay, 12 June 1923.

[3] Now Roorkee, a city in the Indian state of Uttarakhand. It is also known for Roorkee Cantonment, one of India's oldest, and has been the headquarters of the Bengal Engineer Group (Bengal Sappers) since 1853.

Army in January 1852: in India in 1853. When in the Kashmir Survey I married your mother in Budrawar[4] in May 1858. She was not a common woman but of good birth – land owners in the district of Poonch in the outer hills – not a Kashmiri. We were married under Mahomedan rites by a Mullah from Budrawar, her mother was present and all my Survey Camp. To all intents and purposes I was then a follower of the Prophet. Your mother's name was Kudji or Kudikji. She was a good wife with me and if there is anything now left that I can look back on and love, it is her memory and the way she looked after me. She went with me into Baltistan when I was doing very hard work on the Great Glaciers there. When your father was born at Sealkote[5] on 15th March 1859,[6] he was known as Edward or Eddy: the other names you give Henry, Hastings, Godwin are new to me and must come from the Milners. I must put you right here. You mention Mr Godwin-Austen, Deputy Commissioner in Berar,[7] Retired. He is my own brother,[8] I took him to India and got him into the uncovenanted Civil Service.

Knowing Messrs Tweedie's[9] address is very likely due to a sister of yours coming home as nurse with sick soldiers during the war. She went to Tweedie and wrote to me but I was not in Town and did not see her but it lead *(sic)* to letters passing with your mother who I am glad to hear is still alive.[10] I left all she knew in her keeping, the change has now come and I tell you all I know.

I am glad you have a good husband and are happy. I shall always be glad to hear of you. I liked India, the climate suited me better than this. I was very strong and stood both heat and cold. I wish I was there now, although it is no longer the country it was and never will be the same again. Lloyd-George *(sic)* and Montagu[11] did their best to ruin it.

It is something to think that your father had fifteen children and that you are the fourth daughter Florence Naomi Milner. You must give me a complete list of

---

[4] Budrawar: a town in the border area of Indian Kashmir and Himachal Pradesh, just south of Dharamsala, the seat-in-exile of HH the Dalai Lama.

[5] Sealkote: now Sialkot, a city situated in the north-east of the Punjab province in Pakistan near the Chenab river. It was formerly the winter capital of Kashmir. The city is about 125km/78 miles north-west of Lahore and only a few kilometres from Jammu in India.

[6] Almost certainly a slip for 1857.

[7] Berar Province, known also as the Hyderabad Assigned Districts, was a province of British India, ruled by the Nizam of Hyderabad; it was administered by the British after 1853, although the Nizam retained formal sovereignty.

[8] Harold, Haversham's youngest sibling. The other parts of his career are described in Chapter 3.

[9] Tweedie's: at that time and for several decades before and after, the Godwin-Austen family solicitors. Still a prominent London legal firm.

[10] Theresa died in 1958, aged 96.

[11] In June 1917, Lloyd George offered Montagu the office of Secretary of State for India. He had strong views that a further round of reform was inevitable in the government of India. Montagu believed this process should be taken forward and coupled this plan with a declaration that Britain's ultimate goal was Indian self-government. This was, however, to be in the far-distant future and Montagu's reforms were meant to swing India behind the war effort and quieten unrest for the next decade or so.

Montagu's intentions were embodied in the 1919 Government of India Act. However, it seemed initially to stimulate violent protests, rather than produce calm. Montagu urged the Indian Government to moderate its response to the violence and asked for a stronger condemnation of General Dyer after his massacre of demonstrators at Amritsar in 1919. He took a strong interest in foreign policy. In particular, he insisted the government should not provoke Indian Muslim opinion by partitioning Turkey, the only Muslim power of that period.

In March 1922 Montagu authorised the publication of a message from the Viceroy, setting out India's wish to revise the Treaty of Sèvres that had been imposed on Turkey in 1920. This severely embarrassed the Foreign Secretary in the middle of negotiations on this subject. Montagu had already challenged Cabinet policy in Egypt and Kenya and Lloyd George was finding him an increasingly obstreperous colleague. He took the opportunity to force him to resign. Montagu's career was over, and when he lost his seat in the 1922 election he retired from politics, an embittered man. He took a number of jobs in the City and died in London on 15 November 1924. Adapted from: www.liberalhistory.org.uk.

them. I think I have answered all and must conclude being
Yours affectly.
Sd – H H Godwin-Austen
Lt-Col

PTO

I am still doing work for the Survey of India connected with work I did in Kashmir in 1863.

1923
88
1875
20
1855[12]

Godwin-Austen's unwisely accommodating letter on the one hand displays his natural kindliness and wish to be reassuring. The idea of 'claiming kin', however, was not a new one, and his wariness can be seen of potentially lifelong difficult consequences of the kind perhaps best described in Hardy's *Tess of the D'Urbervilles* (1891): the innocent Tess is sent to 'claim kin' from her rich relations (who are not really related to her at all). This action releases a sequence of disasters which leads to the death of Hardy's greatest tragic heroine. Godwin-Austen's wariness made him anxious and more innocent than he really was: he tried a tactic of reassurance which backfired: Naomi Florence replied speedily and with equally foolish, blundering enthusiasm.

Sangor C.P.[13]
29 Cantonments
3-7-1923

My dear Grandpa,

I was glad indeed to get your letter of the 12th June 23. I am glad beyond words to know that our relationship is without flaw and that dear old Dad was really your legitimate son. If I have given you any pain in asking you to make this disclosure, which you have long held sacred, forgive me; but you have cleared the atmosphere of doubt which was weighing so heavy on me. You can understand our situation when I say that the common belief in India in most quarters is that Eurasians and Anglo-Indians cannot trace back their legitimate ancestry. That it is so in some cases I have no doubt myself and to this end I sought proof of mine: its *(sic)* the legal right of all: the humbled native knows exactly who he is and whence he came then why ought not we.

Now like all women I am just screaming with curiosity to know whether Grandma was really so beautiful a girl that you thought her worthy of you and the great renouncement you made for her sake – giving up your religion? You are, after all, not the only Englishman whose sojourn in India was attended with romance of this nature but where you were honourable – others were not. I do not blame you, nor do I the others, for after all we are the

– 2 –

children of Circumstances. I am a fatalist and believe he only trod the path

---

[12] These figures are written in pencil, almost certainly Godwin-Austen's handwriting. They are evidently an attempt, albeit an inaccurate one, to calculate Eddie's age.

[13] Perhaps Sangrur in the Punjab? Sangor is referred to as an army base in the 'right' area of Pakistan in British Army records for 1939.

aligned for us: the longer he lives, the greater the vissicitudes *(sic)* we pass through, this belief becomes more pronounced, with every cross *(?)* incident in life. I can sympathise with poor old Omar Khayyam[14] when he says, 'Ah Love! Could you and I with Fate conspire etc' and yet he might have been able to remould his life if he only did not drink so much wine.

It was so nice of you to have given me so complete a domestic history of your life in India. I feel very moved to hear that you still cherish the memory of the woman you loved. One rarely meets people so.

Now I must relate to you the little incident which led up to the finding of my Grandpa. Upto *(sic)* November last I was in ignorance of your existence: My mother should never have told me and I should never have asked but Fate decreed otherwise. I had a little gold coin of Vijnanagar Dynasty[15] given me by a friend together with its history. It so happened that this history was

- 3 -

published under the signature of Mr Godwin-Austin *(sic)*, Dy. Commr. *(?)* of the district in which it was found (Geotwal[16] I think).

While at Cossipore[17] (Bengal) I happened to meet a Mr Morris Carr, who was about sailing for England, at my Sister Eva's house, and has volunteered to take my collection of coins with him to have them valued and sold at Christy's[18] *(sic)* if the prices were appreciable. My brother-in-law Willie Lithgow, who is a Foreman *(?)* in the Govt. Shell Factory, on reading the name said that Mr. Godwin-Austin *(sic)* was a relation of mine and when I protested grew vehement in his assertion but beyond that he could not tell me anything interesting or convincing. However he sowed the seed of inquisitiveness in me: this and the few scraps of information I gleaned after coming to Sangor led me to take the action I did of writing to you direct. Yes, a letter of Rose's from England to Mother gave me your solicitor's address. Rose was very anxious to see you but when you referred her to your solicitors she lost heart & thought you did not care to see her. I am sorry you did not see her. She is the most unselfish sweet girl I have seen. She was very pretty once but keeps very sickly and thin due to the hard strain of private nursing. She and Dorothy live in a flat in Bombay and have a good practice. Irene was with them till last November and was married on the 19th. November.

- 4 -

Mr Carr, to whom I have entrusted my collection of coins, lives at 51, St John's Road, Erith, Kent which adjoins Surrey. I should now not like to lose that lucky little gold coin for it was the means of finding you.

Yes, you are quite right – India is no longer the country it was and never will be the same again. You who are many thousands of miles away can paint the picture of its ruin from the short laconic paragraphs in papers. We have the living fact before our very eyes and have to wallow in the political mire churned up by the Loyd George-Montague[19] *(sic)* collaboration. England has already killed the Goose that laid the golden Egg! The majority of high class Indians say themselves that they

[14] Omar Khayyam (b.1048, Neyshapur, Iran, d.1123), was a Persian polymath, mathematician, philosopher, astronomer and poet.

[15] The Vijayanagara Empire was a South Indian empire based on the Deccan Plateau. Established in 1336 by Harihara I and his brother Bukka Raya I, it lasted until 1646 although its power declined after a major military defeat in 1565 by the Deccan sultanates.

[16] Unknown.

[17] Cossipore (also spelt Cossipur, Kashipur) is a part of north Kolkata, formerly known as Calcutta, in the Indian state of West Bengal.

[18] Christie's, the London auctioneers.

[19] Lloyd George: British PM; Montagu: Secretary of State for India.

> were never born to rule: and they will never be able to do so disinterestedly owing to their inherent distrust of one another. For this reason they want the Englishman, but they want something else along with him: that something they cannot clearly define. It may be the Swaraj:[20] it might be anything. The fire of Political unrest was kindled by Mr Kerr Hardy[21] *(sic)* and has since been kept alight
>
> – 5 –
>
> by the unfortunate behaviour of some of our members in the Commons. To sum up – the English have got a fright and are prepared to do anything the Indian demands and he is also content to stand impotently by while the Indians commit the most atrocious deeds. The Prince of Wales[22] visit to India for instance. The papers said very little, but he saw and knew more than is good to reproduce on paper. The Kohat tragedies[23] are something to set us thinking. In my opinions of India I wish to be fair. There are faults on both sides. The administrator of today is of a distinct & different class of man compared with the pioneers or builders of India.

Naomi Florence then went on to give a list of the Milner siblings' birthdates and other pieces of objective family interest, before closing the letter:

> In conclusion we wish you a happy birthday. May God spare you for many years. I will be 38 on the 5th of July – a day before your birthday.
>
> Goodbye with love from Bert and me Yours affectly H Barclay[24]

In her understandable enthusiasm after Godwin-Austen's misfired attempt at reassurance, Naomi Florence had rushed into the wrong conclusions about Eddy's legitimacy and the consequent nature of the lack of family contact. We know that her enthusiasm had got the better of her rational sense because she had failed to take account of Godwin-Austen never having attempted to meet Eddy, despite them living not very far apart in northern India for a couple of decades. (And his unwillingness to get in touch with the sister who previously attempted to contact him should have been a strong-enough hint of his reluctance to develop contacts.) Even if her letter's subtext is a hope for a deathbed recognition of Eddy's legitimacy – a forlorn hope – it is more melodramatic than well-judged. On 6 October 1923, only 57 days before his death, Godwin-Austen, by now it seems horrified by the turn events had taken, replied in a tone which had altered from calculated warmth to a forbidding distancing:

> Dear Mrs Barclay,
>
> I have your letter of 12th September 1923, which is written in a very affectionate way, but I am obliged to write and point out that you cannot and must not call me 'Grandfather'.[25] My son, your ancestor, was with his mother Kudikje or Kudoj's *(sic)* consent, adopted by Mr Milner, and I and all concerned owe everything to Mr and Mrs Milner's bringing up. With my assistance he was educated at Roorkee, went there and passed out well into the Department of Public

---

[20] Swaraj can mean generally 'self-governance' or 'home-rule' (*swa-* 'self', *raj-* 'rule') but the word usually refers to Mahatma Gandhi's concept for Indian independence from foreign domination.
[21] James Keir Hardie (1856-1915) was a Scottish socialist and Labour leader, and was the first Independent Labour Member of Parliament.
[22] The future Edward VIII.
[23] Riots in 1921.
[24] This signing-off line is written at the top of the letter's first side, as Florence Naomi had not left herself sufficient space at the end of the last page.
[25] A remarkable volte-face from his previous letter, where his acknowledgement of Naomi Florence as his 'grand-daughter' is underlined!

Works in which he did distinguished service. To you I am in no way a Grandfather. You are related to me, and you can call me 'My dear relative' but better Colonel Godwin-Austen'.[26] I have a son by marriage, living in this country. My position as grandfather is occupied there.[27]

I thank you for all the kind things you say of me & the Civil List Pension which brings my name into notice.[28] I always did my duty and worked hard when in the Survey and made my name as an Explorer and Naturalist.

I shall send you a photo of myself which you ask for, very likely by next mail.

With kind wishes to you all. I am

Yours affectly.

H H Godwin-Austen

Lt Col.

It is a muddled letter: apart from the memory lapses and repetition, his rejection of a close emotional and indeed legal tie is a brutal contradiction of his assertion in his previous letter; the switch to a more amiable tone towards the letter's conclusion would have hardly lessened the shock that Naomi Florence must have felt. As for any decent and kindly person, the problem he was faced with was really a dilemma: his marriage to Kudikji was a sexual union, and although it had a legal basis in Islamic and Raj law, Eddy had been doubly unfortunate, being born outside a marriage which was also a mixed one – he was, for ever, illegitimate. How do you reconcile natural warmth of feeling towards those with whom you are connected by blood but whose connection with you the law refuses to recognise? For a man in his 90th year, Godwin-Austen made a strong attempt at squaring this particular circle, but his success was, at best, uneven.

THE RETURN OF NAOMI FLORENCE

For most members of the family, that position has been where the tale ends. There is, however, a twist. Much later, when she believed she was approaching death, Naomi Florence wrote a long and rambling letter to the President of the Royal Geographical Society, Major General Francis James Rennell Rodd, second Baron Rennell KBE, CB, JP (1895-1978), on 3 September 1945. It begins melodramatically and continues eccentrically (and also contrasts unfavourably with Annesley's requests for help written in 1919):

Dear Sir,

Before I close my eyes in death[29] and while I am able to write, I send you a copy of my grand-father's (Lt Col Godwin-Austen) letter, being the truth about the second-highest mountain in the world, named after him. I have the original falling to bits, which as you will understand, I must keep all my life. If you look up Who's Who, you will not find this entry of his first marriage but he married, in 1861, Pauline Chichele-Plowden (* shown as his first wife),[30] daughter of Major Plowden, whose heirs are now in possession of the property and heirs of Godwin-Austen. I presume, my father was disinherited, as his mother was a non-Christian & not a European and due to the family's prejudice against such a marriage.[31] I want to leave this on record

---

[26] He misses out the first inverted comma before 'Godwin-Austen'.

[27] Rather disingenuous: Arthur and Betty's marriage had been childless since their wedding in September 1897.

[28] For the successful but belated outcome of Annesley's negotiations, see Chapter 19.

[29] In fact, she died three years after writing the letter, in August 1948, aged 63.

[30] *Inserted in the text in Florence's handwriting immediately above 'Chichele-Plowden' *(sic)*.

[31] An almost entirely fictitious assertion.

and do not seek to oust the present heirs. We cannot dodge our Fate! My father died 17th February 1917, after a very hard & *(illegible)* life (of mastoid abcess *(sic)*). He struggled from the lowest rung of the ladder to Asstt: Engr. P.W.D. Berar.[32] When he retired he was Divisional Engr: Local Funds Berar Division. My eldest brother is in the P.W.D. as Asstt: Engr: in Central Provinces, he has a son, nearing completion of education (doing B.Sc in Bombay) in Nami Jah *(?)* and 4 girls. He married late and is on the verge of retirement with these children's futures to face ...

The letter continues for about another hundred words in this vein, before coming to the point:

I have appealed to Lord Wavell,[33] June 1944 & Lady Linlithgow,[34] 1936 for some little*[35] help, but have not been successful, yet criminals, on the Indian frontier are paid to keep the Peace – a pitiful case like mine is not considered worth-while helping ...

The letter's inaccuracies begin by suggesting some mystery surrounding K2/Mt Godwin-Austen. There has been much discussion about the name, but there is no mystery: the letter therefore comes across as somewhat self-indulgently self-dramatising. The Godwin-Austen letter to which Naomi Florence refers appears to have been lost. If she is referring, somehow, to the letter of June 1923, then it is not easy to understand how she extracts any reference to K2 from it.

The inaccuracies are also misguided and unfair. Haversham himself was well-enough known for his amiable and considerate attitude towards the native Indians (or Baltis or Kashmiris) he worked with (see Chapters 10 or 11, for example); there is not a jot of evidence that Annesley or Godwin (as the heirs) entertained the kind of prejudices to which Florence refers, and I can vouch for the attitudes referred to not being any characteristic of the many Godwin-Austens in the present composition of the family. Sadly, it seems more likely that Florence had nursed her resentment against Haversham because of the violence of his rebuff to her in his letter of October 1923 and has allowed this resentment over 22 years to grow disproportionately. There is, in any case, no basis for her criticism: Haversham's marriage to Kudikji was merely a matter of form, concluded by a mullah who was happy to go along with it: it was legal, but Edward was born outside it, therefore illegitimate, and unable, in English and Raj law, to inherit. Again, this unhappy letter continues to describe further family financial problems, but one has to wonder at Florence Naomi's state of mind at this point, in thinking that powerful people could be moved to help those who have simply done their job, and, it seems, done it well, but not in any conspicuous or exceptional way. As she nears the latter part of her letter, however, she returns to her main theme of her family's connection with Godwin-Austen:

My father never kept in touch with his father & none of us were told of this 'skeleton in the cupboard' till I accidently *(sic)* came to know & made enquiries. The finding of the Secret is another interesting story. Too long to relate. I feel like a Sherlock Holmes about this case. Needless to say my mother was furious about

---

[32] See n7, above.

[33] Field Marshal Archibald Percival Wavell, 1st Earl Wavell GCB, GCSI, GCIE, CMG, MC, PC (1883-1950) was a commander of British Army forces in the Middle East during World War II. He led British forces to victory over the Italians, only to be defeated by the German army. He was the penultimate Viceroy of India from 1943-7. He was a supportive commanding officer to General Sir Alfred Reade Godwin-Austen KCSI, CB, OBE, MC (1889-1963), 'Uncle Reade'.

[34] The wife of Victor Alexander John Hope, 2nd Marquess of Linlithgow KG, KT, GCSI, GCIE, OBE, PC (1887- 1952), who served as Governor-General and Viceroy of India from 1936 to 1943.

[35] *Against the asterisk, Florence writes the following vertically in the left-hand margin: *to restore part of my husband's commuted pension that has long since been paid back to Gov: (20 years are over) and grant him the full pension of 100/- p.a.*

> these discoveries & were my father alive he would have swallowed me up. I was then married (1923) & did as I liked about it on this story and the truth would never have been known. The pity of it is that my Indian grand mother, *(sic)* who brought my grandfather fame & who was such a brave courageous heroine, should not share in the glory of her *(sic)* in these achievements, just on account of racial and religious bias – her name is dead to the World. One does not like to unfold the story because of the present successors. That is the reason of so late a disclosure but you must admit it is essential. My father was born in Mahomedanism: was Church of England in adoption to Milners who put him in St Peter's College (orphanage), Agra where he became Roman Catholic. My people did not like these disclosures because of this, for they both were bigoted R.C. My mother's mother was Rebecca Harding, daughter of Capt Harding, R.N.*[36] and related to the Viceroys Hardinge of Lahore[37] fame. Lord Hardinge of Lahore who was made thus on account of good work, for three *(?)* lives. See References.[38] I think I have given quite a history of my ancestry, so will conclude Yours truly
>
> (Mrs) Florence Barclay

The unravelling of the 'Secret' family connection is described above in Naomi Florence's letter to Godwin-Austen of 3 July 1923. It is perhaps significant that she gives the word a capital letter. But the idea of any other kind of 'Secret' has to be set against the hundreds of books and websites devoted to K2 which never refer to such a matter, on the grounds that no genuine 'secret' about K2 could remain hidden from the thousands of people who have researched its history and/or attempted to climb it. A Freudian interpretation may even be possible, by identifying Godwin-Austen with the great peak, K2, itself; the 'Secret' is simply a symbolic one: that K2's effective discoverer had an unacknowledged illegitimate child.

This final part of the letter seems to me an odd assessment of the nature of Haversham's mountaineering achievements, and not just from a Western perspective: Kudikji was either dead or separated from Haversham during his greatest Karakoram, Tibetan and Bhutanese expeditions (as well, of course, as his six seasons in Burma), although accompanying Haversham on his first great Baltistan expedition in 1860 undoubtedly required considerable hardiness and some courage, no doubt underpinned by her otherwise unrecorded love for her Englishman.

It is worth restating at this point that it was not just Kudikji's name that was 'dead to the World' but, outside a smallish company of scholars, Haversham's name also; the aim of this biography is to bring Haversham's name to general recognition, as well as the name of those, such as Kudikji – who sounds like a fascinating and admirable character – into focus through a clear and accurate assessment of their significance.

Sadly, Naomi Florence is mistaken in believing she was making a 'disclosure': Haversham's affair with Kudikji, and their son, must have been an open secret at the Royal Geographical Society since at least the late 1870s, and I suspect much earlier, as it is usually cited as the only plausible reason for Godwin-Austen's Founder's Gold Medal being delayed until 1910, after the death of the strait-laced President, Sir Clements Markham (see chapter 20).

Naomi Florence's letter was well intentioned, but its self-regard, inaccuracy and lack of balance, to my mind, achieved the opposite of her intended effect.

The sudden reappearance of the ghost of Kudikji was a shock to Godwin-Austen's system very near the end of his life. We cannot say that his death was hastened by it, but it undoubtedly exercised a powerful effect on him in his final few months.

---

[36] *Another vertical addendum has been made in the left-hand margin, further describing her mother's background.

[37] Field Marshal Henry Hardinge, 1st Viscount Hardinge, GCB, PC (1785-1856) was a Governor-General of India. No evidence is advanced in support of Florence's claim.

[38] Unclear what is being referred to.

CHAPTER TWENTY-SIX

LAST CALL

Final Years

ONE OF THE HAPPIEST PARTS OF Godwin-Austen's last two decades was the time spent working, in a genteel version of Tolstoyan labour, in the various gardens at Nore.[1] Fresh air, gardening and healthy walks over the Surrey Hills and the Downs, particularly over the Austen lands near Shalford and the especially beautiful walk along the South Downs from above Washington past Chanctonbury Ring, not far to the south of Nore, are mentioned frequently and with evident appreciation in the letters of this period.

AILMENTS

But these pleasurable activities were not enough to prevent increasing infirmity. As has been suggested above,[2] worsening physical health accompanied the decline in Godwin-Austen's financial wellbeing, although he carried on serious work until just a few weeks before his death. References to illness begin to crop up with sharper insistence in his letters from about the time of the beginning of the First World War and, perhaps more tellingly, from not many months after his wife's death. In 1914,[3] he complains that:

> I have not been away from here (Nore) all this month & kept in most of it with a cold which I had to take care of when it developed into a cough – I have been able to work in the Garden Room a good deal more than I have been able to for a long time.

In 1915,[4] when he records various pains, it seems to be not so much the physical discomfort which concerns him, but rather the possibility that his work might have to be curtailed:

> On the 13th I strained my back working in the Wild Garden, & got stiffer and stiffer during the Shell Work, & on Monday the 22nd when I left off to go to tea at 5.30 it was so painful I went to bed & have been there up to now ... I had been

[1] For Nore's history see Chapter 19.
[2] See Chapter 25.
[3] Letter to Harmer, 31 January 1914.
[4] Letter to Harmer, 24 March 1915.

> looking forward to being present tomorrow at the Royal Soc. when Günther's pictures will be presented & bring a few old friends but that is not possible. I shall have to keep very quiet, even sitting up in bed tells.

Only a few days later, he varies his theme:[5] '*I have kept free of colds I am glad to say but very sore heels from chilblains.*' A more worrying symptom appears in the later stages of the war:[6]

> Since first week *(sic)* in December I have had to keep indoors. I got influenza and spent some days in bed; it left a bothersome cough so I could not get out many days & even now to go to town wd not be advisable ... We have had a great deal of influenza about here; I hope you have escaped.

The following year, a new kind of physical trouble, with accompanying anxiety, emerges:[7]

> Now I am laid up – I met with a nasty accident from treading on an untied bootlace wh. threw me off my balance. I had a heavy fall on my side. This is the 10th day – am very stiff & the pain is very great at times especially after I move at all – I can just sit up & write, & how long I shall be in this state is uncertain. Muscles all much bruised.

The tenseness caused by the pain comes through very strongly in his terse and contracted expression. Almost a fortnight later, his condition has not improved any:[8]

> Writing in bed so excuse pencil. Pain is so great when I walk I cannot get down to the Garden Room to get out or do anything. My Diary[9] is in a drawer & I cannot trust anybody to look it up – they wd upset everything in it. And my sister is very shortsighted and can never find anything ... This accident is a great nuisance, & will take time for muscles to get all right again.

Pain, frustration and anxiety about consequent lost income seep from every phrase of this letter. By late summer 1918 a new source of agony puts in its first appearance:[10]

> I am sorry I cannot report much progress (*in the classificatory work for which he is being paid*). Having got over the effects of the fall, sciatica is giving me much trouble & is at times extremely painful, preventing my going out and moving about as I wd like to do.

The extraordinary frustration felt by the first Westerner to tramp the greatest glaciers, break the world high-altitude summit record and sweat his way through some of the planet's densest jungles can readily be imagined. But it is not just immobility, a more or less unknown concept in his life up to that point, or suffering from acute pain, for example, after breaking his femur in Calcutta,[11] to which he seemed to be immune; in all these references to his illnesses, Godwin-Austen never mentions his age (he was 83 when he wrote the letter above). For him, age appears to be an irrelevance instead of something to come to terms with. The inevitable physical decline brought

---

5 Letter to Harmer, 30 March 1915.
6 Letter to Harmer, 12 January 1917.
7 Letter to Harmer, 2 June 1918.
8 Letter to Harmer, 15 June 1918.
9 Not a personal journal, but Godwin-Austen's notes on time spent on his work, in order that his monthly accounts for the British Museum would be correct.
10 Letter to Harmer, 5 August 1918.
11 See Chapter 5.

on by old age is for him not so much painful in itself or because it immobilises him but is mentally agonising through preventing him from carrying out his scientific work.

By October 1918, although the end of the war was in sight, Godwin-Austen reported a mild improvement to Harmer, '*I am not free enough from rheumatism to get to town, although I can get out for a short walk now.*'[12] The fall alluded to above moved his nephew Annesley to request the Secretary of State for India to increase his pension:

> He has recently had an accident, a fall, which has seriously shocked his system, and incapacitated him from work.[13]

This sad story of increasing infirmity went on. Six months later, Godwin-Austen told Harmer – evidently one of the few people he confided in after Jessie's death – that he had been '*much troubled with lumbago.*'[14] No sooner had he got rid of this irksome back pain when a cold snap brought it all back again; the previous day (to this letter being written) had been spent in bed. He hoped for milder weather to bring him some relief. The warmer weather failed him, however, and he confessed to Harmer[15] later that summer:

> For the last three months I have been bullied with lumbago, followed by a bad back, so that sitting up to a microscope was often impossible. I am glad to say I am much better and got a nice long walk this morning.

It did not require too much more work to induce more pain, however, particularly given his intense coupling of physical with mental activity ...

> Just using the saw on some old rose bushes produced the present attack *(of backache)* ... I am busy on a big shells paper. I have got together the finest collection of this genera in the World. Nothing to touch it any where.[16]

While this letter concludes self-pityingly, '*but my travels now are ended*' – perhaps an unconscious echo of Prospero's line, 'Our revels now are ended' in *The Tempest*, 4.1? – his famous application to disciplined scholarly work was defined: '*I read and use the microscope 6 hours a day.*'

A theme of Godwin-Austen's letters from about the time of his wife's death to his own 11 years later is, unsurprisingly, the distant past. He reminisced about pleasant holidays with his wife, '*Much of interest in Jersey. A month there with my dear wife in Bouley Bay will ever be remembered & her enjoyment at that time.*'[17] And his memories return to Kashmir, where some of his great early work was accomplished, '*The whole season of 1858, after the winter at Umritsur* (Amritsar) *with my regiment was spent in the foothills of Budrawar & Kishtwar, to the head of the Wurdwan.*'[18]

By 1922, with only just over a year of his life left, we learn in a letter Godwin-Austen receives from Harmer of a further, quite serious fall; the theme remains the same, 'I am extremely sorry to hear you have had an accident which confines you to your bed but I trust that a few days *(sic)* rest will put you right.'[19]

Godwin-Austen's remaining affliction was severe deafness: from its sudden, accidental

---

[12] Letter to Harmer, 30 October 1918.
[13] Letter from Annesley, at Smithbrook Manor, Cranleigh, to the Secretary of State for India, 26 November 1918.
[14] Letter to Harmer, 5 May 1919.
[15] Letter to Harmer, 1 August 1919.
[16] Letter to Sir Aurel Stein, 25 March 1920.
[17] Letter to Sir Aurel Stein, 29 September 1920.
[18] Letter to Sir Aurel Stein, 25 December (sic), 1915.
[19] Letter from Harmer to Godwin-Austen, from c/o General Somerset, Higher Brixham, Devon, 1 September 1922.

beginning in 1853 on the day of General Godwin's departure from Burma (see Chapter 6), this new form of frustration had returned, to become steadily worse over the last 20 years of his life. One story describes him rushing into the garden at Nore during a German air-raid to listen, with his ear-trumpet upraised, to the defence guns sounding off from the nearby Great Western Railway cutting, and declaring that he had not heard guns like those since Delhi in the 1860s.[20] While having to use his ear-trumpet was necessarily a great handicap, he was described, characteristically enough, as using it 'with unruffled calm'.[21]

Godwin-Austen's last letter in which he mentions his health was written just under six months before his death. Writing to his friend and fellow malacologist,[22] Buller Newton, his mind is evidently less well-focused, his handwriting lacks its former vigour. Taking refuge in a remarkable past becomes a source of comfort:

> I hope you are well and that I may see you when summer comes and I can get up to town. I am fairly well. But cannot get out much more ... but I get through a great deal of work, just lately for the Indian Survey in Kashmir to settle a dispute between that Gov't and Tibet, both claim the same valley, & I was laying down the boundaries between the two countries in 1863.

This letter is by no means the only one to conclude in this way; the boundary work which Godwin-Austen carried out in the earlier part of his career evidently meant a great deal to him – after all, boundary work is itself a form of classification, the intellectual process which had provided the momentum for most of his thinking life.

Illness and financial strain continued to take their toll. A further demand on Godwin-Austen's emotional stamina of course arose from his longevity: many members of his family and, more so, his friends, died before he did. Captain Scott and his fellows endured a pathetic death, suffused in imperial propaganda, in 1912; his wife died in a way which caused great distress for both, in 1913; Haversham's sister Elfrida died in 1914; after the war, his sister Caroline died in 1922, and Sophia in 1923, not long before him. About the same time, he commented generously on the death of Colonel Worthington Wilmer (1838-1923), a fellow malacologist who saw action in the Indian Mutiny, including the Relief of Lucknow, and is said to have collected before retiring to the Isle of Wight over 40,000 shells, many from the Andaman Islands:

> I knew him for many years & have a large number of letters from him, all of great interest. I got to like him as everyone wd who came across him, but as I cd never get to Ryde, I never saw him in his home as you did ... Well, he is gone and will be missed. There are not many like him for he was all round so good and useful a man.[23]

The kindness of the comment is much more typical of Godwin-Austen, the man who kept a steady mind through life's tribulations, than the irascibility sometimes ascribed to him, although at this late stage in his life, Godwin-Austen may well also have been thinking subconsciously about how others will soon assess *him*.

One of the last, and related, matters on which Godwin-Austen pronounced, only a few months before his death, was the technical development and ethics of mountaineering. He was interviewed by Kenneth Mason, one of the widest-ranging mountaineer-surveyor-explorers of his day and a future Professor of Geography at Oxford; in the interview Mason found him:

---

[20] Letter from Marjorie W Cass to Dr Richard Godwin-Austen, cited in *A Rich Inheritance*, by Rita Ghey, privately published, Québec, 1992, Appendix pX.
[21] *Ibid*, Appendix pX.
[22] Letter, 16 June 1923, to R Buller Newton, member of the Council of the Malacological Society of London.
[23] Letter to Buller Newton, 16 June 1923.

> In his ninetieth year still enjoying dissecting snails and other unpleasant creatures, examining them under a microscope, and preparing exquisite coloured sections of their internal organs.

He was, continued Mason:

> A very forceful old man who insisted that new-fangled gear such as clamps, ice-axes, ropes etc should never be used by any self-respecting mountaineer.[24]

Godwin-Austen was horrified to hear of pitons, oxygen and, worst of all, crampons, and emphatically said so. This reactionary pronouncement did not deter Mason, who evidently responded to Godwin-Austen at this stage in his life with a combination of respect and gentle irony, from describing Godwin-Austen in *Abode of Snow*,[25] his magisterial survey of Himalayan mountaineering, as 'a great explorer and probably the greatest mountaineer of his day'.

THE RGS PICTURES

Such memories of life and pronouncements in the public arena have to be contrasted with the distressing correspondence with Florence Barclay in private, as the final significant scenes in Godwin-Austen's life played themselves out. Just before such events, however, near the end of 1921, lapses in his mental grasp led to an awkward but lengthy exchange between him and the Royal Geographical Society. The story began as early as 1916, when he expressed the following wish to Secretary Hinks:

> I have a framed watercolour of K2, made from original sketches taken on the south side of the Baltoro Glacier. This I shall have much pleasure in leaving to the Royal Geographical Society, together with another W.C.[26] of the head of the Punmah Glacier, Baltistan.
>
> While on this topic of sketches, I have quite a large collection commencing when I first went to India in 1852 up to 1876 – they are in pencil, pen & ink, sepia, and watercolour, the largest are 16" by 12", of course among these sketches some are slight and unfinished but they convey a good idea of the place & if they are not artistic they are accurate in detail. The Total number is 740 ...
>
> It is an extremely interesting collection, its value considerable from this point of view, that many of the places depicted made so many years ago are now civilized & improved out of recognition, others are in parts of the country even now most difficult to visit, while native costume has sadly changed from the picturesque old style to an uncouth European mixture.
>
> ... I hope when the weather improves you will come down here and look over it.
>
> ... My memory and eyesight are both very good & I can work all day long *(on writing up his journals)*. Mr Munn of Messrs Arnold & Coy. was particularly anxious I should take up an account of my Survey life in Kashmir, it would interest the Alpine climbers but until the war ends (because of paper and publishing restrictions) there is little use putting a new iron in the fire.
>
> I told Stein[27] I thought the RGS the best and fittest place for such a collection

---

[24] Reported by Phillimore.
[25] Kenneth Mason, *Abode of Snow*, Diadem, 1987; facsimile of the OUP edition, 1955.
[26] W.C. meaning watercolour.
[27] Among Sir Aurel Stein's enormous collection of papers there is an undated, but almost certainly 1921 list of the 'missing' works, typed out by Godwin-Austen himself 'from memory' and running to over 150 items.

> ... Free Trade has made me a poor man & the war does not improve that position[28] so I should not be justified in presenting this work of my life to any Department or Society.

Having supposedly bequeathed a number of his sketches to the RGS, an act of generosity not followed through in his will, Godwin-Austen then became convinced that they had been stolen. He wrote an accusatory letter to Arthur Hinks, the RGS Secretary, a notoriously prissy and pedantic mathematician (who played an influential, if often negative part in the setting up of the early Mount Everest expeditions and who was therefore well-known to Godwin-Austen).

Hinks replied speedily and, by his standards, remarkably diplomatically:[29]

> The recollection of our hall porter and of the assistant to the map room that the box was wrapped in the wrapper Mrs Romer[30] brought, and the big portfolio of sketches in 'brown paper', as it was impossible to send it without covering. As I have already explained, the mistake seems to have arisen from the fact that although your letter to me and my instructions to the assistant map curator referred to the box, he erroneously took it as referring to the whole of the sketches in our possession.
>
> It is quite impossible that it could have left this Society at some other time and in some other way. If it did not go with Mrs Romer it is here now, but search has failed to disclose it. We are, however, looking again most carefully.

Godwin-Austen remained dissatisfied by Hinks's explanation, and put the matter in the hands of Scotland Yard:

> On the 19th August 1921 I went to the Society to look at my sketches. They were not to be found but after waiting some time in the Chief Clerk's office he came back with a sheet of paper on which was a statement written in pencil, without any signature or date, that Mrs Romer had taken the portfolio away when she came up for the oak box the previous year. Before I left I asked to see the room where the portfolio had been kept and was last seen. I was shown the Council Room which is near the entrance and open to every visitor.
>
> On the 23rd August, Mrs Romer being in town, I took her with me to the Society and we saw Mr Allen. I had not seen him before. At my request he went for and came back with the above-mentioned sheet of paper, and after my asking him, more than once, acknowledged that he was the writer. On Mrs Romer seeing it, she took the paper and wrote across it that she had never seen the portfolio – never heard of it. This is my firm belief. I had not mentioned it to her, for I did not want it, neither did I mention it to the Secretary when writing to him before I left for London.
>
> As the Secretary was away on leave, we left the Society's rooms: there was no use entering into a discussion with one of the staff, and the matter was left to be taken up by him on his return. However, on 26th August we find Mr Allen writing to the hall porter to obtain information as to Mrs Romer having taken the portfolio, how it got into her taxi-cab, etc. He certainly should have remembered this after the statement he had written. It is a very discreditable case of carelessness on the part of a great Society ... [31]

---

[28] He is still selling off land as late as 1919 (advertisement, *Surrey Advertiser*).

[29] Letter to Godwin-Austen from Arthur Hinks, 1 November 1921.

[30] It is not clear who Mrs Romer was or more precisely how she fits into the narrative. There was, however, a Romer who was a Fellow of the Society at this time, an Egyptologist.

[31] Letter from Godwin-Austen to AE Lawrence, Assistant Commissioner, New Scotland Yard, 22 March 1923.

The investigation rumbled along for well over a year, but nothing had been resolved by April 1923,[32] when Godwin-Austen wrote to one of his fellow malacologists:

> Yours is a most kind letter, most sympathetic, and of that it is a pleasure to get, for I have been terribly worried over this loss, more than I can tell you, it is never out of my thoughts and prevents my doing work, as I used to. There is no doubt the sketches have been stolen. Sir Thomas Holdich came to that conclusion some time ago. Publicity is the only thing & I am glad that Scotland Yard and my Solicitor are working on it. The loss of sketches made between 1852 and 1858 renders my journals useless because they illustrated all I was making. Nearly all geology. All the time I was in Burma I devoted to Geology – when very little was known of that of India ... The large panoramic sketch of Prome (lost) was wonderful to compare with the Geol. Map of Pegu made years afterwards by Theobald ...

The letter then rambled onwards, not uninterestingly however, about Godwin-Austen's early geological work with his father, '*from the age of 12*'. He followed his story to its conclusion shortly after:

> ... as I passed on afterwards to the Punjab to join my Regiment ... I always knew what ground I was treading. It all makes the loss of these sketches more bitter to bear. But no more now. I wonder I am as well as I am, but the worry tells.

This whole incident may be imaginary. Godwin-Austen's worried insistence suggests a kind of obsession with something which exists more in his imagination than in external reality. Haversham's present portfolio contains, in a large artist's folder (with some of the best having been professionally mounted and framed) numerous sketches, quite possibly '740', and fully-realised watercolours of Burma, Kashmir and the Punjab (and an astonishing variety of other subjects, geographical, human, natural and technological); the K2 and Punmah Glacier paintings are there. While it is impossible to conclude that these are not the same sketches as he refers to in these letters, it is hard to believe they are not identical and that he had simply forgotten where he had stored them at Nore (a large, ten-bedroomed house) and filled the vacuum of this forgetfulness with some other RGS memories. At any event, the problem (if it was one) had not been resolved by the time of his death only eight months after his letter to Buller Newton.

DEATH, BURIAL AND OBITUARIES

The shock and emotional exertion required to deal with the sudden renewal of connections with his illegitimate son's descendants might have hastened Godwin-Austen's death. Despite several claims[33] that he died in 1924, Godwin-Austen actually died, at Nore, on 2 December 1923, aged just short of 89 years and five months. In the only surviving letter I know of by his sister Beatrice, who evidently looked after him during the last ten years of his life, his decline and death are described:

> It was rather sudden in a way. I had noticed during the summer and autumn he was getting very feeble. In October, his memory failed so much. He tried to stay up as long as he could and was in the Garden Room writing in the morning and the afternoon. That was on the 9th of November. On the 10th he said he would stay in bed. It was a cold day but he was out again. He was not ill. He really had nothing the matter with him, he was just three weeks in bed, not a pain or an ache

[32] Letter from Godwin-Austen to Buller Newton, 7 April 1923.

[33] Mason, *Abode of Snow*; Curran, *K2*; Keay, *Where Men and Mountains Meet*, for example.

> all that time. Then he said it was freezing cold and he was quite worn out. His mind failed at the last. He passed away so gracefully and looked so serene and beautiful in his last long sleep. It seemed so lonely at first without him. I have looked after him for 10 years. His collection of shells and his life long work is *(sic)* going to the Hist. Museum. A man from the Museum has been down and got it all together.[34]

The funeral, at St Mary's church, Shalford at noon on 8 December, was reported in detail by the local newspaper:

> A service in the Parish Church was conducted by the Rev David Green (vicar of Bramley) assisted by the Vicar of Shalford (the Rev. Hamilton Jackson). The hymns sung were 'On the Resurrection Morning' and 'Lead Kindly Light'. At the graveside, the Rev. Hamilton Jackson officiated.

The family mourners, led by Arthur and his wife, were those whom one might have expected. The others were listed as:

> The Earl of Midleton,[35] Sir Charles and the Hon. Lady Cook, Sir George F Bonham, Bart,[36] Sir Herbert and Lady Jekyll, Miss Jekyll, Sir Charles Metcalfe,[37] Col. Sir Thomas and Lady Holditch *(sic)*, Lady Victoria Rowe,[38] Admiral and Mrs Copland Sparkes,[39] Brig.-Gen and Mrs Dawson,[40] Mrs H Ricardo, Col. Crosthwaite, Major Williams Wynn,[41] Mr L Benson,[42] Mr and Mrs WF Rawnsley,[43] Mr and Mrs Theodore Pim,[44] Miss Sivier, Mr A Thorburn,[45] Mr GA Franks, Mr G Musgrave, Mrs FC Longbourne,[46] Mrs Ingham, Mr and Mrs EL Rowcliffe, Mr H Rowcliffe,[47] Miss Gulbourn, Mrs Montgomery, Miss Lightfoot, Mr LCW Phillips,[48] Mrs Weston, Mr J Weston, Mr W Saree, Mr J Godman,[49] Mr EH Everett, Miss Haydon, Mr HA Hayes, Mr WP Trench, Mr ES Bates, Mr H Sherburn, Mr WH Messenger, Mr FH Elsley (representing Surrey Archaeological Society) and Mr F Lasham (President, Guildford Natural History Society).

---

[34] Letter from Beatrice Godwin-Austen to Sir Aurel Stein, 19 February 1924.
[35] William St John Fremantle Brodrick, 1st Earl of Midleton, KP, PC (1856-1942), commonly known as St John Brodrick, was a British Conservative Party statesman.
[36] Sir George Francis Bonham, 2nd Bt, (1847-1927), succeeded to the title of *2nd Baronet Bonham* on 8 October 1863. He graduated from Exeter College, Oxford. He was Secretary of the Legation to Lisbon between 1885 and 1890. He was also a JP.
[37] Son of a former Governor-General of India.
[38] Lady Victoria Isabella Liddell (1841-1935). Owner of Thorncombe, near Hascombe.
[39] The inheritor of Woodhill Manor, near Bramley, Surrey.
[40] Brigadier-General Sir Douglas Frederick Rawdon Dawson, GCVO, KCB, CMG (1854-1933). He attended Eton College, and then joined the Coldstream Guards in 1874. He saw service with the Egyptian Campaign of 1882, where he fought at the battles of Mahuta, Kassassin, Tel-el-Kebir and the capture of Cairo. In the Nile Expedition of 1884-5, he was part of the Guards' Camel Corps, was mentioned in despatches, and saw action at the Battle of Abu Klea. He was then appointed a Military attaché from 1895 to 1901, posted to Austria-Hungary, Serbia, France, Belgium and Switzerland, before returning to the United Kingdom as Master of the Ceremonies to Edward VIII from 1903-7 and then as a comptroller in the Lord Chamberlain's Department. After the war, in 1920, he was appointed State Chamberlain until 1924.
[41] Sir Robert William Herbert Williams-Wynn, 9th Bt, 1862-1951.
[42] Possibly the son of WF Benson (d.1887), Godwin-Austen's old friend and fellow-malacologist.
[43] Poetry and travel-writing editor.
[44] Theodore Pim of Snowdenham Hall, Bramley; possibly of Dublin origin. He died in 1929.
[45] Archibald Thorburn, 1860-1935, ornithologist and ornithological painter.
[46] Wife of Brigadier FC Longbourne, possibly of the West Surrey Regiment, who fought on the front line in the First World War.
[47] Fellow Hascombe landowners.
[48] Owner of Unstead Park, near Bramley.
[49] Possibly the writer of *Rambles of a Naturalist*, concerning Indian natural history.

While those listed seem mostly to represent local landowners, military men and Conservative politicians, the 'many scientists' referred to in one of the obituaries hardly received a mention.

The obituaries stressed Godwin-Austen's good character:

> Of a most attractive disposition and singular charm of manner, Godwin-Austen was much beloved by his contemporaries, most of whom he had long out-lived.[50]

The malacologists' obituary (written by 'BBW', presumably BB Woodward, Godwin-Austen's successor as president of the Malacological Society from 1901-3), described him as:

> Modest, unassuming and inordinately grateful for any little service rendered to him, he endeared himself to all who came to know him and will be sorely missed by his many friends as well as the whole malacological world.

It is Sir Thomas Holdich, one of his oldest friends, to whom we turn for the final assessment.[51] Cleverly and poignantly referring to Haversham as 'Nestor', the wise elder statesman of Greek legend who is sometimes a little too lengthily concerned to describe his past achievements, Holdich reminds us that Godwin-Austen was the oldest member of the Royal Geographical Society and the last of that band of Indian malacologists who, through the hardest of work in the field, laid the foundations of all future science in that subject. Having then described his career, Holdich concluded:

> *(he)* undoubtedly started subsequent scientific investigations in that system *(relating geology, palaeontology and malacology)* of the very greatest interest and value. He seemed to live in the bones of an extinct fauna[52] that must once have lived and grazed over a vast tract of northern India.
>
> I found him lately still as keen as ever on those scientific investigations which have made his name great in many learned societies, and which brought together a large number of leading scientific men to join at the old Shalford church in honouring the last call of this most eminent geographer.

Officialdom, however, even in the persona of an old and valued friend, was no respecter of the dead. Only three days after Haversham's death, Sydney Harmer wrote to Annesley to enquire about the concluding of the classifying of Godwin-Austen's collection, a process which was still advancing on a muddled *ad hoc* basis of a peculiarly English kind, having been initiated by Godwin-Austen no less than a quarter of a century beforehand:[53]

> You are no doubt aware of the fact *(that Godwin-Austen)* made an arrangement with the Trustees of the British Museum, as long ago as July 1904, for the presentation of his collection to the Museum. He wrote to me on August 28th, 1922, to explain his wishes with regard to the disposal of the Collections of Indian Shells in his room at Nore. He specially emphasised the importance of having them handled by someone experienced and trustworthy, and went on to express his wish that this matter should be placed in the hands of Mrs M Malcolm when the time came. As she was at that time just starting for Canada I wrote to her on the subject,

[50] *Nature* 112, 946, 29 December 1923.

[51] *Proceedings of the Royal Geographical Society*, 1923-4.

[52] *The Times* obituary comments 'his picturesque little bungalow in Dehra Dun was always littered with the bones of extinct creatures which once browsed among the folds of the Himalayan foothills.' (8 December 1923).

[53] Letter from Harmer to Annesley Godwin-Austen, 5 December 1923.

> expressing the hope that she would come to see me on her return to England in the spring. She did not do this, and I have no idea where she is now. Are you able to give me her present address, so that we get into touch with her?

This correspondence bumbled along until a mutually acceptable conclusion was reached. Godwin-Austen's vast collection can still be found in the Research Section of the Natural History Museum, South Kensington.

Perhaps the most remarkable of all the Austens and Godwins, Henry Haversham was buried in Shalford village graveyard, its most distinguished occupant. In Shalford's St Mary's church, next to many of his ancestors, he has a plaque which reads:

> In honoured memory of LT COL HENRY HAVERSHAM GODWIN-AUSTEN FRS:FRGS:FZS of Shalford Surrey born July 6 1834 died Dec 3[54] 1923 Explorer. Surveyor. Scientist. In appreciation of his enterprise and skill in exploring and surveying the N W Himalayas – especially in the vicinity of the Kara-Koram range. He was awarded the Founders Gold Medal of the Royal Geographical Society. His family and friends place this memorial here. *mens aequa rebus in arduis*[55]

Godwin-Austen was one of Victorian, Edwardian and Georgian Britain's widest-ranging explorers and ablest natural historians. He was an astonishingly productive and accurate surveyor, one of the greatest early mountaineers, a highly accomplished artist, a distinguished scientist in several disciplines and a human being of sensitivity, modesty and kindness. While it was right that the various scientific bodies of which he was a prominent member gained vastly in knowledge from the publication of his research, it is fitting that his extraordinary achievements should now reach a wider public and that his greatness, as a polymath among polymaths, an adventurer among adventurers, becomes fully appreciated.

---

[54] A slip for 'December 2nd'.

[55] Adapted from Horace's *Odes*: 'a steady mind in life's tribulations'. It was also the motto of Viscount Hardinge and of Warren Hastings.

MAPS

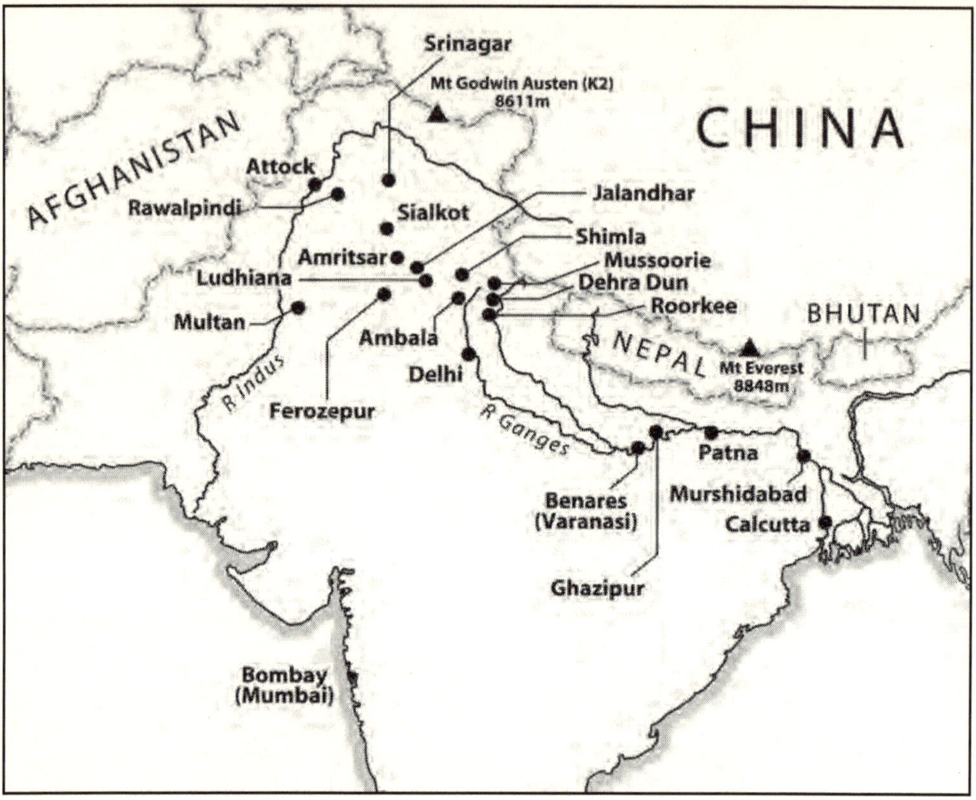

1. British Northern India in the mid-19th century. See Chapter 7.

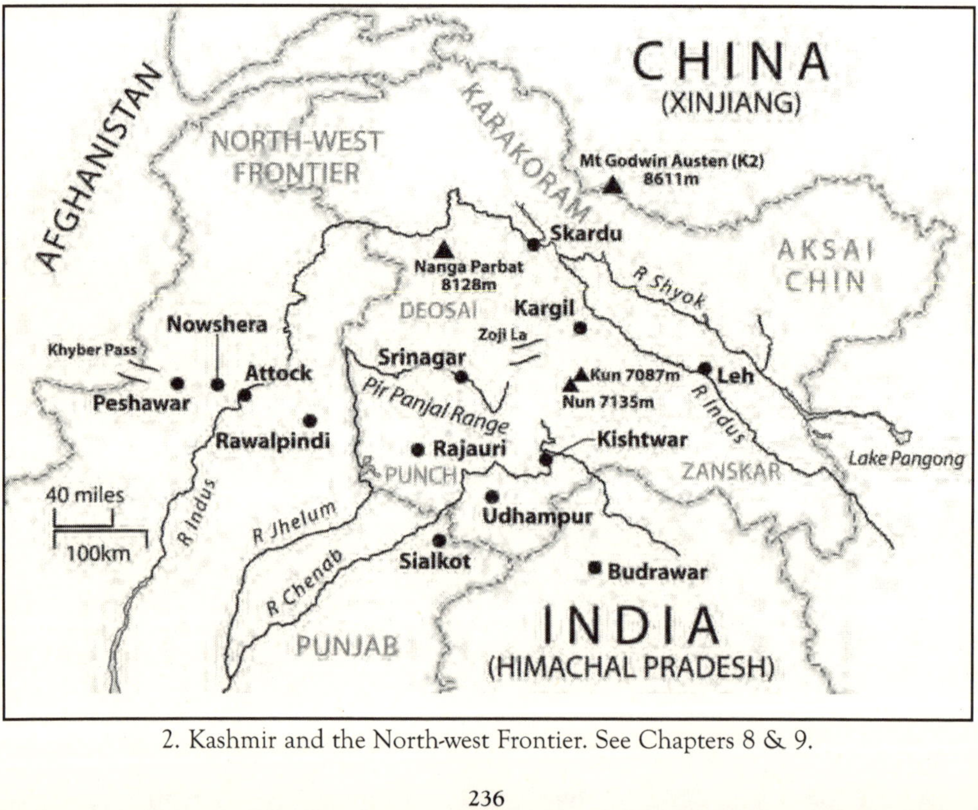

2. Kashmir and the North-west Frontier. See Chapters 8 & 9.

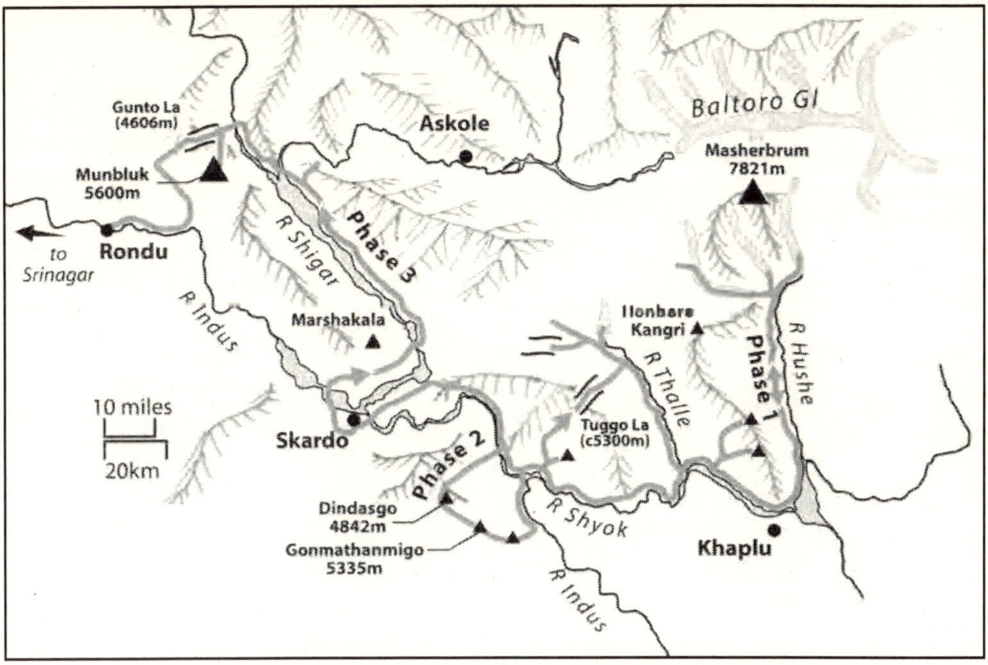

3. The First Karakoram Expedition, 1860. See Chapter 10.

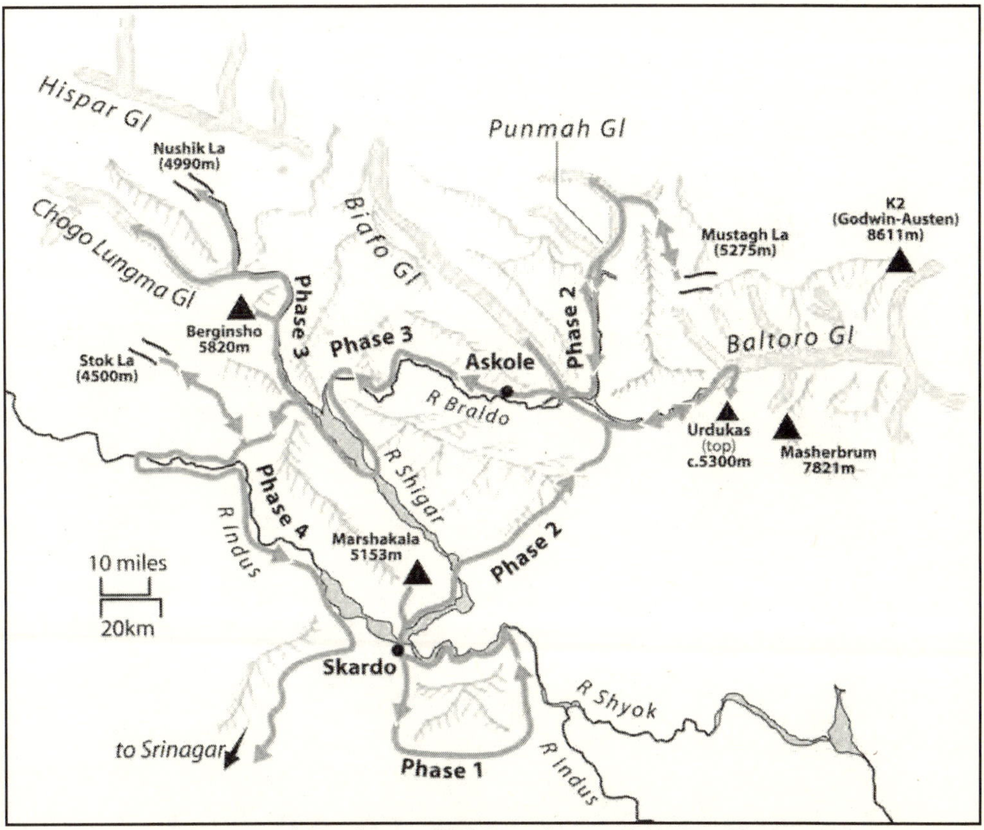

4. The Second Karakoram (K2) Expedition, 1861. See Chapter 11.

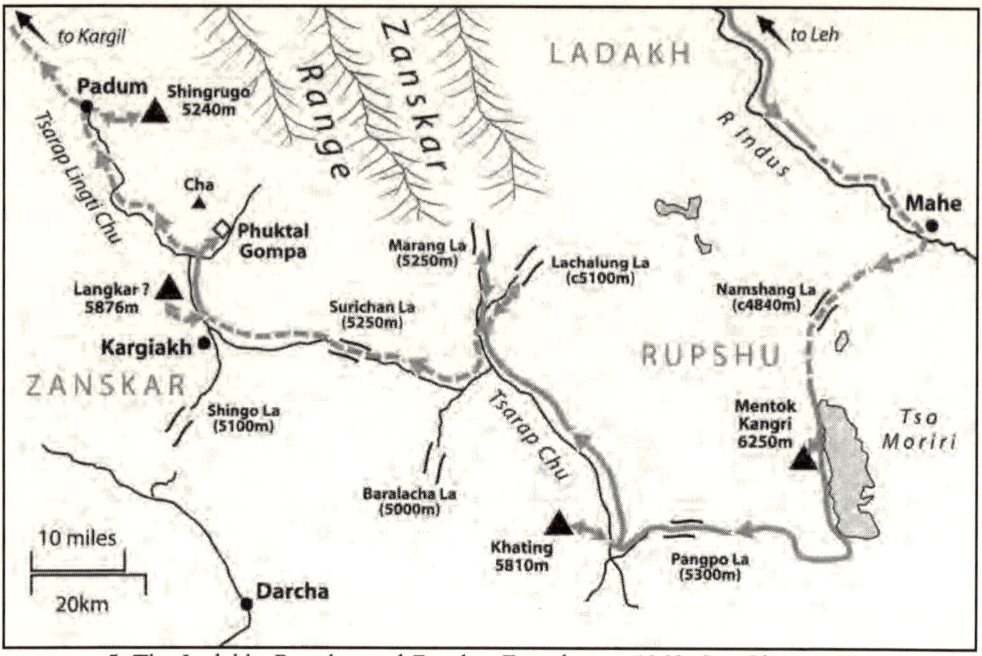

5. The Ladakh, Rupshu and Zanskar Expedition, 1862. See Chapter 12.

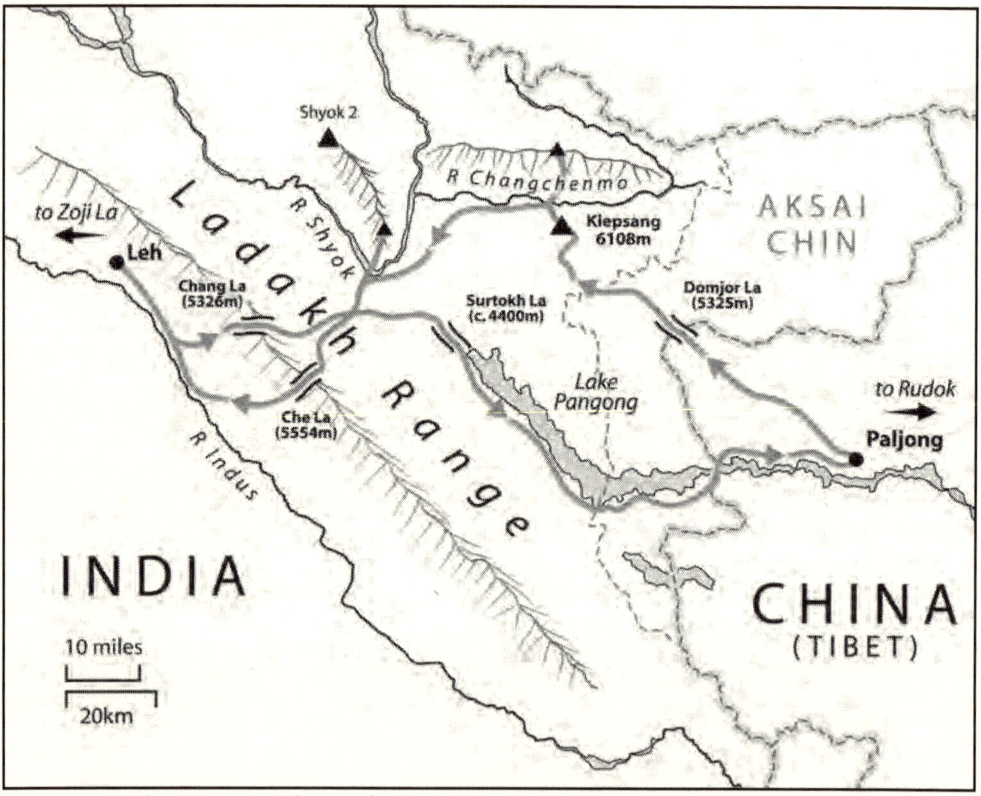

6. The Western Tibet and Aksai Chin Expedition, 1863. See Chapter 12.

7. The Bhutan Expeditions, 1864-5. See Chapter 13.

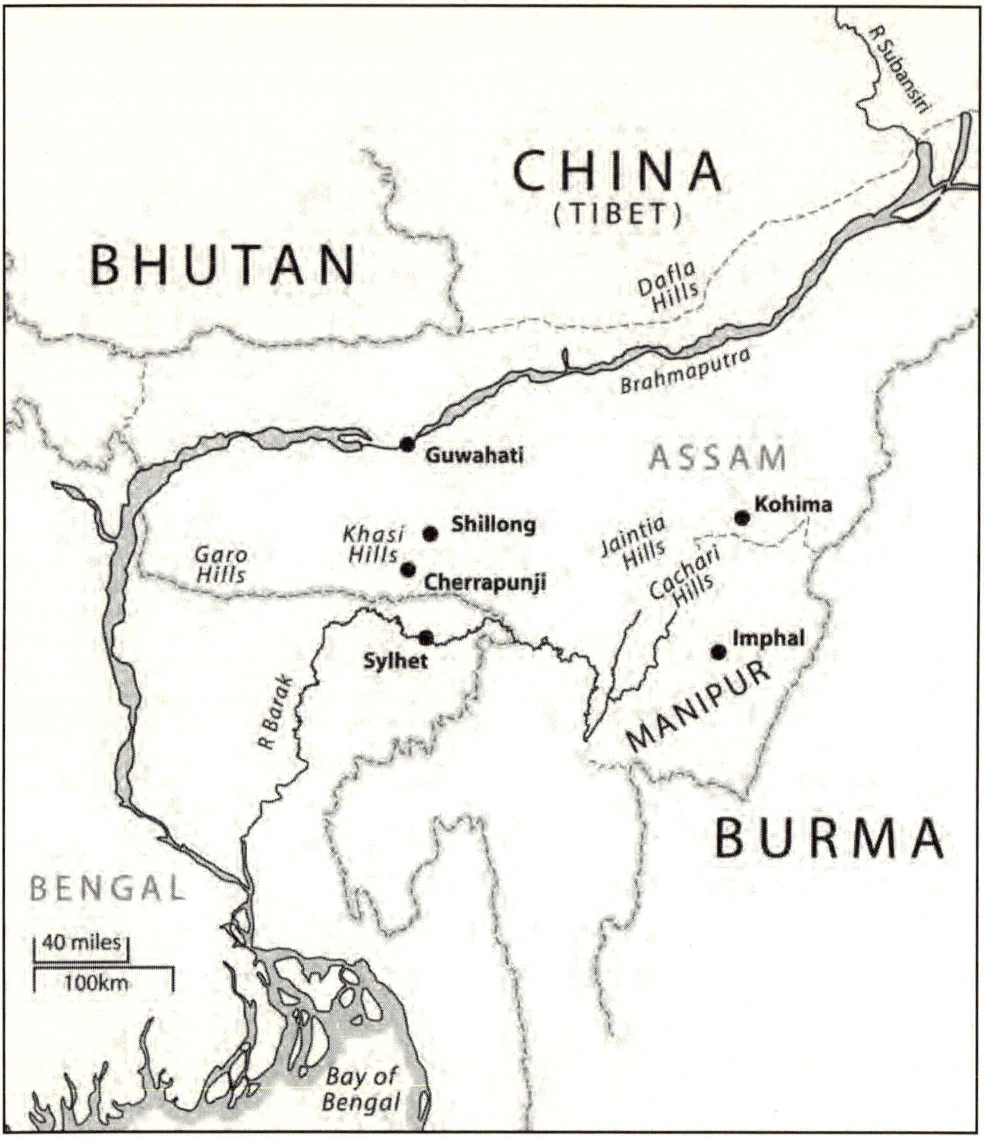

8. Northern Burma and Assam. See Chapters 15 & 16.

APPENDICES

APPENDIX A

SUMMARY OF THE LIFE OF HAVERSHAM GODWIN-AUSTEN

1834: 6 July: born eldest child of Robert Alfred Cloyne Austen and Maria Elizabeth Austen, née Godwin, at Ogwell House, nr Newton Abbott, Devon.
1841: Registered for Sandhurst by grandfather, Colonel, later Major-General, Godwin.
1845-8: probably at the Royal Grammar School, Guildford.
1848-51: Sandhurst; December 1851: gazetted Subaltern in 24th Foot.
1852: Chatham to Calcutta (via Cape Town) by sea with his regiment.
15 November: breaks thigh in riding accident.
1853: To Burma by ship; catches smallpox. Second Burmese War concluding. Promoted to Ensign. Mapped distributaries of the Irrawaddy Delta. Travels with General Godwin, September, to Simla, northern India. Death of General Godwin.
25 October: attends General Godwin's funeral, Simla.
1854: Serving in Sialkot, Punjab. Meets painter William Carpenter.
17 October: name-change to Godwin-Austen by Special Licence from Queen Victoria.
1855: Stationed at Peshawar; likely beginning of the affair with Kudikji.
1856: Meets Lumsden (Kashmir Survey) in Peshawar; surveys Kuram Valley.
27 May: gazetted Lieutenant.
1857: Indian Mutiny. Posted to Kashmir Survey, Srinagar.
15 March: son, Edward, born to him and Kudikji; survey work in Manganwar, Kajnag, Jhelum Valley, Pir Panjal, Kishtwar areas; Winter 57-58: in Amritsar, guarding disarmed native regiment.
1858: Surveying in Jammu area, then Budrawar, Punch; surveying in Nun-Kun area, Zanskar; May: marries Kudikji; surveys Chandra Bagha and Chata Dhar Ranges.
November: beaten up by villagers at Udhampur; hospitalised in Sialkot.
1859: Son adopted by Milners.
April: invalided out to England; elected FRGS.
1860: May: elected member of the Alpine Club.
June: returns to Survey at Srinagar. First Karakoram expedition, accompanied by Kudikji; Winter recess at Dehra Dun; triangulation of Valley of Kashmir; probable death or disappearance of Kudikji.
1861: 5 April: marries Pauline Chichele Plowden; measures K2 from Deosai Plateau; second Karakoram (K2) expedition; near-approach to Mustagh La; explores glaciers to K2 and fixes its height and position; explores to Nushik La and Skoro La.
1862: Expedition via Zoji La to Rupshu, Ladakh and Zanskar; Father awarded the Wollaston Medal for Geology.
1863: Lake Pangong and Aksai Chin expedition.
September: son, Arthur, born Srinagar.
November: despatched to join Eden Mission in Darjeeling; thanked by Government of India and Viceroy for his service.

1864: January 1: Eden Mission to Bhutan.
March: storm on Tegong La; reaches Punakha; returns through Bhutan by outward route and recesses at Mussoorie.
October-November: at sieges of Dalingkote and Chamurchi, Cooch Behar.
1865: Surveys Bhutan Duars.
May: retires to Dehra Dun (ill); six months' sick-leave.
1866: Sick-leave continues; geological work in Siwalik Hills.
November: takes charge of Topographical Party No 6, Khasi Hills, Assam.
1867: Khasi Hills; instructs young surveyors.
1868: Burrail Range; Lushai tribe attack South Cachar tea-plantations; joins General Nuttall in South Cachar with British punitive force.
November: gazetted Major.
1869: January: caught in Cachar earthquake; Burrail Range survey until May.
1870: Health 'seriously broken'; begins two years' furlough in England.
1871: Probable death of wife, Pauline, in Calcutta.
1872: October: returns to Burma via Brindisi and Suez; continues survey of Burrail Range on Irrawaddy-Brahmaputra watershed; obstructiveness by Raja of Manipur.
1873: Completes map of previous work; takes leave in England for 'urgent private affairs'; elected to British Ornithological Union.
1874: Returns to India; Dafla (Assam) expedition; eastern Himalayan peaks fixed; survey as far as Dunsiri River, Akka Hills; some work and collecting in the Naga Hills.
1875: Dafla Expedition: Subansiri River mapped.
1876: Sick leave in England; winter: assists in arranging the collections in the new Indian Museum, Calcutta.
1877: re-arranges bird collection in Indian Museum; geological work in the Siwalik Range.
June: retires from Indian Army with rank of lieutenant-colonel.
1878: Finishes *Birds of Assam*; undergoes course in biology with T H Huxley; begins work on *The Land- and Freshwater Molluscs of India*; offers to sell bird collection to British Museum; elected Fellow of the Geological Society.
1880: Recovers health.
3 June: elected Fellow of the Royal Society.
1881: 1 June: marries Jessie Robinson, daughter of Examiner of the House of Lords; honeymoon probably in Swiss and Italian Alps.
1883: President, Section E (Geography) of the British Association.
1884: Death of Robert Alfred Cloyne; inherits the estate.
1888: Attempt at Royal Geographical Society by General Walker to have K2 renamed Mount Godwin Austen; accepted by meeting, refused by RGS Council; visits Devon.
1890: Poor farming prospects.
1891: Long talks with Sir Martin Conway before his Karakoram Expedition.
1892: Resigns membership of the Alpine Club.
1893: First extended stay at Nore, Hascombe, nr Godalming, as financial prospects worsen.
1894: Bhutan article rejected by Sir Clements Markham.
1895: Tries to sell bird collection; financial worries; until 1897, Vice-President of the Zoological Society of London.
1897: Placed on list of Extraordinary Members of the British Ornithological Union. Until 1899, President of the Malacological Society.
1898: 22 September: files for bankruptcy.
1899:10 February: Presidential Address to the Malacological Society; auction in Shalford of personal effects and of Shalford House contents.
1900 (onwards): frequent visits to Shell Gallery, Natural History Museum, South Kensington.
1902: 3 January: bankruptcy discharged.

1904: Tries to sell his collection of Mollusca.
1905: President of Alpine Club regrets not seeing Haversham in town very often.
1906: Visits Orkney Islands, possibly other parts of Scotland.
1907: Attends Alpine Club Jubilee Dinner.
1908: President of the Conchological Society until 1909; publishes *The Fauna of British India*, Vol I.
1910: Awarded the Founder's Gold Medal of the Royal Geographical Society.
June: Scott of the Antarctic spends his last night in England at Nore; visits St Ives; gives up ornithology.
1912: Made Honorary Member of the Alpine Club.
1913: Death of Jessie, probably of stomach cancer.
1914: Publishes *The Freshwater Molluscs of India*; outbreak of First World War. Row lasting until 1921 with Gude about errors in Gude's edition of the *Mollusca of India*.
1915: In pain from chilblains and strained back; Zeppelin drops bomb over St Catherine's, Guildford.
1916: Records good haymaking; visit from Sir Aurel Stein.
1917: 19 February: death of his son by Kudikji; first entry in *Who's Who*; awarded Barclay Memorial Medal by Academy of Bengal Studies; flu.
1918: Bad fall; Annesley Godwin-Austen attempts improvement to pension with RGS and Government of India; Armistice.
1919: Lumbago; various political comments.
1920: Restores the Great Bell of the Shwesandaw Pagoda, Burma.
1921: Gives his opinion about forthcoming Mount Everest expedition.
1922: Overnight visit from Kenneth Mason; comments on mountaineering ethics recorded; an 'accident', also sciatica, rheumatism, deafness.
1923: Correspondence with daughter of son by Kudikji.
2 December, dies at Nore; buried at St Mary's Shalford. Various obituaries.

APPENDIX B

DATES, ROUTES AND ASCENTS ON THE MAJOR EXPEDITIONS

Items with 'Mumm' in brackets after them are from Arnold Mumm's entry for HH Godwin-Austen in the Alpine Club Register, 1857-63. Heights given first in feet show Haversham's original measurements.

Part One: the five 'great' Expeditions

1. Karakoram 1860

Walk-in of 20 days from Srinagar to Skardu 'by the usual route'.

First Phase: 5-28 August

Ascents

- Bianchu, 15,990ft/4874m, 'above Skardu'
- Kon Ton (height unspecified, above valley north of Kiris: Mumm)
- Tunggo La (pass) over a ridge into the Thalle Valley, 'over 16,000ft/4877m'
- Trigonometrical Station on Monlong, 14,905ft/4543m, (Mumm)
- Lanzopheuma (aka Kanchinokla), 16,820ft/5127m
- Chungoksigo, 18,843ft/5743m
- Thulle Pass and Tusserpola Passes, (heights unspecified: Mumm)
- Ridge crossed into the Hushe valley, height unspecified but c.3000+m
- about 5800+m/c.19,000ft on south face of Masherbrum.

Glaciers surveyed

- Two hanging glaciers on descent from Kiris La to Upper Thalle Valley
- Atoser Glacier
- Gondokhoro Glacier (in part)
- Masherbrum Glacier (in part)
- Serac Glacier (in part)
- Ating Glacier.

Second phase: early September to c20 September 1860.

Ascents

- Gonmathaumigo, 17,500ft/5334m
- Dindasgo, 15,882ft/4841m (Mumm)
- Thurigo (height unspecified).

Third phase: c20 September to unspecified date, probably early to mid-October.

Ascents

- failed attempt at an unspecified height on 'peak above Kuardo' (Marshakala, 5183m, climbed by Haversham in 1861)

- Marpo Chungi, 15,400ft/4694m
- Busper Peak, 14,940ft/4554m
- Gunto La, 15,107ft/4605m
- Munbluk, c.18,373ft/c.5600m

Glaciers surveyed
- Gunto La area.

2. The Second Karakoram, or 'K2' Expedition, 7 July – c28 September 1861.
Ascents (peaks and passes) on the first phase:
- 'Several peaks in the neighbourhood of Skardu.'
- Thyarlung, 16,844ft/5134m
- the Burji La, 15,878ft/4840m
- Shigar Luggo pass, height unspecified, probably around 3963m/13,000ft
- Mashkulla (Marshakala), 16,919ft/5157m
- Thalanka, (height unspecified) (Mumm)
- Skoro La, 16,644ft/5073m
- 'point at 2500ft above the Biafo Glacier', 12,645ft/3854m
- point on ridge above a Punmah Glacier Camp, height unspecified, probably about 4000m/13,123ft
- Drenmung Ridge, opposite Nobande Sobande Glacier, c.4572m/15,000ft
- Mirego? ('a fine peak above Skeenmung camp') 18,342ft/5591m
- turning-point just short of Mustagh Pass, 17,301ft/5273m.

Glaciers surveyed
- Burji La and Shigar Luggo areas
- Skoro La summit area
- Skoro La Glacier
- Biafo Glacier (part)
- Panmah Glacier
- Choktoi Glacier (mouth)
- Nobande Sobande Glacier (part)
- Lower Skamri Glacier (part)
- Chiring Glacier.

Ascents on the second phase
- the K2 viewpoint on a spur of Masherbrum above Urdukas, height unspecified, but '1000ft ' then 'some steep rock' then 'another slope' then 'another 1000ft of elevation' then 'another sharp push' above the Baltoro Glacier, probably c.5640m/18,504ft
- a 'low knob 5 miles (8km) above Korophon', height unspecified
- 'mountain above Chokpojong', height unspecified, c.5300m

Glaciers surveyed
- Biaho Glacier
- Baltoro Glacier (lower part only).

Ascents on the third phase
- Summit c900m/3000ft above Doko, possibly Berginsho, 5820m/19,094ft
- spur bounding lateral glacier of Kero Loombah (Lungma) Glacier, 'until Trans-Indus 2 and Trans-Indus 4 were visible', probably c4500m
- Nushik La, 4990m
- spur above Chogo Loombah (Chogo Lungma) Glacier, height unspecified

- a 'fine knob' 600m/2,000ft above the Chogo Lungma Glacier, height unspecified but probably c.5000m.

Glaciers surveyed
- Hanging Glacier above Doko
- Kero Lungma Glacier
- Hispar Glacier viewed
- Chogo Lungma Glacier.

Ascents on fourth phase
- pass to Upper Tormik Valley (height unknown) (Mumm)
- near approach to the Stok La, c.4500m
- Kutchi Bore La, c.3000m
- a 'low spur' above Basho, height unspecified, probably c3000m
- Alumpi La, 15,200ft/4633m; Dore La, c.4250m; Raj Diangan (Tragbal) Pass, 11,950ft/3642m.

Glaciers surveyed
- Tormik Glacier (part)
- Alumpi La area.

3. Rupshu, Ladakh and Zanskar, early July – early September, 1862 (start and finish dates unspecified).

Ascents on the walk-in from Srinagar to Leh:
- Zoji La, 3528m/11,575ft. Probably Haversham's second return crossing of this pass. Almost certainly re-crossed on the return.

Ascents
- Mata Peak (Mentok Kangri), 6283m/20608ft (Haversham's highest ascent)
- 'Red Knob' (south end of Tso Moriri), 17,000ft/5182m
- 'near Peak P', 18,000ft/5486m
- Pangpo La, 17,000ft/5182m (crossed three times in one week)
- 'Peak 16,500ft', (5029m)
- Kato, 19,000ft/5791m
- Lachi-lunglango (pass in a side-valley), 16,630ft/5069m
- Marang La, 17,000ft/5182m
- 'Survey Peak', 19,000ft/5791m
- Surichan La, 18,000ft/5486m; Lankar, 19,273ft/5874m; Shingrugo, 17,208ft/5245m.

4. Lake Pangong and the Changchenmo Basin, 14 July – September, 1863.

Preliminary walk-in: via the Zoji La, 3528m/11,575ft, from Srinagar. Almost certainly the return route from Leh.

Ascents
- Chang La, 17,470ft/5325m
- Surtokh La, c.4400m
- Lake Pangong TS station, 13,931ft/4246m
- Tuggu Nuggu, 'the most eastern eminence' on the south shore of Lake Pangong, height unspecified, probably about 3963m/13,000ft
- a 'low pass' above Pal, northern shore of Lake Pangong, height unspecified
- a 'limestone mountain east of Pal', height unspecified

- Dingo La, 16,270ft/4959m; 'small hill' above Dingo La, c.5260m/17,257ft
- Pa Luen Pass, height unspecified, probably c.5182m/17,000ft
- Deenjor La, 17,465ft/5323m
- Unnamed summit south of Deenjor La, 20,240ft/6169m
- Kiepsang, 20,035ft/6107m
- Kiung Gang La, 17,259ft/5261m
- Ororotze La, 18,050ft/5502m
- Kay La, 18,256ft/5564m
- 'peak of about 19,000ft (5791m) which overlooked the Shyok Valley'.

5. The first Bhutan Expedition, 1 January – April 1864

Passes

- Pango La
- Tule La
- Tegong La, 12,200ft/3719m
- Cheli La, 3780m
- Bie La
- Dochu La, 3116m (all repeated on the return journey).

Part Two: Other Areas involving Exploration and Ascent.

1. Kashmir

a) May 1857: Kajnag Hills on the north-west side of the Valley of Kashmir. These hills rise to around 4000m, with a few just exceeding that height.
Haversham mapped a portion of the valley of the River Jhelum (another major tributary of the Indus).
He was said, by Phillimore, to have helped to produce the 'first map of the Kuram Valley', on 'reconnaissance work by Lt Peter Lumsden and others'.

b) September-December 1857: Kishtwar area.

c) February-May 1858: Jammu area, to survey 'that part of the Maharajah's territory adjoining the Sealkote district, together with part of the outer sandstone range': a southern extension of the Pir Panjal, around Udhampur and south of the Chenab. These hills rise to over 4000m, although, very regrettably, we do not know how many of them Haversham climbed. Described as 'very intricate and difficult ground'.

d) May-end of June 1858: Budrawar area, south of his Kishtwar area of 1857, and lying between the Chenab (also known as the Chandra Bagha) and the Chata Dhar Range. Probably via the Indrahar Pass (4350m) north of Dharamsala.

e) July-early August 1858: Kishtwar, then the Maro-Wardwan River as far as 'Ainshin' (now more usually 'Inshan'). Until 23 August 1858: in Srinagar.

f) From 23 August – end of October 1858: eight days to reach Ainshin. Zoji La. Near approach to then bypassing Nun and Kun (also known as Ser and Mer) from the east then north, exiting probably at Kargil and probably continuing as far as Mulbekh Gompa, Ladakh. (The visit to Mulbekh Gompa could have been in July 1858, however.)
'Commencing a fresh piece of country in the Low Hills', near Udaipur.

No records have been left of Haversham's ascents in this area, although there is one allusion to the surrounding country reaching 15,000ft/4572m (in a way which suggests he was measuring at or near that height). Over two seasons, several ascents must have been made, and high passes crossed in the Kajnag, Pir Panjal, Punch, Chata Dhar and Western Ladakh ranges, from anywhere between c.2000m (below which, because of forestation, good surveying would have been unlikely) and c.4500m.

2. Bhutan Duars

This is all relatively low-lying country. No records remain of any ascents in it by Haversham. They would, in any case, have amounted to not more than a few hundred metres at most.

3. Burma, Assam and Munipur

a) Season 1866-7: no records remain. The Khasi Hills appear to rise to about 1000m asl, in general, although one rock dome appears to reach c.1646m/5,400ft.

b) Season 1867-8: no records remain. The Jaintia Hills rise to a plateau at about 4000ft/c.1219m asl. The Shillong range rises to 6,450ft/c1966m.

c) Season 1868-9: Naga Hills and South Cachar. Burrail Range. No records remain other than the mention of Mahadeo, 'a high peak', south of the Cachar earthquake epicentre; Haversham ascended it at least twice. See under Season 1872-73 below.

d) Season 1869-70: Garo Hills. No records of ascents remain.

e) Season 1872-3: records are not comprehensive.

- Some minor ascents, e.g. '1500 feet above a village'
- Jotsama (Jotsomah, modern spelling): 'a stiff ascent', over 2400m/c.7874ft asl
- Japvo: 3014m/c 9,888ft just south-west of Kohima
- Ascents along the watershed near Kohima, such as the Napoo Ridge on 14 January 1873
- 17 January: Khunho, a peak of indeterminate height, 2500-2800m asl in the Kohima area
- Telizo, probably 2485m/8153ft, on the northern perimeter of Manipur
- Ridge named Kopamedja
- 20 February: Munghri, height unknown
- 2 March 1873: Nunshigung, solo; height unknown
- 30 March: 'climbing in the Shiroifurar Range'; location unknown, but south of Imphal, probably close to the watershed. Ascent of about 1000m
- Pundengroo (second ascent after Beavan): height unknown; probably in the Shiroifurar Range.
- Marangthang peak: 1389ft (c423m)
- March/April: various unspecified spurs and ridges.

f) Season 1874-5: Dafla Hills, Assam: no records remain, although it is known that Haversham was in the Subansiri and probably the Dunsiri River areas. He does say that during this survey, the heights of 'thirty-seven' peaks were determined, although it is not known how many were established by him. The Dafla Hills rise to c.4708m in the Upper Subansiri, although most hills in the area Haversham is likely to have surveyed are in the 1500–3000m asl range.

Part Three: Distribution of heights of known ascents (peaks and passes) on the five major expeditions, 1860-4.

Summits

above 6000m: 4
5500-5999m: 8
5000-5499m: 11
4500-4999m: 6

Plus two mountains mentioned by Mumm, height unknown. From their situation, almost certainly between 4-5000m.

Other Survey stations established at

above 5500m: 3
4500-5499m: 2
4000-4499m: 2
3500-3999m: 1

Passes of specified height crossed

above 5500m: 2
5000-5499m: 9, plus one recrossed twice, so 11.
4500-4999m: 6
4000-4499m: 1
3500-3999m: 5, plus one recrossed five times, plus two (probably) recrossed once, so 12.
3000-3499m: 7, plus four (probably) recrossed once, so 11.

Plus one pass of unspecified height near Lake Pangong: c.4000m
Plus three other passes, probably in the 3500-4000m range, but of unspecified heights.

Glaciers surveyed: 23.

APPENDIX C

SURVEYING

Part One: Areas surveyed by Godwin-Austen.

The following figures are Godwin-Austen's own, from an Appendix to his Abstract of Military Service.

1857	Kashmir	1/2 inch scale	1,310 square miles
1858	Kashmir	1/2 inch scale	2,368 square miles
1860	Baltistan	1/4 inch scale	2,250 square miles
1861	Zanskar	1/4 inch scale	1,300 square miles
1862	Baltistan	1/4 inch scale	3,700 square miles
1863	Ladakh	1/4 inch scale	3,300 square miles
1864	Bhutan	1/4 inch scale	2,160 square miles
1865	Bhutan Duars	1/4 inch scale	2,000 square miles
1866-7	Khasi Hills, Assam	1/4 inch scale	249 square miles
1867-8	Jaintia Hills, Assam	1/2 inch scale	411 square miles
1868-9	Naga Hills & South Cachar	1/2 inch scale	98 square miles
1869-70	Garo Hills	1/4 inch scale	515 square miles
1872-3	Naga Hills & Munipur	1/4 inch scale	1,580 square miles
1874-5	Dafla Hills, Assam	1/4 inch scale	1,090 square miles

Total of square miles surveyed: 22,331

Part Two: The Art and Science of Surveying

An article (which Haversham would have edited) in the 1883 edition of *Hints to Travellers*, still accessible on the open stacks of the Royal Geographical Society, tells us what the well-equipped surveyor requires. First, he will need a knowledge of trigonometry and astronomy. His equipment should comprise:

1) A sextant of 6 inches radius to read to ten seconds of arc
2) A mercurial artificial horizon
3) A watch with a second hand
4) A graduated compass
5) A lantern with oil and a wick
6) A thermometer and (water-) boiling apparatus
7) A pocket-size aneroid barometer
8) Drawing instruments such as compasses, pens, protractor, etc
9) Stationery
10) Pen-knives, paints and brushes.

In addition, the following are desirable:

1) Plane-table
2) Telescope
3) Theodolite: this is a movable telescope mounted within two perpendicular axes, the horizontal or trunnion axis, and the vertical axis. When the telescope is pointed at a desired object, the angle of each of these axes can be measured with great precision, typically on the scale of seconds of arc. By measuring horizontal and vertical angles, calculating the height and distance from one 'station' to the next is a matter of fairly simple trigonometry.

In Burma and Assam, Haversham also used the heliotrope, an instrument which contains a mirror to reflect sunlight over great distances to mark the positions of participants in a land survey. Although all this equipment sounds very cumbersome, Haversham breezily remarks that 'a plane-table can always be carried by a native of the country. In India, the guides often carried the stand (i.e. tripod). It is seldom that a European is called on to carry anything in a tropical country.' This sounds patronising, but as quick reflection shows, local porters and pack-animals today are still used to carry personal equipment, other than daysacks, and all other food and equipment for most visiting Western climbers and trekkers. It is a morally ambivalent position: this arrangement helps the economy and income of poor people and struggling countries, but it also perpetuates the sahib-servant relationship. And Haversham's watercolours bear out his opinion convincingly.

In practice, apart from the boiling-point thermometer to measure altitude, the plane-table is the instrument which Haversham valued most. A plane table (called a plain table before 1830) is a device used to provide a solid and level surface on which to make field drawings, charts and maps. The early use of the name *plain table* reflected its simplicity and plainness rather than its flatness. It consists of a smooth table surface mounted on a sturdy base. The connection between the table top and the base permits one to level the table precisely, using spirit levels, in a horizontal plane. The base, a tripod, is designed to support the table over a specific point on land. By adjusting the length of the legs, one can bring the table level regardless of the roughness of the terrain.

To use it, a plane table is set over a point and with the help of the spirit levels is brought to precise horizontal level. A drawing sheet is clamped to the surface and a rule with a telescopic sight can then be used to construct a line on the drawing that is in the direction of the object of interest (a 'ray'). By using the rule as a surveying level, information on the topography of the site can be directly recorded on the drawing as elevations. Distances to the objects can be measured directly or by the use of stadia marks in the telescope of the rule. (Stadia marks, also called stadia lines or stadia hairs, are crosshairs on the reticle (a grid or pattern placed in the eyepiece of an optical instrument), in this case of a theodolite or other surveying instrument and used to establish scale or position.) The table can be protected with a waterproof sheet. In size, the table is recommended to be 2ft 6in (approx 76cm) x 2ft (approx 61cm), and should be made of seasoned deal; a traveller should carry two. The tripod should be adjustable.

Better than using a prismatic compass with the plane-table (states Haversham), is to gum a fresh sheet of paper by its edges to the plane-table, and on some convenient scale, say 1inch to the mile, the route can be sketched by back and forward rays (i.e. lines drawn in the direction of any convenient distant object), and setting the plane-table (i.e. orientating it in relation to observed natural features). A far wider area of country can be sketched in by this method than by the use of the prismatic compass, which can cramp the observer.

While setting up the board (presumably by the 'coolies'), the traveller is recommended to spend his time botanising, geologizing or collecting objects of Natural History, the results of such searches being written up in the evenings.

To train a surveyor, Haversham recommends the following method. Give the learner two counties in the south of England, to survey at 4 miles (c.6.44km) to the inch (2.54cm) using only a plane-table, a sight-rule and a compass. Two or three points of latitude and longitude, of prominent positions about 15 or 20 miles (24km or 32km) apart, should be projected on the table. Selecting Surrey and Sussex, and supposing the traveller has his camp at Dorking, and his line of route west towards Guildford and Farnham, an excellent 8 mile (12.87km) base could be formed by St Martha's Hill (see Chapter 5) and Leith Hill (the highest point in South-east England). Proceeding to Leith Hill, the edge of the sight-rule is laid upon this base-line and directed to St Martha's Chapel. A part of the top of the hill is selected through the telescope where some spot most advantageous to see the stations (measuring points where the plane-table is set up) that are to be afterwards laid down on the plane-table can be found. That having been done, the most conspicuous points for further extension, such as Hindhead, Hascombe Hill, Booker's Tower at Guildford, a point on the Chalk Downs above Wotton, Box Hill and Reigate Beeches should then be selected, with some conspicuous far points, for example on the South Downs near Worthing. Then the compass is placed near the side of the plane-table and when steady, pointing due north, a line is drawn round the (compass) box and the name of the station recorded, then repeated two or three times for accuracy. Rays must then be taken to church-towers, villages, conspicuous trees, etc, and their names recorded in the Field Book. Symbols are very useful along the pencilled ray-lines: thus, a circle for the top of a hill, = for a village, + for a church etc. A hill or mountain top should have its outline sketched on the plane-table. Before leaving Dorking, the traveller should then ascend Box Hill and, setting his plane-table by the ray he took from Leith Hill, should fix the selected station above Reigate, and the intermediate points and clumps of trees on the same chalk ridge, as well as a number of villages to the south and south-east, to give him further information on (say) the source of the River Mole. On his march to his next camp near Albury, he should ascend to the points he would wish to select above Wotton, and thence he would get a great deal of detail in the valley he was about to pass through; whenever he could see the tower on Leith Hill, St Martha's Chapel and the top of Box Hill, he would set up the plane-table by compass and get his position by the intersection of rays from these points...

The extension of this process, taking in the Hog's Back, Crookesbury Hill, Caesar's Camp and the Fox Hills is then described. One final checking device is mentioned: should the original base not be absolutely accurate, the traveller can on some convenient level ground fix the ends of a base and measure the same, or obtain by astronomical observations the latitude and longitude of two or three positions.

In summary, all that is needed (Haversham concludes) in carrying out this kind of work (and it was the work which surveyed India from south to north in one of the greatest enterprises of its kind in history), is to be careful: the edge of the sight-rule must be very true and parallel to the line of back and forward sights; these must be perpendicular so that the hair in the front sight must intersect a distant object both at top and bottom, to get rid of the collimation error: collimated light has rays which are nearly parallel, and therefore will spread slowly as it propagates. The word is related to 'colinear' and implies light that does not disperse with distance (ideally), or that which will disperse minimally (in reality). A perfectly collimated beam with no divergence cannot be created because of diffraction, but light can be approximately collimated by a number of processes, for instance by means of a collimator. Collimated light is sometimes said to be focused at infinity. Thus as the distance from a point source increases, the spherical wavefronts become flatter and closer to plane waves, which are perfectly collimated.; the table must be set up level; the pencil must be sharp and held perpendicularly, so that the ray drawn is at the same distance from the edge of the sight-rule as the station dot. If this is not attended to, a good deal of error may creep in. Finally, the triangles should be as symmetrical as possible, these angles being not less than 30° and not much over 90°.

Combining the plane-table and the theodolite therefore produces a map: the plane-table shows where features are in relation to each other; the theodolite provides their relative heights and distance, though it seems to me amazing that work of any kind of accuracy, where winds are howling, mist is swirling and breathing is arduous, was achieved at all.

BIBLIOGRAPHY & SOURCES

Part One: Godwin-Austen's Personal Bibliography[1]

1. Books and Bound Works

1. *Description of a Mystic Play as Performed in Ladak, Zaskar &c.*, contained in *Journal of the Asiatic Society of Bengal*, part 1, no.2. Baptist Mission Press, Calcutta, 1865.
Ten mounted arched-top albumen prints of Ladakh ghost dancers, mounted two to a page.
2. *On the Garo Hills*, Royal Geographical Society, London, 1873: rare.
3. *Exploration survey work*, Royal Engineers Institute, 1894.
4. *Hints to Travellers*, 5th Edition, ed, Douglas Freshfield and J Knox Laughton, Royal Geographical Society, Stafford, London, 1894.
5. *The Fauna of British India, including Ceylon and Burma: Mollusca.* 4 vols, Taylor and Francis, London, 1908-15.
Vol. I: *Testacellidae and Zonitidae*, by WT Blandford and H. H. Godwin-Austen.
Vol. II: *Trochomorphidae-Janellidae*, by GK Gude.
Vol. III: *Land operculates, (Cyclophoridae. Truncatellidae. Assimineidae. Helicinidae)*, by GK Gude.
Vol. IV: *Freshwater Gastropoda and Pelecypoda*, by HB Preston.
6. *The Land and Freshwater Mollusca of India*: including South Arabia, Baluchistan, Afghanistan, Kashmir, Nepal, Burmah, Pegu, Tenasserim, Malay Peninsula, Ceylon, and other islands of the Indian Ocean. Supplementary to Messrs. Theobald and Hanley's *Conchologica Indica*, Taylor & Francis, London, November 1920.
7. *The Mountain Systems of the Himalaya and Neighbouring Ranges of India.* 5pp. 2 folding maps. Extract from Proceedings of the Royal Geographical Society. Bound in card covers. Date unknown.
8. *Report of the Socotra land and freshwater shells collected by Professor Balfour. Report of the Fifty-first meeting of the British Association for the Advancement of Science held at York in August and September 1881. Vol 51 (Appendix)*: pp194-7, John Murray, London, 1882. (See also under (iv) Malacology and Conchology, Nos 23 and 24.)

2. Scientific Papers and Letters for Publication

Abbreviations:
Ann Mag Nat Hist: Annals and Magazine of Natural History
Brit Ass Rep: British Association Report
GJ: Geographical Journal
JAIGBI: Journal of the Anthropological Institute of Great Britain and Ireland
JASB: Journal of the Asiatic Society of Bengal

[1] The items in each section below are arranged chronologically.

J Moll Stud: Journal of Molluscan Studies
J Roy Anthropol Inst: Journal of the Royal Anthropological Institute
JRGS: Journal of the Royal Geographical Society
n.sp.: new species (one)
n.spp.: new species (more than one)
PRGS: Proceedings of the Royal Geographical Society
Proc As Soc Beng: Proceedings of the Asiatic Society of Bengal
Proc Mala Soc: Proceedings of the Malacological Society (of London)
PZS: Proceedings of the Zoological Society (of London)
QJGS: Quarterly Journal of the Geographical Society
Rep SGI: Report(s) of the Surveyor-General of India
RGSJ: Royal Geographical Society Journal
Sel Rec Gov Ind: Selection from the Records of the Government of India
Ser: series
Vol: volume

(i) Mountaineering, Exploration and Surveying

1. *On the Garo Hills*, Royal Geographical Society, London, 1873. Note: this work also exists in rare bound form.
2. Geographical Names, *Nature*, Vol 47, Issue 1211, p245, 1893.
3. The Nomenclature of the Karakoram Peaks, Correspondence in the *RGSJ*, Vol 1, p177, 1893.
4. Exploration in the Mustagh Mountains (Discussion), Godwin-Austen, Colonel; Biddulph, Michael; Freshfield, Douglas, *GJ*, Vol 2, no.4, pp299-303, October 1893.
5. Peak K2 (Godwin-Austen), HH Godwin-Austen, *GJ*, Vol 3, no.5, pp431-2, May 1894.
6. Exploration Survey Work, Lieutenant-Colonel HH Godwin-Austen FRS etc, *Professional Papers of the Corps of Royal Engineers*, Vol XX, Paper VI, 1894.
7. The Map of Kashmir, HH Godwin-Austen, *GJ*, Vol. 15, no.6, pp657-9, June 1900.
8. The Survey of the Mustakh Ranges, Correspondence in *RGSJ*, 1903.
9. From Srinagar to the Sources of the Chogo Lungma Glacier (Discussion), Godwin-Austen; Wahab, Colonel; Freshfield, Douglas W; *GJ*, Vol 25, no.3, pp265-8, March 1905.
10. Peak K2 Godwin-Austen: the Duke of the Abruzzi in India, *Surrey Advertiser*, 11 September 1909.
11. Discoveries in Southern Tibet (Discussion), Clements Markham; Thomas Holdich; Colonel Godwin-Austen; Henry Trotter; Martin Conway; Douglas Freshfield; Captain Rawling; TG Longstaff; Major Lenox-Conyngham; Sven Hedin, *GJ*, Vol. 33, no.4, pp416-34, April 1909.
12. Explorations in the glacier tributaries of the Shayok River, Kashmir territory, *Nature*, Vol 17, March, 1910.
13. Exploration of the Karakoram Range, *Nature*, Vol 83, Issue 2120, pp469-70, 1910.
14. Glacier Exploration in the Eastern Karakoram (Discussion), Sir Francis Younghusband; Douglas Freshfield and T G Longstaff, *GJ*, Vol 35, no.6, pp653-8, June 1910. Also in *Nature*, 16 June, 1910.
15. "Saltoro" and "Baltoro", *GJ*, Vol. 41, no.2, pp75-176, February 1913.
16. Expedition to the Karakoram and Central Asia (Discussion), de Filippo, Filippi; Younghusband, Francis; Godwin-Austen, Colonel; Neve, Arthur; Holdich, Thomas, *GJ*, Vol 46, no.2, pp99-105, August 1915.
17. To the Editor of the *Surrey Advertiser*: The Ascent of Mount Everest. The importance of Surveyors on the forthcoming Mt Everest Reconnaissance Expedition, February 1921.

(ii) Geology, Glaciology and Physical Geography

1. On the Lacustrine or Karéwah Deposits of Kashmere, *Quarterly Journal of the Geological Society*, Vol 15; issue 1-2; pp221-9, 1859.

Note: This article appears to have been reissued in *Nature*, Vol 53, pp247-8, 16 January 1896.
2. Notes on the Valley of Kashmir, *JRGS*, Vol 29, pp30-7, London, 1859.
3. On the glacier Phenomena of the valley of the Upper Indus, Report of the 32nd Meeting of the British Association for the advancement of Science, UK. Probably given in 1863.
4. The glaciers of the Muztagh Range, *PRGS*, Vol 34, pp19-56, 1864.
5. Geological Notes on Part of the North-western Himalayas, *QJGS*, Vol 20, pp383-388, February 1864.
6. Notes on the Sandstone Formation near Buxa Fort, Bhutan Dooars, with plate, *JASB*, Vol 34, p106, 1865.
7. On the Carboniferous Rocks of the Valley of Kashmir with notes on the Carboniferous Brachiopoda by T Davidson and an Introduction and Résumé by RAC Godwin-Austen, FRS, *QJGS*, Vol XXI, pp492-3 (abstract), 1865. Full article: *QJGS*, Vol XXII, pp29-35, 1866.
8. Notes on the Pangong lake District of Ladakh from a journal made during a survey in 1863, *JRGS*, Volume 37, p347ff, 1867.
9. Notes on Geological features of the country near the foot of the hills in the Western Bhootan Dooars, *JASB*, Vol XXXVII, part ii, pp117-23, 1868.
10. (Extract from) Narrative report of Captain HH Godwin-Austen, in charge of no.6 Topographical Party, Cossyah and Garrow Hills Survey, *Sel Rec Gov Ind*, Vol LXXI, pp140-7, 1869.
11. Memorandum as to Geology and Distribution of Tribes, *Sel Rec Gov Ind*, Vol 74, pp69-73, 1869.
12. Geology of the Garo boundary survey, *JASB*, 1869.
13. Notes from Assaloo, North Cachar, on the great earthquake, January l0th, 1869, *Proc As Soc Beng*, pp91-9, 1869.
14. Earthquake in the Cachar hills. Extracts from letters, *JRGS*, Vol XIII, pp703-23, 1869.
15. Notes to accompany a Geological map of a portion of the Khasi hills near Longitude 91 E, *JASB*, Vol XXXVIII, part ii, p127, 1869.
16. Memorandum as to the Geology of the Jaintia hills and as to the distribution of the tribes, *Sel Rec Gov Ind*, Vol LXXIV, pp69-73, 1869.
17. Notes on the Geology and Physical features of the Jaintia hills, *JASB*, Vol XXXVIII, part ii, pp 151-6, 1869; *Proc As Soc Beng*, pp64-5, 1869.
18. On the Garo hills, *JRGS*, Vol XLIII, 146, (Geological Appendix, pp42-6), 1873.
19. Geology of parts of Naga hills and Manipur, *Rep SGI*, 1872-3, pp79-84, 1874.
20. Notes on the Geology of part of the Dafla hills, Assam; lately visited by the Force under Brigadier-General Stafford, *JASB*, Vol XLIV, part ii, pp35-41, 1875.
21. On the evidence of a past Glacial action in the Naga hills, Assam, *JASB*, Vol XLIV, part ii, pp209-213, 1875.
22. Remarks on Himalayan glaciation, *Proc As Soc Beng*, p4, 1877.
23. On the post-tertiary and more recent Deposits of Cashmir and the Upper Indus Valley, *Brit Ass Rep*, p589, 1880.
24. Presidential address to Section E, *Brit Ass Rep*, 1883, pp576-89. Also in *PRGS*, new series, Vol V, pp610- 25; (Also in *Nature*, XXVIII, pp552-8, 1883.)
25. The Mountain systems of the Himalaya and neighbouring Ranges of India, *PRGS*, New Series, Vol 6, pp83-7, 1884. (See also under 'Books', 7, above.)
26. On the New Railway-Cutting at Guildford:–The Pleistocene Sands and Drift-Gravels observed there, *QJGS*, Vol 40, issue1-4, pp599-613, 1884.
27. Observations on certain Tertiary Formations at the south base of the Alps, in North Italy, *QJGS*, Vol 40, pp855-64, 1884.
28. Correspondence: The Nomenclature of the Karakoram Peaks. *GJ*, Vol 1, p177, 1893.
29. Bhutan and the Himalayas East of Darjeeling, *The Scottish Geographical Magazine*, Vol X, pp635-40, 1894.

(iii) Malacology and Conchology

1. Figures of the species of Diplommatina Benson hitherto described as inhabiting the Himalayas, Khasi Hills and Burma with some additional notes from Darjiling and the Khasi Hills, *JASB*, part ii, p83, 1868.
2. Description of new species of Diplommatinidae from the Khasi Hills, *JASB*, Vol 39 (2), pp1-9, plates 1-2, 1870.
3. Description of the species of Alycaeinae known to inhabit the Khasi Hill ranges, *JASB*, Vol 5 (2), pp87-93, plates 3-5, 1871.
4. Description of new Land and Freshwater Shells from the Khasi, North Cachar, and Naga Hills, NE Bengal, with one plate, *PZS*, pp514-8, 1872.
5. Description of nine new species of Alycaeinae from Assam and the Naga Hills, *JASB*, Vol 43 (2), pp145-50, *pl* 3, 1874.
6. Description of five new Species of Helioidoe of the subgenus Plectopylis, with remarks on all the other known forms, with two plates, *PZS*, Vol 73-4, pp608-13, November 1874.
7. Supplementary Notes on the species of Helicidoe of the subgenus Plectopylis, *PZS*, pp43-4, January 1875.
8. Descriptions of new Species of Mollusca of the Genera Helix and Glessula from the Khasi Hills and Munipur, with one plate, *JASB*, Vol 44, part ii, 1875.
9. Descriptions of four new species of Mollusca belonging to the family Zonitidoe from the NE Frontier of Bengal, with drawings of Helicarion Gigas and of a variety of the same, with 2 plates, *JASB*, pp4-7, 1875.
10. Descriptions of New Operculated Land shells belonging to the genera Craspedotropis, Alycoeus & Diplommatina from the Naga Hills and Assam. One plate, *JASB*, Vol 44, part ii, pp7-10, 1875.
11. On the Cyclostomacea of the Dafla Hills, Assam, *JASB*, Vol 45 (2), pp171-84, 1876.
12. On the Helicidoe collected during the expedition into the Dafla Hills, Assam, with one plate, *JASB*, Vol 14, part ii, pp311-8, 1876.
13. Description of shells from Perak and the Nicobar Islands, *PZS*, pp732-40, 1879.
14. On the capreolus of Lister (or the spermatophore) of some of the Indian species of the Helicidoe, (ad.Zonitidæ), *British Association*, 1879.
15. Descriptions of two new species of Plectopylis a subgenus of the Helicidoe, *Ann Mag Nat Hist*, pp1-2, August 1879.
16. On new species of the genus Plectopylis of the Family Helicidoe, with one plate, *JASB*, Vol 17, part ii, pp1-3, 1879.
17. On the land-molluscan genus Girasia of Gray, with remarks on its anatomy, and on the form of the capreolous of Lister, *PZS*, March 1880.
18. Notes on and drawings of the animals of various Indian Land Mollusca (Pulmonifera); 8 plates, *JASB*, Vol 49, part ii, pp151-58, 1880.
19. On the Anatomy of Ferussacia groviano *(?)*, Risso, from Mentone, concluding with a note on the classification of the genus and its allies by G Nevill, CMZS, *PZS*, 14 December 1880.
20. Descriptions of the animal of Durgella Christianoe, a species of Land-shell from the Andaman Islands, *Ann Mag Nat Hist*, November 1881.
21. Notes on and drawings of the animals of various Indian Land Mollusca (Pulmonifera), no.2: with one plate, *JASB*, Vol 51, part ii, pp67-72, 1882.
22. On the Land Mollusca Genus Durgella, WT Blanford; with notes on its anatomy and description of a new Species, with two plates, *Linnean Society's Journal*, Vol 15, 1881.
23. On the land-shells of the island of Socotra collected by Prof I Bayley Balfour.[2] Part I. Cyclostomaceae, *PZS*, pp251-8, 1881.

---

[2] Sir Isaac Bayley Balfour (1853-1922) was a Scottish botanist. He was the son of John Hutton Balfour who was also a botanist.

24. On the land-shells of the island of Socotra collected by Prof I Bayley Balfour. Part II. Helicaceae, *PZS*, no.52, pp801-12, 1881.
25. Report on the Socotran Expedition, *British Association*, 1882.
26. On a fossil Species of Camptoceras, a Freshwater Mollusc from the Eocene of Sheerness-on-Sea, *QJGS*, 38, pp218-21, 1882.
27. Shells from Kashmir, *Nature*, Vol 25, pp385-6, 23 February 1882.
28. Report on Mr Everett's collection of Land snails found in the Cave Earth – Borneo, *British Association*, probably 1882 [3]
29. On the freshwater shells of the island of Socotra collected by Prof I Bayley Balfour. Part III, *PZS*, pp2-8, 1883.
30. On some Land Mollusca from Burmah with descriptions of some new species, *PZS*, April 1888.
31. Descriptions of new Land-Shells from the Andaman and Nicobar Group of Islands in the Bay of Bengal, *Ann Mag Nat Hist*, July 1888.
32. On a collection of land-shells made in Borneo by Mr A Everett, with Descriptions of supposed new Species, *PZS*, pp332-53, plates 35-9, 1889.
33. Description of a supposed new species of Helix from near Moulmain, Tenasserim, *Ann Mag Nat Hist*, February 1889.
34. On Supposed new species of Land Mollusca from Borneo belonging to Genera Opisthostoma and Diplommatina, *Ann Mag Nat Hist*, September 1890.
35. On a collection of land-shells made in Borneo by Mr A Everett, with Descriptions of supposed new Species, part 2, Zonitidoe & Helicidoe, *PZS*, January 1891.
36. Description of Helicarion Whiteheadi, n.sp, *PZS*, 1891.
37. On new species and varieties of the land-molluscan genus Diplommatina from the Garo, Naga, and Munipur Hill-ranges, Assam, *PZS*, pp509-20, 1892.
38. Description of a new species of Helix of the subgenus Plectopylis (P fultorci) Khasi Hills, *Ann Mag Nat Hist*, October 1892.
39. On a supposed new species of Rhiostoma from Borneo, and notices of Two other species of Shells from Palawau, *Ann Mag Nat Hist*, July 1893.
40. On some new species of the land-molluscan genus Alycoeus from the Khasi and Naga Hill Country, Assam, Munipur, and the Ruby Mine District, Upper Burma; and on one species from the Nicobars, *PZS*, pp592-5, 1893.
41. List of Freshwater and Brackish-water shells collected in Palawan, *Ann Mag Nat Hist*, VI, Vol XII, 1893.
42. On the Molluscan genus Paryphanta and on the Anatomy of P Hochstetteri, Pfr, *J Moll Stud*, Vol 1, pp5-9, October 1893.
43. New Species of Cyclopherus and a Spiraculum from the Khasi and Naga Hills, Assam, with Col R Beddome.[4] *Ann Mag Nat Hist*, June 1894.
44. List and distribution of the land mollusca of the Andaman and Nicobar islands with descriptions of some supposed new species, *PZS*, pp438-57, 1895.
45. Notes on Trochonanina and other Genera of Land Mollusca, with reference to the Generic Position of Martensia Mozambicensis and Other Species, *J Moll Stud*, Vol 1, pp281-7, July 1895.
46. Notes on, and drawings on the animals of various Indian Land Mollusca (Pulmonifera). *JASB*, Vol 14, part ii, pp131-6, 1895.
47. On the structure and Affinities of some New Species of Mollusca from Borneo, in conjunction with WE Collinge, Assistant Lecturer and Demonstrator in Zoology and Comparative Anatomy, Josiah Mason College Birmingham, with four plates, *PZS*, no.16, pp241-50, 1895.
48. Description of a supposed new species of Land-Mollusca of the genus Parmarion from Pulo

[3] Everett's expedition to Borneo took place in 1879. The paper is undated in Godwin-Austen's résumé, but is next to an entry for 1882; this fits with the known date of the expedition.
[4] Godwin-Austen rather surprisingly mistypes the name in his list as 'Beddouce'.

Lat, an island off the South East Coast of Borneo; one plate, *Ann Mag Nat Hist*, pp434-7, December 1895.
49. Notes on the Genus Euplecta of Semper, with descriptions of Supposed New Species from Ceylon, *J Moll Stud*, Vol 2, pp173-8, April 1897.
50. Untitled Paper: Ordinary Meeting of the Malacological Society, *J Moll Stud*, Vol 2, 30 November 1897.
51. On a collection of land-shells made in Borneo by Mr A Everett, with descriptions of supposed new species, *PZS*, pp22-47, plates 2-6, 1897.
52. On Philalanka, a new Sub-genus of Endodonta, with descriptions of two new species from the Indian Region, *Proc Mala Soc*, April 3, pp11-3, April 1898.
53. Untitled Paper, Ordinary Meeting of the Malacological Society, Friday 12 May, 1899.
54. Address of the President to the Malacological Society, 10 February 1899, *Proc Mala Soc*, part 5, July 1899.
55. Title Uncertain, *J Moll Stud*, Vol 3, pp333-4, October 1899.
56. On the Anatomy of Hemiplecta Floweri, EA Smith, from Perak, Malay Peninsula; with Notes on some other Eastern Genera, *J Moll Stud*, Vol 4, pp31-6, March 1900.
57. Intercalation on Anatomy, Fauna Hawaiienisis, Vol ii, 1900.
58. On the Anatomy of the Helix Ampulla of Benson, and its Generic Position in the Ariophantinæ, *J Moll Stud*, Vol 4, pp187-90, July 1901.
59. On the Anatomy of Helix Politissima, and on its position in the Ariophantinæ, *J Moll Stud*, Vol 4, pp261-3, October 1901.
60. On Helix Basileus, Benson, from Southern India: Its Anatomy and Generic Position, *J Moll Stud*, Vol 5, pp248-52, October 1902.
61. On the Anatomy of two Land Mollusca (Helicarion) Willeyana & Helicarion Woodwardi. N.spp. From New Britain & Lifu Loyalty Islands. Collected by Dr A Willey FRS in 1895-7, *Proc Mala Soc*, Vol V, part 6, pp296-8, 1902.
62. On the Anatomy of two Land Molluscs (Helicarion) Willeyana and H Woodwardi, n.spp, from New Britain and Lifu, Loyalty Islands, collected by Dr Arthur Willey, FRS, in 1895-7, *J Moll Stud*, Vol 5, pp296-9, April 1903.
63. Further Description of the Animal of Damayantia Carinata, Collinge, showing its similarity to D Smithi, Collinge and Godwin-Austen, with remarks on this Genus of Issel, Collingea of Simroth, and Isselentia of Collinge, *J Moll Stud*, Vol 5, pp311-6, June 1903.
64. Morphological observations on genera of the families Cyclostomidae and Helicidae. Morphological observations on species belonging to the families Cyclostomidae and Helicidae, pp161-73. 10 figures. Plain plate XIII A; Morphology of Guillainia and Lothidion. Depicts 3 species; artist Henry Haversham Godwin-Austen. In, *Natural History of Sokotra and Abd-el-Kuri*, Editor: HO Forbes. *Bulletin of the Liverpool Museums (Special Bulletin)*, 1903.
65. Further descriptions of the animal of Damayantia Carinata, Collinge, showing its similarity to D Smithi, C & GA, with remarks on this Genus of Issel, Collingea of Simroth, & Isselentia of Collinge, *Proc Mala Soc*, Vol 5, part 5, read 13 March, 1903.
66. On the Genus Eurystoma of Albers (Type Vittata, Müller), its Anatomy and Reference to Other Indian Species, *J Moll Stud*, Vol 6, pp48-50, March 1904.
67. Note on Damayantia Smithi and Godwin-Austen, *J Moll Stud*, Vol 6, p133, September 1904.
68. On the Extension of the Genus Macrochlamys to the Island of Mauritius, *J Moll Stud*, Vol 6, pp319-21, September 1905.
69. On a Species of the Land Molluscan Genus Dyakia from Siam, *J Moll Stud*, Vol 7, pp93-6, June 1906.
70. On the generic position of Benson's Helix Hyba and the similarity of its anatomy to that of Khasiella vidua, WT Blanford, *Ann Mag Nat Hist*, Ser 7, Vol 20, July 1907.
71. Description of nine terrestrial mollusca from South Africa, including anatomical descriptions of two proposed new genera Afrodonta, M & P and Peltatus, G-A: joint paper with James Cosmo Melvill and John Henry Ponsonby, *Ann Mag Nat Hist*, Ser 8, Vol I, February 1908.

72. Dispersal of land shells by the agency of man, *Proc Mala Soc*, Vol 8, pp146-7, October 1908.
73. On the animals of genera and species of Mascarene Land mollusca belonging to the family Zonitidae, collected by ME Dupont, *Ann Mag Nat Hist*, Ser 6, Vol II, November 1908.
74. Description of a new slug from Tibet, collected by Captain FH Stewart, IM Service, with the details of its anatomy, *Records of the Indian Museum, Calcutta*, Vol 2, part iv, no.42, December 1908.
75. Descriptions of the Animals of Two Land Shells from Perak, Skeat Expedition in the Malay Peninsula, 1899-1900, *J Moll Stud*, Vol 8, pp365-8, September 1909.
76. The importance of the animal in the Land Mollusca, shown by certain evolutionary stages in some Genera of two Zonitidoe: Presidential Address, read at Annual Meeting, 16 October 1909; *Journal of Conchology*, April 1910.
77. Some remarks on the determination of Genera and Species, Presidential Address: Annual Meeting of the Conchological Society, 15 October 1910.
78. Description of Oxytes Beddomei, n.sp, from Upper Burma, *J Moll Stud*, Vol 9, pp327-8, June 1911.
79. A review of South African Land-Mollusca belonging to the family Zonitidoe, Part i, *Ann Mag Nat Hist*, Ser 8, Vol ix, p122, January 1912.
80. A review of South African Land-Mollusca belonging to the family Zonitidoe, Part ii, *Ann Mag Nat Hist*, Ser 8, Vol ix, p569, May 1912.
81. A review of South African Land-Mollusca belonging to the family Zonitidoe, Part iii, *Ann Mag Nat Hist*, Ser 8, Vol xiii, p449, May 1914.
82. Zoological Results of the Abor Expedition, 1911-12; Mollusca ii: Zonitidae and Helicidae, *Records of the Indian Museum, Calcutta*, Vol 8, pp359-63, March 1914.
83. Zoological Results of the Abor Expedition, 1911-12; Mollusca iii: Cyclophoridae (in part), *Records of the Indian Museum, Calcutta*, Vol 8, part vii, no.39, August 1915.
84. Zoological Results of the Abor Expedition, 1911-12; Mollusca vi: Zonitidae, *Records of the Indian Museum, Calcutta*, Vol 8, part ix, no.46, August 1916.
85. Zoological Results of the Abor Expedition Mollusca vii: Cyclophoridae in part, *Records of the Indian Museum, Calcutta*, Vol 8, part 10, no.47, pp547-59, April 1917.
86. The Land Mollusca collected on the Island of Barkuda in the Chilka Lake, *Records of the Indian Museum, Calcutta*, Vol 13, part vi, no.21, December 1917.
87. Zoological Results of the Abor Expedition Mollusca viii; Macrochlamys (in part), *Records of the Indian Museum, Calcutta*, Vol 8, February 1918.
88. New species of slug-like molluscs belonging to the family Zonitidae from the Dawna Hills, Tenasserim, collected by Dr FH Gravely,[5] *Records of the Indian Museum, Calcutta*, Vol 20, part iii, no.18, August 1918.
89. Zoological Results of the Abor Expedition Mollusca ix, *Records of the Indian Museum, Calcutta*, Vol 8, part xii, November 1918.
90. On the supposed occurrence of the Miocene genus Fossarulus in India, *Records of the Indian Museum, Calcutta*, Vol 16, pp209-13, 1919.
91. Indian Land Mollusca, *Nature*, Vol 108, (2708), pp106-8, 1921.
92. A list of molluscs with localities, mainly from India, but including specimens from Thailand, Solomon Islands, Lord Howe Island, and Ceylon: origin uncertain.
93. Morphological observations on genera of the families Cyclostomidae and Helicidae: origin uncertain.

(iv) Ornithology and Lepidopterology

1. A List of Birds Obtained in the Khasi and North Cachar Hills, *JASB*, Vol 39, January 1870.

---

[5] Dr Frederic Henry Gravely (1885-1965) was an eminent arachnologist, entomologist, zoologist and student of archaeology, who conducted pioneering research and wrote extensively on various subjects during his tenure at the Indian Museum, Calcutta, and the Government Museum, Madras.

2. Second List of Birds obtained in the Khasi and North Cachar Hills ranges, including the Garo Hills and country at their base in Mymensing and Sylhet Districts, *JASB*, July 1870.
3. Third list of birds obtained in the Khasi and Garo hill ranges, with some corrections and additions to the former lists, *JASB*, Vol 41 (2), pp142-143, 1872.
4. Fourth List of Birds principally from the Naga Hills and Munipur including others from the Khasi, Garo and Tipperah Hills, with six plates, *JASB*, Vol 43, 1874.
5. Description of ten new birds from the Nágá Hills and Munipúr Valley, NE frontier of Bengal. Probably in *Ann Mag Nat Hist* or *JASB*, 1874a.
6. Description of a new Sibia from the Naga Hills, northeast frontier, Bengal, *Ann Mag Nat Hist* (4) Vol 13, pp160-1, 1874b.
7. Descriptions of some supposed new Species of Birds, *Ibis*, Vol 17, Issue 2, pp250-3 (with Arthur, Viscount Walden), April 1875.
8. Description of a Supposed New Actinura from the Dafla Hills, *Ann Mag Nat Hist*, November 1875.
9. Description of a Supposed New Suthora from the Dafla Hills and a Minla from the Naga Hills with remarks on Pictornis (Crysomma) alti rostri of Jerdon, *Ann Mag Nat Hist*, July 1875.
10. Fifth list of birds from the hill ranges of the north-east frontier of India, *JASB*, 1876.
11. List of the birds collected on the expedition into the Dafla Hills, Assam, together with those obtained in the adjacent Darrang, especially from Baridua bordering Terai, *JASB*, Vol 45 (2), pp64-85, 1876a.
12. Description of a new Garrulax from the Naga Hills, *Ann Mag Nat Hist*, November, 1876.
13. Description of three new species of the genera Pellorneum and Pomaiorhinus lately collected in the neighbourhood of Sadiya, Assam by Mr Ogle, *JASB*, Vol 46 (2), pp41-5, 1877.
14. Untitled: *Ann Mag Nat Hist*, December 1877.
15. Sixth list of birds from the hill ranges of the North-East Frontier of India, *JASB*, Vol 47 (2), pp2-25, 1877.
16. Description of Two Supposed New Birds from Eastern Assam, *Ann Mag Nat Hist*, 1877.
17. Note on the female Lophophorus Selateri from Eastern Assam, *PZS*, 1879.
18. On Specimens of the Male and Female of Phasianus humiæ, from Munipur, with a Description of the latter, *PZS*, Vol 50, Issue 4, pp715-8, 1882.
19. Movement of Sea-Gulls with a Coming Change of Weather, *Nature*, Vol 60, Issue 1560, p491, 1899.
20. On the genus Mysoria, Godwin-Austen, *Records of the Indian Museum, Calcutta*, Vol 16, p431, 1919.

The following note should be added, on the birds named after Godwin-Austen.

The Brown-capped Laughingthrush, *Garrulax austeni* is named after him. The Streak-throated Barwing was earlier called Austen's Barwing as was the Snowy-throated Babbler, which was referred to as Austen's Babbler. The Naga wren-babbler, *Pnoepyga chocolatina Godwin-Austen and Walden*, from the Manipur area, was named after him and its co-discoverer. A race of Assam Laughingthrush is called *Trochalopteron C godwini* as is a race of the Streak-breasted Babbler. The South Assam race of the White-browed Fulvetta is called *Alcippe v austeni* after him.

(v) Ethnology and Anthropology

1. Description of a Mystic Play, as performed in Ladak, Zaskar &c, *JASB*, 1864. (See also under 'Books', no.1.)
2. On the System Employed in Outlining the Figures of Deities and other religious drawings as practiced in Ladak, Zaskar &c, *JASB*, 1864.
3. A Vocabulary of Balti and Kashmiri, *JASB*, 1866.
4. On the Stone Monuments of the Khasi Hill Tribes and on some of the peculiar rites and customs of the people, *JAIGBI*, 1871.

5. Some Monuments of Certain Naga Tribes, *JAIGBI*, 1874.
6. Further notes on the rude stone monuments of the Khasi Hill Tribes, *JAIGBI*, 1874(?).
7. On the Rude Stone Monuments of Certain Naga Tribes, with Some Remarks on Their Customs, etc., *JAIGBI*, Vol 4, pp144-7, 1875.
8. The Ruins at Dunapar (sc. Dimapur) on the Dunsin R, Assam, *JAIGBI*, Vol 4, 1875.
9. On the Garo Hill Tribes, *JAIGBI*, Vol 4, 1875. (See also under 'Mountaineering and Exploration' no.1.)
10. Exhibition of a celt found at the Khasi Hills at Shilong, Probably in *JAIGBI*, 1877.
11. Notes on stone implements from the Khasi Hills, and the Banda and Vellore districts, *Proc As Soc Beng*, p158, 1879.
12. On the discovery of some worked flints, cores and flakes from Blackheath near Chilworth and Bramley, Surrey, *J Roy Anthropol Inst*, Vol 13, pp137-43, 1884.
13. The Barisal Guns and Similar Sounds, *Nature*, Vol 53, Number 1368, pp241-64, 1896.

(vi) Zoology

1. Dr. Stoliczka's Collection of Mammals, *Nature*, Vol 13, pp106-7, 9 December 1875.
2. The Squirrel Crossing Water, *Nature*, Vol 23, pp340, 10 February 1881.

(vii) Miscellaneous

1. Remarks on the Plane-table, in Hints to Travellers, eds J Knox Laughton, D Freshfield and HH Godwin-Austen, 5th Edn, *RGS publications*, Stafford, London, 1883.
2. The Earthquake, *Nature*, Vol 30, pp145, 12 June 1884.
3. Ocean Swells, *Nature*, Vol 30, pp487-88, 18 September 1884.
4. To the Editor of the *Surrey Advertiser*,[6] A Defence of the Sanitary Committee's Secretary, 25 August 1894.
5. To the Editor of the *Surrey Advertiser*, Concerning the Recent Controversy surrounding the first Parish Council Elections, 15 November 1894.
6. To the Editor of the *Surrey Advertiser*, The Shalford Drainage Question, 2 March 1896.
7. To the Editor of the *Surrey Advertiser*, Wanton destruction by Hunters & Poachers of local fauna, written 27 April 1899, published 1 May 1899.
8. To the Editor of the *Geographical Journal*, The Map of Kashmir, Nore, April 1900.
9. The Survey of the Mustagh Range, Letter to the Royal Geographical Society, Nore, 7 November 1903.
10. The History of Upper Assam, Upper Burmah, and North-Eastern Frontier (uncommissioned review), *Nature*, August 1915.
11. Distances at which Sounds of Heavy Gun-firing are Heard, *Nature*, Vol 96, pp173, 14 October 1915.
12. To the Editor of the *Surrey Advertiser*, Agricultural Difficulties: the Necessity for Subsidies, 20 January 1917.
13. Future Treatment of German Scientific Men, *Nature*, Vol 102, pp64-5, 26 September 1918.
12. To the Editor of the *Surrey Advertiser*, Housing Problems for Large Families, 1 November 1919.
13. 'Momiai', *Nature*, Vol 106, Issue 2660, p241, 1920.
14. A Lost Collection of Indian Sketches, *Nature*, Vol 111, pp845-46, 23 June 1923.

(viii) Maps

The Irrawaddy Delta, 1853.
Kuram Valley, Kashmir, 1856.
The Pir Panjal, The Kajnag, The Chata Dhar Ranges; The Muro-Mirwan-Valley, all 1857-8.

---

[6] Although six letters to the Surrey Advertiser are cited here, it is very likely that he had many more published.

The southern Karakoram, 1860.
The northern, central and western Karakoram, 1861.
Rupshu, Zaskar and Ladakh, 1862.
The Lake Pangong and Chanchengmo Regions, 1863.
Western Bhutan from Darjiling to Punakha, 1863.
The (Western) Bhootan Dooars, 1865.
The Khasi, Naga, North Cachar, South Cachar and Munipur Hills, 1867-74.
The Dafla Hills, 1875.

A number of sketch-maps and Surrey estate-maps also exist.

Part Two: Record of Works and Websites consulted

1. Books and Articles (other than those cited in Godwin-Austen's personal bibliography).

Baker, Ian, *The Heart of the World: A Journey to the Last Secret Place*, Souvenir Press, London, 2007.
Barnett, Corelli, *The Desert Generals*, George Allen & Unwin, London, 1960.
Baumer, Christopher, *Traces in the Desert: Journeys of Discovery Across Central Asia*, IB Tauris, London, 2008.
Beckwith, Christopher I, *Empires of the Silk Road*, Princeton University Press, 2009.
Bell, Sir Charles, *Tibet Past and Present*, Book Faith India, Delhi, 1998.
Bingley, Lieutenant-General Sir Alfred H, KCIE, CB, K St J, *The Austens of Shalford and their Kinsfolk*, Bibliobazaar, reprinted from hardback original, dated 1936.
Burrard, Sir Sidney, *The Himalayan Ranges and Godwin-Austen's Map*, letter to the Editor of the *G J*, 27 September 1929.
Chan, Victor, *Tibet Handbook*, Moon Publications, Chico, California, 1994.
Coldstream, John, *Dirk Bogarde, The Authorised Biography*, Phoenix, 2005.
Conway, WM (see below), *The Crossing of the Hispar Pass*, *GJ*, Vol 1, no.2, pp131-8, February 1893.
Conway, WM (see below), *Peak K2*, *GJ*, Vol 3, no.6, p527, June 1894.
Conway of Allington, Lord (Sir Martin Conway), *The Autobiography of a Mountain Climber*, Jonathan Cape, London, 1933.
Curran, Jim, K2, *The Story of the Savage Mountain*, Hodder & Stoughton, London, 1996.
Darwin, Charles, *The Origin of Species* and *The Voyage of the Beagle*, Everyman, Random House, New York, 2003.
David, Saul, *Victoria's Wars*, Penguin, 2007.
Dierden, Margaret, Various privately-published essays in Local (Surrey) History.
Fairley, Jean, *The Lion River, The Indus*, Allen Lane, 1975.
Filippi, Dr Filippo de, *The Expedition of HRH The Duke of the Abruzzi to the Karakoram Himalayas*, lecture at the Royal Geographical Society, November 21, 1910.
Fitzpatrick, Dennis, Non-Christian Marriage, *Journal of the Society of Comparative Legislation*, New Series, Vol 2, pp359-87, 1900.
French, Patrick, *Younghusband*, HarperCollins, London, 1994.
Freshfield, Douglas, On the Glaciers of Kangchenjunga, *RGSJ*, Vol 20, p236, London, Date unknown.
Ghey, Rita, *A Rich Inheritance*, Imprimeries Quebecor, Québec, 1992. (Privately printed.)
Godwin, Major-General Henry, *Burmah. Letters and Papers written in 1852-53*, edited and privately printed by Robert Alfred Cloyne Godwin-Austen, Bosworth, London, 1854.
Godwin-Austen, Dr Richard, *Seizing Opportunities*, The Memoir Club, Durham, 2008.
Graan, David, *The Lost City of Z*, Simon & Schuster, London, 2010.
Gyurme, Dorje, *Tibet Handbook*, Footprint Handbooks, Bath, England, 1999.
Halifax, Lord, *Memorandum*, unpublished, 25 November 1853.
Harvey, Andrew, *A Journey in Ladakh*, Jonathan Cape, London, 1983.

Herrligkoffer, Karl M, *Nanga Parbat*, Elek Books, London, 1954.
Hopkirk, Peter, *The Great Game*, OUP, 1990.
– *Trespassers on the Roof of the World*, OUP, 2001.
Johnson, Robert, *Spying for Empire*, Greenhill Books, London, 2006.
Jordans, Bart, *A Trekker's Guide to Bhutan*, Cicerone, Milnthorpe, Cumbria 2005.
Keay, John, *When Men and Mountains Meet*, John Murray, London, 1977.
Keay, John, *The Gilgit Game*, OUP, 1979.
Keay, John, *The Great Arc*, HarperCollins, London, 2000.
Kick, William, *Chogo Lungma Glacier, Karakoram*, *GJ*, Vol 122, pp93-6, March 1956.
Knight, EF, *Where Three Empires Meet*, Ali Kamran, Lahore, 1986.
Laurie, William FH, *A Narrative of the Operations at Rangoon in 1852*, Smith, Elder & Co, London, 1853.
Lines, Maureen, *Beyond the North-West Frontier*, Oxford Illustrated Press, 1988.
Lonely Planet Guide to Bhutan, Lonely Planet Publications, Hawthorn, Victoria, Australia, 1998.
Lonely Planet Guide to India, Lonely Planet Publications, Hawthorn, Victoria, Australia, 1981.
Lonely Planet Guide to Pakistan, Lonely Planet Publications, Hawthorn, Victoria, Australia, 1998.
Lonely Planet Walking Guide: Trekking in the Indian Himalaya, Lonely Planet Publications, Hawthorn, Victoria, Australia, 1997.
Lonely Planet Travel Survival Kit: Kashmir, Ladakh & Zanskar, Lonely Planet Publications, South Yarra, Victoria, Australia, 1985.
Longstaff, Tom, *This My Voyage*, John Murray, London, 1950.
Mason, Kenneth, *Abode of Snow*, Diadem Books, Seattle, 1987. (Facsimile edition of original.)
Mason, Kenneth, *Autobiography*, MSS, unpublished, Special Collections, Bodleian Library, Oxford. Date not given but from internal evidence probably c.1970.
Messner, Reinhold, *All 14 Eight-thousanders*, transl Audrey Salkeld, Crowood Press, Marlborough, Wilts, 1999.
Mumm, Arnold, *The Alpine Club Register 1857-1863* (Vol I), Edward Arnold & Co, London, 1923.
Nelles Guides: Northern India, Nelles Verlag GmbH, München, 1990.
Peale, SE, *The Nagas and Neighbouring Tribes: JAIGBI*, Vol 3, pp476-81, 1874.
Perrin, James, *Shipton & Tilman*, Hutchinson, London, 2013.
Pevsner, Nikolaus & Nairn, Ian, rev. Bridget Cherry, *Surrey*, in *The Historic Buildings of England*, Yale University Press, New Haven & London, 2002.
– *Sussex*, in *The Historic Buildings of England*, Yale University Press, New Haven & London, 2003.
Phillimore, RH, *Historical records of the Survey of India*, Dehra Dun, India, 1945-8.
Proceedings of the Royal Society of Medicine, Vol 38 (3), pp135-40, 1945.
Rohatgi, Pauline and Partlett, Graham, assisted by Shirley Imray and Pheroza Godrej, *Indian Life and Landscapes by Western Artists*, Chhatrapati Shivaji Maharaj Vastu Sangrahalaya (CSMVS), formerly the Prince of Wales Museum, Mumbai, date unknown.
Russell, Scott, *Mountain Prospect*, Chatto & Windus, London, 1946.
Shipton, Eric, *The True Book about Everest*, Frederick Muller, London, 1955.
Stearns, RP, *Agriculture adaptation in England, 1875-1900*, Agricultural History, Vol 6, no.2, April 1932.
Steel, Lieutenant EH, *On the Khasia Tribe, Transactions of the Ethnological Society of London*, Vol 7, pp305-12, 1869.
Steele, Peter, *Eric Shipton: Everest and Beyond*, Constable, London, 1998.
Stein, Sir Aurel, *Correspondence with HHG-A*, Microfilm, John Radcliffe Science Library, Oxford.
Sturley, DM, *The Royal Grammar School Guildford*, Governors of the Royal Grammar School and DM Sturley, Guildford, Surrey, 1991.
Suffern, Canning and Purdon, William, *On the Trigonometrical Survey and Physical Configuration of the Valley of Kashmir, PRGS*, Vol 4, no.1, pp31-3, London, 1859-60.
Teltscher, Kate, *The High Road to China*, Bloomsbury, London, 2007.
Unsworth, Walt, *Everest*, Allen Lane, London, 1981.

Waller, Derek J, *The Pundits*, University of Kentucky Press, 1990.
Ward, Michael, Bhutan: A *Short History of Exploration*, MSS, Royal Geographical Society Archives, unpublished.
Watson, Nigel, *The Royal Grammar School, Guildford: An Illustrated History*, James & James, London, 2004.
Watt, George, *The Aboriginal Tribes of Manipur, J Roy Anthropol Inst*, Vol 16, 1887.
Waugh, Sir Andrew (Surveyor-General), Survey of India, *General Reports of the Great Trigonometrical Surveys*, Synoptical Volumes for 1858-59, 1864-65, 1878-79. Published by the Government of India.
Wheeler, Stephen, *Obituary: Ney Elias, C. I. E., Geographical Journal*, Vol. 10, no.1, pp101-6, Blackwell Publishing on behalf of The Royal Geographical Society, July, 1897.
Woodthorpe, RG, *Notes on the Wild Tribes Inhabiting the So-called Naga Hills, on our North-Eastern Frontier of India*, Part I, *JAIGBI*, Vol 11, pp56-73, 1882.

2. Websites

(i) Online resources, books and articles.

Allan, Nigel J.R, *Karakorum Himalaya, Sourcebook for a Protected Area*, www.lda.ucdavis.edu.
Black, Charles Edward Drummond, *A memoir on the Indian surveys*, 1875-1890, Google books.
Burke, John, *A genealogical and heraldic history of the commoners of Great Britain and Ireland enjoying territorial possessions or high official rank: but uninvested with heritable honours.*, Google books.
Chicheley Plowden, Walter Francis Courtenay, *Records of the Chicheley Plowdens A.D. 1590-1913; with four alphabetical indices, four pedigree sheets, and a portrait of Edmund, the great Elizabethan lawyer (1914)*, www.archive.org.
Godwin-Austen, Henry Haversham, *Land and freshwater Mollusca of India, including South Arabia, Baluchistan, Afghanistan, Kashmir (1882)*, www.archive.org.
Growse, FS, *Mathura-Brindaban – The Mystical Land Of Lord Krishna*, Google books.
Hobson-Jobson: *A Glossary of Colloquial Anglo-Indian Words and Phrases*, www.amazon.com.
Jina, Prem Singh, *Ladakh: the land and the people*, Google books.
Laurie, Col WFB, *Our Burmese Wars: Relations with Burma , being an abstract of military and political operations 1824-25-26 and 1852-53, with various local, statistical, and commercial information, and summary of events, from 1826-79, including a sketch of King Theebau's progress*, www.archive.org.
Marshall, Julie G, *Britain and Tibet 1765-1947: a select annotated bibliography of British relations with Tibet and the Himalayan states including Nepal, Sikkim and Bhutan*, Google books.
Nautilus, The, A Quarterly Journal devoted to the interests of Conchologists, www.archive.org.
Ogden, ICK, *Younghusband*, www.archive.org.
Pollak OB, *The Origins of the Second Anglo-Burmese War (1852-53)*, www.jstor.org, 1978.
Waller, Derek J, *The Pundits: British Exploration of Tibet and Central Asia*, Google books.
Williams, Chris, *A companion to nineteenth-century Britain*, Google books.

(ii) Institutions, Societies and Other Search Engines.

Websites for societies will generally provide access to their own in-house magazine, periodical or scholarly journal.

www.army.mod.uk
(Sandhurst)
www.billbuxton.com
(very extensive mountaineering bibliography)
www.bl.uk
www.bodley.ox.ac.uk
www.britannica.com
www.british-history.ac.uk

www.britishmuseum.org
www.burkes-peerage.net
www.conchsoc.org
www.copac.ac.uk
(unified academic catalogue)
www.debretts.com
www.earth.google.co.uk
www.exploringsurreyspast.org.uk google.co.uk/scholar
www.glosters.tripod.com/offzdiedm.htm
www.google.co.uk/books
www.google.co.uk/images
www.google.co.uk/maps
www.google.co.uk/scholar
www.google.co.uk/web
www.guildford.gov.uk
(for Guildford Museum)
www.guildford-institute.org.uk
www.imperial.ac.uk
www.jstor.com
www.kcl.ac.uk
www.lyellcollection.org
www.malacsoc.org.uk
www.nationalarchives.gov.uk
www.nationalarchives.gov.uk/a2a/
(unified local archives)
www.nhm.ac.uk
www.nhm.ac.uk/tring
www.oxforddnb.com
(Dictionary of National Biography)
www.prm.ox.ac.uk
(Pitt Rivers Museum, Oxford)
www.questia.com
www.rgs-guildford.co.uk
(Royal Grammar School Guildford)
www.rgs.org/OurWork/Collections/Catalogue+Search/CatalogueSearch.htm
(Royal Geographical Society)
www.royalsociety.org
www.rrw.org.uk
(Royal Regiment of Wales)
www.surreyarchaeology.org.uk
www.surreycc.gov.uk/surreyhistoryservice
www.thebrookingcollection.com
www.thepeerage.com
www.oup.com/whoswho
www.vam.ac.uk
www.weyriver.co.uk/theriver/industry_8_papermaking.htm
www.wikipedia.org
www.worldclimate.com

(iii) Other Websites

www.britishbattles.com/zulu-war/isandlwana.htm
www.fleurdelis.com/meanings.htm

www.gillandersindia.com
www.greatindianoutdoors.com
www.kashmirlehadv.com
www.kolkatabirds.com
www.manenvis.nic.in
www.measuringworth.com
www.rootschat.com
www.vedanti.com
(cultural history of Assam)
www.visiblemantra@googlemail.com
(Mandala interpretation)
www.1911encyclopedia.org

3. Maps

(i) Online

www.earth.google.co.uk
www.google.co.uk/maps
www.mastertopo.com
www.reise-know-how.de
www.stanfords.co.uk
www.tibetmap.com

(ii) Hard Copy Maps

Bhutan Himalaya, 2001, 1:390,000, Nepa Maps, Gondoni
Indian Army Map Service, Corps of Engineers:
Anantnag: 1:250,000
Leh: 1:250,000
Palampur: 1:250,000
Shyok: 1:250,000
India North, with Special Map of Zanskar/Ladakh: 1:1,500,000, Nelles Maps, Nelles Verlag
Indian Himalayan Maps Series: all 1:200,000, Leomann Maps:
Jammu and Kashmir: Sheet 1: Srinagar, Kolohoi Glacier & Kishtwar Area
Sheet 2: Kargil, Zanskar and Nun-Kun Area
Sheet 3: Leh, Zanskar, Markha and Nubra Valley
Sheet 9: Rupshu, Tso Moriri, Pangong Tso
Himachal Pradesh: Sheet 4: Chamba, Dhauladhar Passes, Pangi Valley, & Western Lahaul
Sheet 5: Kullu Valley, Parbati Valley & Central Lahaul
Karakoram: Sheet 4: Siachen, Rimo, Saser Kangri
Karakoram Highway, The, 1:1,000,000, John Conway
Ladakh Zanskar, Carte de Randonée, 1:350,000, Editions Olizane, Geneva
Ladakh & Zanskar, Nord, Brog-yul, Nubra, Shyog, 1:150,000, Editions Olizane
Ladakh & Zanskar, Centre, Indus, Zanskar, Pangong, 1:150,000, Editions Olizane
Ladakh & Zanskar, Sud, upper Zanskar, Rupshu, 1:150,000, Editions Olizane
Myanmar Burma, 1:1,500,000, Reise Karten, Germany
Northern India, 1:1,500,000, Nelles Maps, Nellas Verlag
Ordnance Survey of Great Britain: Landranger Series: Sheet 186: Aldershot & Guildford, 1:50,000
Explorer Series: Sheet 133, Haslemere & Petersfield, 1:25,000
Explorer Series: Sheet 134, Horsham, 1:25,000
Explorer Series: Sheet 145, Guildford & Farnham, 1:25,000
Pakistan: 1:1,500,000, Nelles Maps, Nellas Verlag
Pakistan, Northern Area of, Gilgit, Diamar,Skardu, Ghangche, 1:500,000, Professor Manzoom Ali, Pakistan (based on other maps and author's own survey)

Swiss Foundation for Alpine Research, Orographical Sketch Map:
Karakoram Sheet 1: 1:250,000
Karakoram Sheet 2: 1:250,000
Tactical Pilotage Charts, American Ministry of Defense: all maps 1:500,000:
G-7D (Kashmir, Himachal Pradesh, Lake Pangong)
H-8B (Aksai Chin)
H-9A (Himachal Pradesh)
H-9C (Northern India/Nepal)
H-9D (Northern India/Delhi)
H-10A (Tsangpo-Brahmaputra)
H-10D (Bhutan, Munipur, Assam, Nagaland)
ONC G-7 (Baltistan,Tibet, Chinese Pamirs)
Tibet, Nepal, Bhutan, 1:2,000,000, GeoCenter
Tibet: Nepal, Sikkim, Bhutan, 1:2,400,000, Nepa Maps, Gondoni

SOURCES

Family

A.i Godwin-Austen family:

Lt Col Robin Godwin-Austen: private conversations and access to Haversham's private papers (see section Bi).
Dr Richard Godwin-Austen: private conversations; his autobiography, *Seizing Opportunities*, Memoir Club; the Southwell (Notts) archive (see section Bii); General Sir Alfred Bingley's *History of the Austens of Shalford and their Kinsfolk*; Jonathan Godwin-Austen, private conversations; the Papplewick Hall (Notts) archive (see section Biii); Mr Gilly Marsh (Bampton, Oxon): private conversations. Some drawings.

ii Descendants of Edward Milner and Theresa Smith: Mr Patrick Conquest, Australia: genealogical history by email via Mr R Pearce; Rita Ghey: *A Rich Inheritance* (family biography); Mr Robin Pearce, Kings Langley, NSW, Australia: genealogical material (photographs, family trees, email correspondence).

B.i Haversham Godwin-Austen's private papers:

Correspondence, mainly between Haversham Godwin-Austen & Robert Alfred Cloyne Godwin-Austen; some letters between Haversham and Maria Godwin-Austen.
Correspondence (one letter) from Pauline Godwin-Austen to Haversham.
Correspondence between Haversham and Florence Barclay.
Letters of resignation from the Indian Service.
Miscellaneous papers pertaining to the life of General Godwin.
Papers pertaining to the sale and contents of Shalford House.
Papers pertaining to the Austen estates.
Other miscellaneous family papers and correspondence, including a Freedom of Guildford certificate dated 1748.
A Godwin-Austen family pedigree; an Annesley family pedigree.
A family history (drafts) probably compiled by Robert Alfred Cloyne.
A family births, marriages, deaths record containing numerous newspaper clippings.
Abstract of the Record of Haversham's military service.
Memoir on life at Sandhurst.
Untitled memoir on 'The Battle of Bagshot'.
Untitled essay on the education of a British Army officer.
Many completed watercolours of scenes in Burma, the Punjab, Kashmir and the Karakaoram-Baltistan areas.
Miscellaneous newspaper cuttings pertaining to the Godwin-Austen family.

About seven hundred sketches and drawings on miscellaneous subjects.
Family portraits and photographs.

ii Southwell Archive:
Haversham's diaries from 1852-1854
Various watercolours, mostly of European and British scenes; one Karakoram watercolour.
A large pencil portrait of Robert Alfred Cloyne.

iii Papplewick Hall Archive:
Family portraits.
Family pedigree.
Paintings by Gould of birds named after Haversham.
A portfolio of paintings of scenes from the First and Second Anglo-Burmese Wars.
Some watercolours of *gompas* in Baltistan.
Watercolours of Scottish, Devon and Alpine scenes.
Haversham's Kashmir geological large notebook.
Watercolours of Buddhist mandalas, probably by Haversham.
Newspaper cuttings relating to the K2 expedition by Jonathan and Stephen Godwin-Austen.

National Libraries and Archives
i. Royal Geographical Society, Foyle Reading Room and Archives, London.
ii. Natural History Museum, Main Library and Department of Malacology Library, South Kensington, London.
iii. British Library, India Reading Room, London.
iv. Bodleian Library Oxford: Duke Humfrey and John Radcliffe Science Library, Oxford.
v. British Museum, Departments of Rome and Greece; Middle East; East.
vi. Malacological and Conchological Societies' Archives at Wye College, Ashford, Kent.
vii. Victoria and Albert Museum, South Kensington, London.
viii. Alpine Club Library and Archives.

Local Archives and Other Local Sources
i. Guildford Institute, University of Surrey.
ii. Shalford local historian, Margaret Dierden.
iii. Surrey History Centre, Woking.
iv. Surrey Archaeological Society, Castle Gate, Guildford.
v. Guildford Museum, Castle Gate, Guildford.
vi. History of Nore: Mr and Mrs G Kemp (the present owners); Mr Christopher Wells.

Military Archives
Regimental (24th Foot/South Wales Borderers/Royal Regiment of Wales) Archives, Brecon.
Imperial War Museum.

Bibliography of works referring to Haversham
Dictionary of National Biography; *Who's Who* (1917), *Who Was Who*, *Debrett's*, *Burke's Peerage*, Directories of notable and landed families.
Contemporary works eg Mason, Curran, Keay *et al*: see under Bibliography and Sources on page 255.
Correspondence and reviews by others e.g. Huxley, Darwin.

Newspapers and Magazines (other than scholarly journals cited under Bibliography and Sources):
The Auk
The Illustrated London News

The London Gazette
The Surrey Advertiser
The Times

The Royal Grammar School, Guildford

DM Sturley, *History of the Royal Grammar School*, Guildford (Governors of the RGS Guildford, 1981).
The Guildfordian (school magazine, published annually).

LIST OF ILLUSTRATIONS

25. Map of Baltoro and K2 area made by the zincographic process pioneered by Godwin-Austen and General Walker, from Godwin-Austen's original survey map.
26. The junction of the Punmah and Choktoi Glaciers, looking north (1861). *HH Godwin-Austen. Courtesy of Dr Richard Godwin-Austen*
27. Panorama of the Baltoro Glacier from near Concordia, looking west, to show the area of Godwin-Austen's first sighting and measurement of K2. *Courtesy of John Cleare*
28. The site of Godwin-Austen's likely ascent above Urdukas, on the south side of the Baltoro Glacier, to reach the point on a spur of Masherbrum from which the first close sighting of K2 was made. *Courtesy of John Cleare*
29. K2, to the left at the rear, from a top above Urdukas, about 1000m/3281ft above the Baltoro Glacier, looking over the Biange heights. The first close-up sighting of the world's second-highest mountain. (August, 1861) *HH Godwin-Austen*
30. View of the Godwin-Austen Glacier leading to K2, from Concordia. *Courtesy of Jonathan Godwin-Austen*
31. Left: view looking south over Tso Moriri, in the remote country of Rupshu. Mentok Kangri (6250m/20,505ft), on which Haversham broke the high-altitude summiting record, is the snow-covered peak beyond the lake. *Courtesy of Royal Geographical Society, South Kensington, London*
32. Lake Pangong, Tibet. Part of the area traversed by Haversham during the 1863 expedition. *Courtesy of Royal Geographical Society, South Kensington, London*
33. Paro Dzong: administrative, military, religious and legal centre, Bhutan, 1864. *HH Godwin-Austen*
34. A survey map of Punakha Dzong, Bhutan. This kind of illustration shows Haversham's attention to the fort's military significance as well as its architecture. (1864). *HH Godwin-Austen*
35. The first map of Sikkim and Western Bhutan, made by Haversham in 1864 when attached to Sir Ashley Eden's failed diplomatic mission.
36. Godwin-Austen outside Colonel McCulloch's bungalow at Cherrapoonjee (Cherrapunji), 1869. Haversham is aged 35 or 36. His sunken cheek may have been caused not only by the physical demands of a soldier-surveyor's life in a trying climate but also indicates that he had already suffered two bouts of malaria. *Courtesy of Natural History Museum, South Kensington, London*
37. A murong, or Burmese communal hut, from a diary drawing by Godwin-Austen, 1872. *Courtesy of Dr Richard Godwin-Austen*
38. Some sketches and a page entry from Haversham's Burma Diary. *Courtesy of Dr Richard Godwin-Austen*
39. The Château de Chillon and Lake Geneva, looking east. (c.1882) *HH Godwin-Austen. Courtesy of Dr Richard Godwin-Austen*
40. Left: the Matterhorn from the Zinal Valley. (c.1882) *HH Godwin-Austen. Courtesy of Papplewick Hall*
41. Haversham's third wife, Jessie Robinson, (c.1910) from the memoir on her life published (1914) by Gertrude Jekyl (sic) and 'MSW' (Margaret Watts). *Courtesy of V&A images: Victoria andAlbert Museum, London*
42. The postcard from the *Terra Nova,* sent to the Godwin-Austens by Captain Scott and his crew from Dunedin on their way to the Antarctic. Jessie has marked crosses against those who died on the expedition, not long before she herself died. *Courtesy of V&A images: Victoria and Albert Museum, London*
43. A Buddhist Mandala, or sand painting, symbolising the order and ephemerality of the universe. This watercolour by Godwin-Austen is one of several pieces of circumstantial evidence for his conversion to Buddhism late in life. *Courtesy of Papplewick Hall*
44. The Buddhist temple of Burmese design and probably made from Northern Indian sandstone, discovered at Nore by Dirk Bogarde after he bought the house in 1950. *Courtesy of Mr and Mrs G Kemp, Nore, Hascombe, Surrey*

INDEX